"While both popular and scholarly accounts of the Dust Bowl confine it to the 1930s, this careful and authoritative reconstruction of southeastern Colorado provides a much longer time frame for assessing two pivotal processes of the 1940s and 1950s: how farmers adopted a new and largely effective set of soil and water conservation practices and how the region came to depend on a labor regime of migratory workers. Sheflin deftly threads an analysis of the Dirty Thirties together with the broadest questions of postwar agricultural history."

—SARAH T. PHILLIPS, associate professor of history and director of graduate studies at Boston University

"This is a serious and thoughtful history of Colorado agriculture. The way it mixes environmental, political, and labor history is always interesting and sometimes downright poetic. The material on migrant children is important and absolutely fascinating."

—JONATHAN REES, professor of history at Colorado State University—Pueblo

"A boon to scholars and policy makers.... [This is a] carefully researched monograph with copious notes, excellent maps, good photographs, and an extensive bibliography."

—STEPHEN J. LEONARD, *Colorado Book Review*

"Douglas Sheflin's new, exceptionally well-researched study of the legacy of New Deal Dust Bowl policies in southeast Colorado convincingly reveals how the combined work of the Colorado Extension Service, the Social Conservation Service, the Production Management Administration, and Soil Conservation Districts rectified the unsustainable production-first mentality of farmers in the 1920s. As a result, Sheflin clearly illustrates how these policies produced for farmers a federal safety net well beyond the 1930s, especially for those who practiced soil conservation."

—JAMES E. SHEROW, University Distinguished Professor at Kansas State University

D1260400

LEGACIES OF DUST

Legacies *of* Dust

Land Use *and* Labor *on the* Colorado Plains

DOUGLAS SHEFLIN

University of Nebraska Press | Lincoln

© 2019 by the Board of Regents of the University of Nebraska

Parts of chapters 2–5 previously appeared as "The New Deal Personified: A. J. Hamman and the Cooperative Extension Service in Colorado," *Agricultural History* 90, no. 3 (Summer 2016): 356–78.

Library of Congress Cataloging-in-Publication Data
Names: Sheflin, Douglas, author.
Title: Legacies of dust: land use and labor
on the Colorado plains / Douglas Sheflin.
Description: Lincoln: University of Nebraska Press, [2019] |
Includes bibliographical references and index.
Identifiers: LCCN 2018047773
ISBN 9780803285538 (cloth: alk. paper)
ISBN 9781496224996 (paperback)
ISBN 9781496215390 (epub)
ISBN 9781496215406 (mobi)
ISBN 9781496215413 (pdf)
Subjects: LCSH: Agriculture—Environmental aspects—
Colorado—History—20th century. | Agriculture—Social
aspects—Colorado—History—20th century. | Dust Bowl Era,
1931–1939. | Dust storms—Colorado—History—20th century. |
Farmers—Colorado—Social conditions—20th century.
Classification: LCC S451.C6 S54 2019 | DDC 338.109764—dc23
LC record available at https://lccn.loc.gov/2018047773

Set in Garamond Premier Pro by E. Cuddy.
Designed by L. Auten.

To Ingrid, Lou, and Cecilia

CONTENTS

ILLUSTRATIONS

ACKNOWLEDGMENTS

This project has taken several years to complete, and I have accrued many debts over that time. The following is a brief note of thanks to those who have guided, funded, and supported me while I was researching and writing this book.

I received financial assistance from a number of sources. I would like to thank the Department of History at the University of Colorado for several incarnations of dissertation funding. The most important funding came in the form of the Bean Fund Research Fellowship, which paid for a research trip to Fort Worth, Texas. I am also grateful to the University of Colorado Center for the Humanities and the Arts for finding me a worthy Thomas Edwin Devaney fellow and offering significant financial support for a year of research and writing. The Boulder Historical Society provided support for research through the Thomas J. Meier Fellowship, so I thank both the organization and the Meier family for their generosity. Most recently, the Department of History at Colorado State University has been extremely considerate in its financial support for this project, specifically by helping to offset some of the publication costs.

The funding noted above helped me traverse the state and the Southwest in search of relevant sources, and several individuals and staffs guided me during the various legs of that exploration. The Agricultural Archive at Morgan Library at Colorado State University houses an incredible amount of pertinent information, much of which I found only with assistance from Vicky Lopez-Terrill and Linda Meyer, who pulled a number of boxes on the Colorado Cooperative Extension Service. Patricia Rettig, who heads the Water Resources Archive at Colorado State, always knew exactly what I meant even when I failed to explain it well and pointed me toward some fantastic materials. Wendel Cox, senior special collections librarian in the Western History and Genealogy Department at the Denver Public Library, gave me a tour through their collections and helped me brainstorm about sources I had not yet considered. Archivist Rosemary Evetts at Auraria Library similarly talked with me about its various sources on Amache. Finally, David Hays, archivist at Norlin Library on the University of Colorado at Boulder campus, offered a wealth of information and friendly banter.

I also found helpful professionals at the National Archives and Records Administration (NARA) facilities in Denver and Fort Worth. Eric Bittner guided part of my time at the NARA–Rocky Mountain archive outside of Denver and proved especially helpful with the War Manpower Commission records. Marene Sweeney Baker held my hand for a good portion of my initial visit (and a few subsequent trips), so I am in her debt. I had the pleasure of working with many people in Fort Worth during a hot July week researching the Natural Resource Conservation Service (NRCS) records. They have since moved into a new building, but I am certain that their hospitality remains intact at the new site. I also made contact with a number of NRCS employees in Prowers County, including conservationist Susan Hansen, who offered me information about the area and insight into regional history that I could not find in the sources.

I also owe a tremendous debt to those who have helped me better define my questions and scope and challenged my findings, all in hopes of making this a better project. As one might imagine, many of these individuals are (or were) faculty at the University of Colorado. Specifically, I thank Thomas Andrews, Peter Boag, Daryl Maeda, Ralph Mann, and Phoebe S. K. Young. Other faculty members, including Fred Anderson, Virginia

Anderson, Brian DeLay, and Tom Zeiler, had a more indirect impact on this book but nonetheless helped me on my way. Paul Sutter's arrival at the University of Colorado changed this project in dramatic ways. Paul's impact has been tremendous. He has influenced every chapter and every page of what follows. He has offered constructive criticism, abundant commentary, and useful questions, all designed to get me to focus more clearly on both the subject and the process. He continues to be a staunch supporter and considerate listener whenever we meet.

My current and former colleagues at Colorado State University have been fantastic since I arrived in 2012. Thanks especially to Eli Alberts, Ruth Alexander, Tracy Brady, Leisl Carr Childers, Michael Childers, Mark Fiege, Robert Gudmestad, Kristin Heineman, Diane Margolf, Janet Ore, Jared Orsi, Sarah Payne, and Doug Yarrington, who are good people and have offered moments of levity as well as aid in getting me acclimated to life at csu. They have helped make my work life an enjoyable one, which has afforded me more time to work on this project and thoroughly enjoy my time in Fort Collins. Adrian Hawkins read through an early draft of the manuscript and offered penetrating insights that helped reform significant pieces of the manuscript, and I thank him for his effort.

I would also like to thank the folks at the University of Nebraska Press who have shown tremendous patience throughout this process. Bridget Barry contacted me just after I finished my dissertation and suggested that we might find a good way to turn it into a book. I was flattered and a bit overwhelmed but very much appreciated her faith in me. She has responded to every query quickly and compassionately, and she has been a tremendous guide through this process. Brain Cannon and Jonathan Rees read this in its entirety and helped me in several ways, most notably by getting me to refine my focus, tighten my argument, and by giving me confidence in my ability to finish this thing. I very much appreciate their effort and expertise. Thanks as well to Sarah Phillips and Jim Sherow, who offered their time and energy to read the manuscript and supported this endeavor.

I am incredibly fortunate that my siblings and their families have lived near me for the last fifteen years. Family can sustain us in ways that friends cannot, and their unwavering encouragement has fueled me over the past several years. We have all enjoyed our time living so close to each other, but I doubt they can understand how much their presence has meant to me. My parents steadfastly stood by me throughout the process and well before it. We lost my mother while I was completing this project, but I like to think that she would have been tickled by seeing it come to fruition.

I dedicate this book to my wife and our children. We welcomed both of our children, Lou and Cecilia, while I was in the process of researching and writing this book. I think that they'll eventually understand all of this in an unusual way given our relationship and how they've come to understand this project, but I hope that if they sit down to read this that they'll enjoy it. My wife, Ingrid, has been my best friend, my confidante, and my biggest cheerleader. I owe her for more than she will ever realize.

LEGACIES OF DUST

INTRODUCTION

The Dust and Everything After

The dust storm that gathered momentum on April 14, 1935, the day later named Black Sunday, was only the most visible example of the devastating storms that swept across the United States during the 1930s. The storms blew away hundreds of millions of tons of topsoil that had been loosened by scorching heat and plows driven by farmers more concerned with production and immediate economic gains than with conservation. Residents of Baca and Prowers Counties in southeastern Colorado, which sat at the center of the geographical region known as the Dust Bowl, witnessed these storms so frequently during the 1930s that they became almost commonplace. Every storm left residents fighting harder to hold on to their soil, water, families, and their sanity. Images of dilapidated and abandoned farms, clouds of dust enveloping entire towns, and movies like Pare Larentz's *The Plow That Broke the Plains* captured America's attention and brought an outpouring of support from the federal government.

Knowledge of the dust storms and of the residents' suffering have become more prominent in the last decade thanks to popular renditions of the period in documentary films and an award-winning book on the subject by Timothy Egan.[1] Such emphasis has been renewed, but it is not new. Indeed, Dorothea Lange made a name for herself with a picture long thought to capture the downtrodden and devastated visage of a migrant woman named Florence Owens Thompson, whom Lange photographed

FIG. 1. A dust storm moves across Baca County on Black Sunday, April 14, 1935. Photo by J. H. Ward, Library of Congress, Prints & Photographs Division, FSA/OWI Collection (LC-USZ62-47982).

in a migrant camp in California.[2] It has become almost synonymous with the Dust Bowl and the suffering that it caused. John Steinbeck's treatment of the Dust Bowl in his canonical *The Grapes of Wrath* contributed to the growing awareness among Americans that residents on the Great Plains suffered a particular kind of misery. On the one hand, the recent increased attention to the plight of farmers and residents on the Great Plains signals an attempt to better understand their experience and sympathize with their challenges. On the other hand, however, we have, to this point, failed to really understand the Dust Bowl's impact on American agriculture and the people of the Great Plains. It was destructive, scary, and even tragic.

It was also transformational. The Dust Bowl inspired widespread change among farmers and policymakers on the fields of southeastern Colorado and in the halls of Washington DC, because it marked the moment at which most agriculturalists in the region came to terms with the challenges that accompanied farming on the arid High Plains. This book explains that shift and contends that the real legacies of the Dust Bowl only emerged well after the catastrophe had ended.

To put this another way, we cannot understand the Dust Bowl's true impact unless we adopt new ways to recognize what the crisis produced. This requires that we adjust our chronology a bit to better contextualize the Dust Bowl years. The dominant narrative in the vast majority of books written about the crisis only attend to the 1930s and therefore fail to consider anything more than the temporary and immediate response to the calamity.[3] We need to remember that the dust storms eventually stopped and farmers found themselves having to face the reality of trying to put the pieces back together. The worst years ran from about 1932 until about 1938, and those who stayed beyond that six-year period celebrated as rain returned to the countryside in 1939 and 1940. Once the weather turned, the Second World War began, and the Allies ate up everything that plains farmers offered. The war restored farmers' economic stability, covered some of the scars left by drought and depression, and left southeastern Coloradans eager for the postwar world. Over the course of little more than a decade, farmers in southeastern Colorado faced such ecological distress that few could produce enough to live on without federal relief only to return to the same fields to meet unprecedented demand. They helped win the war and then expected the good times to continue. Unfortunately for farmers, they did not. Another drought and renewed dust storms again battered the region in the 1950s, producing the Filthy Fifties and again challenging farmers and policymakers to address the environmental challenges of farming on the arid plains. When the dust storms of the 1950s hit, farmers were better prepared, the government was more willing to help, and the landscape was able to withstand the abuse. In the face of that second series of dusty, drought-filled years, the real legacies of the Dust Bowl became clear.

By recognizing the connection between the 1930s and 1950s, and by attempting to draw that connection by evaluating the Dust Bowl's long-term impact on both agricultural production and the people who fueled it, this book offers an entirely new interpretation of the Dust Bowl. It contributes to our understanding of the catastrophe itself, of environmental history and the history of the American West during the interwar period, and of the decline of family farms on the western Great Plains. It emphasizes the agro-environmental aspects of the disaster by assessing how the disaster influenced land use regimes, namely, by showing how these adaptations played out in three distinct environmental contexts: land, water, and labor. In each case, the gravity of the ecological disaster, coupled with the severity of the economic devastation caused by the Great Depression, compelled farmers and the state to combine their efforts to achieve one primary goal: to keep farmers farming. Those who lived through the Dust Bowl would never forget it, and with the help of the government, they spent the rest of their lives trying to protect themselves from the vulnerability they experienced during the 1930s.

Most obviously, the adaptation that seemed to make the most difference occurred in relation to the land itself, specifically in terms of a widespread adoption of soil conservation methods. Farmers employed conservation in part because the government incentivized it and in part because they successfully created some autonomy for themselves in terms of determining how, when, and where to practice conservation. They had just enough control and enough financial assistance to better manage a more sustainable agriculture, and that combination of federal, state, and local contributions allowed farmers to better prepare themselves against drought. The drought of the 1950s again challenged farmers, but in no way did it prove as destructive as the earlier crisis, because the support system had already been established. The foundation for a more sound agricultural policy, for government assistance programs, resettlement options, and erosion control existed when the dirt started to blow again. In that respect, the interwar period proved vital to the development of a new approach to land use that became more common in the postwar years.[4]

We can also see adaptation in another environmental context in terms of water use and conservation, as farmers incorporated lessons learned during the Dust Bowl by pushing hard for access to irrigation that might provide a buffer against drought and insecurity. It seems obvious, of course, that farmers would attempt to get more water in the face of regular drought, but the ways in which southeastern Colorado farmers attempted to increase their access further demonstrate federal, state, and local cooperation in response to the crisis. Those who had water rights and could rely on irrigation generally adopted water conservation practices to ensure that they took better care of what water they could use. In addition, they convinced the state and federal governments that more water might provide the kind of stability and security to the agricultural economy that could defend them against depression and drought. They received federal and state funding for several water projects designed to capitalize on the Arkansas River and its tributaries, but the two most important examples are the John Martin Reservoir and the Fryingpan-Arkansas Project. Farmers also capitalized on technological innovations to mine for water underground. By using aquifer water for irrigation, farmers in the 1950s who had long been used to dryland farming could rely on some irrigation. This combination of river and aquifer water meant that more farmers had more access to water, and federal acquiescence to their requests for funding further underscores the fact that farmers, policymakers, and officials recognized the region's vulnerability to drought that had become so apparent in the 1930s.

Agricultural labor represents a third environmental context through which we can see the legacies of the Dust Bowl. Indeed, while nearly every study of the Dust Bowl investigates how the catastrophe affected landowners, this book shifts attention to the plight of farmworkers.[5] The Dust Bowl fractured a system of agricultural labor that had emerged around 1900 and remained relatively static through the 1920s.[6] The dip in labor supply and demand during the 1930s meant a break in that system, which proved especially problematic for farmers who faced World War II with an acute need for workers and not many options on how to find some. Franklin Roosevelt's administration responded, providing an array of people to fit that need by importing Mexican braceros and Jamaican guest workers, by providing German prisoners of war, and by establishing work contracts for Japanese American prisoners from the incarceration camp near Granada, Colorado. In that way, farmers in southeastern Colorado capitalized on the increasing role played by the federal government in terms of supplying workers just as they had with land and water. The early postwar years ushered in a new labor regime, since the wartime conglomeration proved unsustainable. Slowly farmers moved away from bracero labor and returned to a combination of migrants from neighboring states and immigrants from Mexico. They increasingly relied on migrant workers, in large part because they could pay them less and because the state constructed an infrastructure that attempted to address migrants' education, health, and employment needs. The state effectively authorized the use of a largely Latino labor force and provided workers and their families with the kinds of services that might keep them coming back to Colorado fields for years to come.

In many ways, then, the state and federal governments allowed for, even facilitated, the shift away from small family farming in southeastern Colorado. The government provided expertise in water and soil conservation, supplied access to water, and even made pools of cheap labor available to ensure that farmers could cut costs and reinvest that money into growing their farms. Indeed, this period witnessed a remarkable trend in the dramatic increase in farm size and decrease in the number of farmers. While the Dust Bowl spurred this shift by forcing out middling and small-time farmers who could not sustain themselves during the crisis, federal policy also largely rewarded big operations to ensure they could survive the lean years. In most cases, this meant pushing for fewer farms on larger acreages—what we might call a series of "anti–Homestead Act" policies. The push for larger farms continued during the war, primarily because some federal officials believed size meant that farmers were more likely to survive lean years and more prone to practice conservation. In other words, this broad transition in American agriculture corresponded with the decline in small farms and the advent of agribusiness, which was in some ways a product of New Deal policy and the Dust Bowl's impact on the Great Plains.

This larger narrative of growing federal intervention, the adoption of soil and water conservation among locals, and the shifts in agricultural labor relies heavily on the records of the Colorado Cooperative Extension Service. These records have been virtually untapped by historians looking to better understand the relationship between farmers and state.[7] Indeed, Cooperative Extension Service employees, specifically the county agents, became vital to this process of instituting policies that reflected the legacies of the Dust Bowl and best ensured that farmers could keep farming in southeastern Colorado. They became the face of federal and state policy.[8] Although agents had been working in rural America since the Smith-Lever Act of 1914 introduced a national Extension program, they only really took root in southeastern Colorado during the early years of the New Deal. This was not coincidental. One of the agents' key tasks was to ensure that farmers abided by federal policy; once they checked that farmers followed regulations or cut production according to policy guidelines (e.g., the Agricultural Adjustment Act and its crop reduction plan), then the agent distributed federal money to practitioners. In this way, the agent worked as intermediary between the federal government and the local farm as he instituted government policy while simultaneously keeping farmers afloat. Consequently, the position became more valuable to locals once they understood that Extension held the purse strings for federal financial assistance.

Furthermore, the agents maintained this position as interlocutor when federal land use policy shifted to emphasize conservation and production reduction in 1935 and afterward, and as a result they represent what we might call agents of continuity. The growing sense that soil exhaustion contributed heavily to the Dust Bowl left federal observers and agents convinced that soil conservation should be central to their efforts to rehabilitate the land. It reflected a dual response to the Dust Bowl and the Depression, as soil conservationists believed that poor land made poor people and that rural poverty and soil degradation were inextricably linked. This premise led to more emphasis on soil and water conservation to stabilize the rural economy and make agriculture more sustainable. Again, the agents toiled on the front lines, offering instruction and expertise as well as machinery and labor, to help farmers plow on the contour, strip crop, or plant shelterbelts to reduce erosion and maintain soil fertility. The formation of soil conservation districts as a result of a state law passed in 1937 represented the key moment in the fight against erosion. The districts relied on agents to coordinate and manage conservation efforts but gave local farmers some autonomy in terms of deciding where and how to focus on conservation. In other words, rather than abide by directives given by the Soil Conservation Service or another federal agency, the agents and farmers cooperated in executing a system of conservation on private lands and did so of their own volition. Moreover, and representative of a broader theme in the New Deal years, the federal and state government provided the district with expertise, training, machinery, and funding, demonstrating an unprecedented commitment to natural resource conservation on the Great Plains. The combination of federal largesse and local control, orchestrated by the county agent, allowed for the New Deal conservation state to mature in the Colorado countryside and to remain foundational to the agricultural economy through the 1950s and beyond.

These legacies become clear when we focus on the state of Colorado and more specifically on the southeastern corner of it. The book utilizes two counties, Baca and Prowers, to show how the Dust Bowl engendered dramatic change to the agricultural economy. The two counties offer two distinct land use patterns within the Dust Bowl region and thus offer prime points of comparison and contrast. While these counties share a border and enjoy roughly the same weather patterns and soil type, Prowers

County farmers have the benefit of the Arkansas River and its tributaries. Consequently, irrigation allowed farmers in Prowers to grow cash crops such as sugar beets and alfalfa and to raise livestock, whereas farmers in Baca focused on growing dryland crops like wheat and broomcorn during much of the period. In addition, water rights, though hotly contested and by no means universally available, helped mitigate the consequences of drought and dust for Prowers County farmers. In the absence of a consistent water supply, farmers in Baca County were more responsive to the soil and water conservation lessons taught by federal agents during the New Deal. The two counties also shared many characteristics. Both experienced out-migration during the Dust Bowl and again during the war as residents followed the promise of better wages and stable employment. Most important, both received an unprecedented level of attention from the federal government with the onset of the New Deal and played an important part in supplying integral wartime commodities like sugar and wheat. In essence, then, we can see how both dryland and irrigated farmers faced the dual crises and better understand how different constituencies utilized federal largesse to survive the worst years.

Despite their differences, farmers in both counties relied on agricultural laborers to produce their most marketable products, specifically sugar beets and broomcorn. The agricultural system in southeast Colorado relied on tenants, migratory labor, and local workers, especially from the Latino community, in the decades leading to the 1930s. The Dust Bowl and the Great Depression marked a point of crisis on Colorado farms, specifically in sugar beet fields and factories, because much of the available labor pool left once individuals understood the bleak prospects for gainful employment and employers had no reason to hire workers when there was no crop in the field. Moreover, Colorado governor Ed Johnson closed the borders of Colorado to migrant labor in 1936, a demonstration of how some white Americans deemed migratory labor a threat to white employment. The situation did not much improve once the drought broke in 1939 and 1940, as most workers found better jobs during the war, which left farmers at a

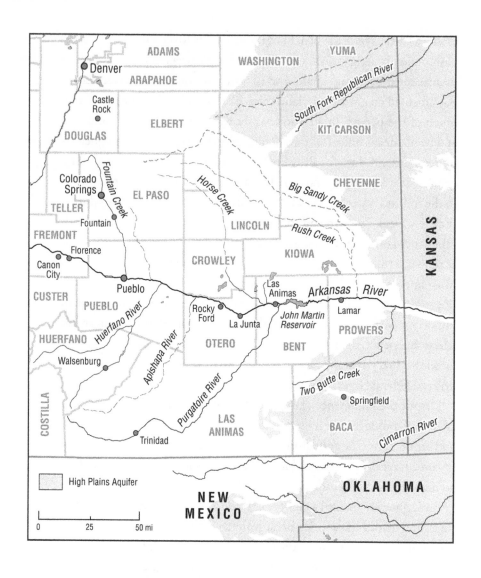

FIG. 2. The southeastern plains of Colorado. Map created by Erin Greb.

point of crisis to produce for war. While the federal and state government supplied them with workers during the war, farmers found themselves scrambling to bring in harvests once the war ended. They again tapped into migrant labor streams, this time capitalizing on Latino labor from Texas and Mexican Nationals who came in as part of the bracero program. By the late 1950s and early 1960s, the influx of migrants had become so pronounced that the state established several programs designed to meet their health, education, and employment needs. Their presence and the ways that Colorado communities responded to them are another legacy of the Dust Bowl, since it really marked a point of departure for the shift from small farms to industrialized agriculture that could only function with an available pool of cheap labor.

By addressing the issues of labor and land use from the 1920s to the 1960s, this book shows how interwar efforts at conservation were successful in mitigating overproduction and combatting soil erosion and how those efforts left a legacy that changed agriculture in the region. The extent to which farmers practiced conservation techniques, though support never became universal, speaks to the effectiveness of New Deal policy. Farmers benefited from federal efforts to provide a new agricultural labor force during World War II, and the availability of that labor force in turn influenced an increase in the scale of farming operations during and after World War II. Despite the fact that farmers felt the impact of such federal policy and set the foundation for the agricultural system that still exists throughout the nation, however, very little attention has been paid to the construction of that basis. The most important legacy of the Dust Bowl is that nothing will ever be the same; it had an indelible impact on the farmers and residents, the migrant and immigrant workers, and the landscape. The Dust Bowl's economic, political, social, and cultural effect on the United States then and since is in fact much larger than we had previously assumed. This book hopes to illuminate why those legacies, those most significant pieces of how people responded to the crisis, matter so much to America in the twenty-first century.

The book proceeds in a mostly chronological format that sometimes combines a thematic approach within a chapter to best explain how these disparate themes relate. There is some overlap between chapters in that they occasionally cover similar periods, but the focus on water or land or labor generally provides for a distinct perspective on each event. By blending the chronological and thematic approaches, the work attempts to emphasize how each environmental context helps us better understand the Dust Bowl while also appreciating that the relationship between these eras best demonstrates how the Dust Bowl impacted the agricultural economy for decades to follow.

Chapter 1 begins by examining Baca and Prowers Counties, their populations, their agricultural economies, and their landscapes in 1929. It gives background history to demonstrate how and why these communities and the environment appeared as they did. By dealing with settlement and initial land use patterns, including the impact of the homesteaders who flooded the region, this chapter presents the typical boom and bust agricultural cycle experienced throughout the plains and the West. The aversion to federal involvement, the reticence to change land use, and the mentality that nature will indeed provide demonstrated farmers' inability to take responsibility for reconciling their methods to reflect the arid environment. The chapter argues that the decades leading up to the Depression demonstrate how economic self-interest dictated how farmers approached the land and resulted in both overproduction and significant personal debt. Thus it connects the counties' trials in 1929 to their histories to provide insight into the place and people.

Chapter 2 traces changes in the region from 1929 to 1934, from the beginning of the Depression to the onset of the dust storms and the initial influx of New Deal programs. It shows how Colorado farmers developed ties to federal officials and New Deal agents, ties which were crucial to how farmers engaged conservation policy and land use reform broached by these relative outsiders. It argues that the Colorado Cooperative Extension Service, and specifically the county agents, facilitated this embrace by working as mediators between the federal and the local. The agents educated farmers on federal subsidy programs, soil conservation, marketing plans, and general home economics. In the process they gained farmers' trust and effectively became the face of the federal interventionism launched under the New Deal. This maturing combination of county agents' efforts and federal largesse gained ground during these early New Deal years and set the stage for continued relations between farmers, agents, and the federal government.

Chapter 3 covers the period from the early New Deal years through the first years of World War II and focuses on the relationship between farmers, agents, and Washington that resulted from the drought and depression. The genesis of the Soil Conservation Service (SCS) marked a transitional point for New Deal policy in the area and is the central event in this chapter. The SCS was the flagship agency persuading farmers to consider soil erosion a significant obstacle to long-term sustainability and economic security. The process of cultivating farmers' support for conservation and adjusting their methods required both education and economic incentive. While the SCS and the federal government offered the economic incentive, county agents largely supplied the education and were hands-on in their approach to reconciling overproduction and mistreatment of soil. They helped coordinate, organize, and lead the soil conservation districts that best represent farmers' embrace of the maturing conservation state. In essence, New Deal conservation policies gained traction in the area and particularly in Baca County because of the positive relationship between agents and farmers.

The fourth chapter focuses more intently on the situation in Prowers County between 1934 and the construction of John Martin Dam and Reservoir in 1948. Like those in Baca County, residents of Prowers County heeded advice about conservation, but they enjoyed irrigation from the Arkansas River and were thus sheltered from the worst of the drought. As such, they were less prone to institute conservation techniques on a grand scale as had been accomplished in Baca County. They instead pushed for the construction of the Caddoa Dam to provide more irrigation for the county—irrigators generally think of more water as the best solution to any problems with production. Like Baca County farmers, Prowers County farmers started to look more favorably at federal involvement. Rather than embrace conservation, however, they asked for federal funding and expertise to build an expansive dam and reservoir along the Arkansas River to stabilize the water levels and provide irrigators with more of the valuable resource. Prowers farmers had utilized the river as a sort of crutch and protection from drought, but the Dust Bowl proved so severe and the Roosevelt administration proved so generous that they fought for and won federal support for dam construction. We thus see an example of how federal intervention helped sustain farmers during lean years and ensured the persistence of agriculture in the region.

Chapter 5 addresses how the Depression and drought affected the labor regimes in place in southeastern Colorado. It focuses on the emergence of a labor force composed primarily of migrant workers and tenants and then illustrates how the drought and the Depression fractured that regime. Workers made their living by managing and working wheat and sugar fields in the area, but economic decline, drought, and the dearth in good jobs meant less opportunity for employment. These workers, mainly poor whites and Latinos, had a difficult time making ends meet, finding work, and staying in the area long enough for the drought to break. Like landowners, some left, and their departure changed the nature of agriculture in the area because the biggest farms, most of which were tied to cash crops like wheat and sugar, no longer had an abundant and available labor pool from which to draw. This chapter also inserts the story of tenants and workers into the broader Dust Bowl narrative. Analyses of the Dust Bowl neglect their presence in the fields, instead focusing on landowners who stayed or white migrants who left the area.

Chapter 6 deals with the influx of contract labor in 1942 and 1943 through the end of the war to show how replacement labor influenced area farmers and agricultural practices. It focuses extensively on the

Extension Service and its Emergency Farm Labor Program, particularly its success in bringing more than 250,000 workers to the state during the war. The increase in labor was a blessing for many farmers in the two counties. Indeed, most workers were warmly received. Farmers had relied on migratory labor and tenants previously, so this transition to braceros, inmates, prisoners of war, and others was not much out of the ordinary, especially in Prowers County with the emphasis on sugar beets and other cash crops. The use of replacement labor was a new chapter, however, as they were paid less, were temporary, and were less tied to the community. The reliance on this kind of worker, provided by federal intervention and federal programs, helped push many large farms onto the path to agribusiness and the employment of cheap workers on a grand scale. Federal and Extension Service intervention during the war, though focused on production rather than conservation, evidenced the agents' continuing role in working with both farmers and federal officials to meet farmers' needs. Only through such involvement could farmers meet demand. The outside workers effectively worked to supply the Allied war effort and helped the farm economy rebound in the process. The agents' continued role affirms their dynamic role in these communities and their importance in sustaining farmers during the war.

Chapter 7 explains in broad terms how farmers in Baca and Prowers Counties responded to the end of the war and then focuses more specifically on how they sought access to more water to protect themselves from the ravages of the Filthy Fifties. That drought was in some ways worse than its predecessor in the 1930s, but most farmers in the region actually fared pretty well. It attends to the rise of using wells to tap aquifer water, the push to conserve precious water among water users, and the public movement to capitalize on federal largesse in order to construct the Fryingpan-Arkansas (Fry-Ark) Project. This entailed the diversion of waters from rivers flowing west down the Continental Divide to the Front Range to meet the increasing demands of urban centers and farmers clamoring for more water. Much like they had with the John Martin Dam, locals spent years building enough state and federal support for the idea until Colorado politicians pushed it through Congress and earned funding. John F. Kennedy's 1962 speech in Pueblo to celebrate the project encapsulates what it meant to many Coloradans—the "right" kind of federal intervention that would allow residents, and especially farmers, the kind of security that they knew they needed to combat aridity. They had been fighting for more water since the 1930s, and the 1950s drought both rekindled their demand and convinced doubters that diverting water would bolster the agricultural economy. In this case, the Dust Bowl produced an active federal government, a thirsty constituency, and a partnership between the two that eventually produced a remarkable project.

The final chapter wrestles with the legacies of New Deal conservation programs in terms of land use and dryland farming. These farmers were generally better prepared for this round of storms because most had become big enough and diverse enough to be financially stable. The large reduction in submarginal land under intensive cultivation meant that less acreage was prone to blowing, resulting in fewer and less intense dust storms. In effect, these newly formed large farms, what some would call examples of agribusiness, were better able to sustain themselves than homestead-style 160-acre farms that had become exposed during the 1930s and almost ceased to exist by the 1950s. This development marked a victory for the New Deal because policymakers had hoped to maintain agricultural production in the area but pushed to expand individual holdings and remove small-scale farmers because they believed that small operations on largely submarginal land were a financial and ecological liability. The chapter also ties back together the issues of land use and labor by looking at how farmers reorganized their workforce in the wake of the war. Farmers constructed a new regime in part by rekindling ties with migrant groups but also by utilizing the bracero program, by recruiting immigrants, and by tapping into resident communities of once-migrant workers. Obviously, farmers hoped to maximize their profit base, and that meant limiting labor costs. As a result, farmers proved less likely to take good care of their workers, contractors provided very little by way of fringe benefits, and both companies and farmers reached out to the most inexpensive workers. These developments induced broad changes in the region, changes that become clearer when we adopt a wider lens to view the ways that drought and depression impacted humans and the environment.

The book concludes with a broader argument about how this story fits with our understanding of the Dust Bowl and of modern agriculture in the West more broadly. As an environmental history, it suggests that we have yet to find a balance between production and sustainability, in large part because so much about farming in the Colorado plains remains out of our control. As droughts continue to ravage the area, and as whispers of another Dust Bowl begin to surface, it becomes clear that folks have a long memory when it comes to the worst ecological disaster in American history.

I

Early Lessons from the Land of Opportunity

Writing in 1929, Baca County, Colorado, resident J. L. Farrand expressed concern over the state of agriculture in southeastern Colorado, which he believed to be dangerously untenable. Farrand cited a dire need in Baca County to "acquaint farmers with all phases of cooperative marketing" so that they could work toward stabilizing prices and become more adept at "analyzing the farm business." Farmers looked for safeguards against continued economic strain, and in most cases this meant that they prioritized production as the best and only path to economic prosperity. Farrand believed that such focus led farmers to sacrifice their land's long-term fertility and therefore its potential for sustained production. He thought they gave short shrift to soil fertility and conservation. For example, he noted that they dismissed the importance of crop rotation in trying to reinvigorate soil's productive capacity by temporarily retiring parts of their acreage. He took it upon himself to offer several demonstrations of suitable soil conservation techniques "to show the efficacy of proper cultural and tillage methods in the prevention of soil blowing and the conservation of soil moisture." Farrand understood that farmers remained most concerned for their economic well-being, and he seemed sympathetic in most respects. But he also knew that productive farming meant protecting soil fertility. Thriving in agriculture meant more than planting, killing weeds, and waiting for rain. To ensure a successful farm, family security, and economic prosperity, farmers needed to value the soil. Convincing agriculturalists of that need often proved a Sisyphean task, however. Farrand clearly appreciated the peril that would result from wanton degradation of the soil, but the intensive agricultural practices that accompanied the "wheat boom" and that made the land vulnerable had become commonplace in the region by the 1920s, and their hold over farmers made Farrand's task incredibly difficult. To promote a shift in land use, Farrand and others who shared his dismay over exploitive farming had to help farmers admit that their survival required them to reconsider their relationship with the land in southeastern Colorado. At its core, this meant recognizing the importance of protecting the soil before farmers could achieve a prosperous and sustainable agricultural economy.[1]

This might not have meant much then or now had Farrand just been jotting this down in a journal or his memoirs. But Farrand was not a passive observer. He served as the Baca County Extension Service county agent in 1929, and his position allowed him to develop an intimate perspective on the process and prospects of farming while requiring him to share his views with county farmers. His Extension training prepared him for his post by instructing him on how to assess land use, to work with local communities, and to extend his knowledge and resources to improve the agricultural economy. He visited local farms, published bulletins, responded to letters and inquiries, and conducted informational demonstration meetings on issues ranging from food preservation to improving the family diet to proper crop rotation. These factors afforded Farrand a unique point of view regarding land use in southeastern Colorado; his pronounced anxiety over the agricultural economy's stability and his distress over exploitative land use proved remarkably prescient. When he extolled the need to adapt land use, he did so with a firm understanding of the region's history and an idea of its rather grim future.

Farrand was certainly not alone in identifying the need to reconsider the fundamental approach of American farmers throughout the country and in the West more specifically. By 1929, concern over the farm economy had already gained steam across the Great Plains as farmers tried to recover from the downward spiral beginning after the Great War, continuing through the 1920s, and becoming even more serious in the 1930s.[2] He found that this economic situation only made the destruction of the landscape more problematic, which gave him a sense of urgency to facilitate a change in land use practices in the region. Little evidence of this shift existed in southeastern Colorado during Farrand's brief tenure, most likely because farmers had not yet faced a crisis severe enough to dispel the myth of abundance that boosters attached to the region in the late nineteenth century.

The first step on the path to identifying the need to adapt their land use patterns, a step that in large part originated because of and reflected the sense of urgency caused by the Dust Bowl, required farmers to combat the generations-old belief that the area, which boosters had labeled the "Valley of Content," was destined to inevitably become an agricultural juggernaut.[3] Not surprisingly, periods of heavy migration into the area coincided with years of unusually wet weather that encouraged farmers to dismiss the area's very real climatic challenges to agricultural production. This willful dismissal, based in their stubborn refusal to acknowledge the historic frequency of droughts across the region and the fact that many migrants moved west during the wet years of the 1880s, left most southeastern Colorado farmers without a healthy and necessary appreciation for the region's environmental constraints. The disinclination to admit such constraints existed became increasingly entrenched during good rain years and when market demand encouraged farmers to maximize production to capitalize on high prices. These two factors emerged simultaneously during the 1910s and resulted in the "wheat boom" that accompanied World War I and that invited farmers to push production at all costs. By 1920 populations in Baca and Prowers Counties had peaked, evidence of the enthusiasm spurred by the wheat boom and more general faith in the future of agriculture in the region. Within the decade, however, much of that fervor had abated. The temporary run on wheat stalled in the early 1920s, the wet years appeared only intermittently, and the brief optimism faded by the late 1920s. Despite some out-migration, most agriculturalists remained committed to farming in that area and hoped that another break might come their way. Even as the Great Depression set in across the country, residents of southeastern Colorado occupied an unusual sort of stasis as they continued to produce sugar beets, wheat, and alfalfa and tried to sustain themselves. It was against this backdrop that Farrand attempted to convince farmers to buy into his idea of conservation and to change their ways. Their stubborn refusal to embrace his advice reflected what had by then become a common obstinacy among southeastern Coloradans who remained devoted to the idea that the next good year stood just around the corner.

The causes of the Great Depression are many and varied; the economic collapse that ushered in more than a decade of depression affected all Americans. While it is easy to assume that the Great Crash of 1929 signaled the beginning of the decline, in truth the crisis proved more complex and more gradual. The extensive depression in the agricultural economy following World War I started the decline for most Colorado farmers, so they had been facing tough times for several years before 1929 and paid little attention to the Crash. As longtime Baca County resident Ike Osteen noted, "Few people in Baca County even knew what the stock market was, and I don't know of a single person in Baca County who killed themselves because of what happened on Wall Street."[4] Indeed, of the two major newspapers in the area, only the *Lamar Daily News* recognized the event, giving it front-page coverage with the title "Billions Lost on Stock Exchange."[5] Even then, the paper contained no evidence of vocal support for or rejection of President Herbert Hoover's reaction to the Crash, and instead both the *Lamar Daily News* and the *Springfield Democrat Herald* emphasized the local and the familiar. In a parochial way, the newspaper coverage of the Depression's first years suggested that national issues mattered little to area residents, most of whom seemed more concerned with their own economic trials.

Despite this localized sense of identity and regardless of whether residents ever realized it, southeastern Coloradans were very much a part of the larger narrative of American history since the early nineteenth century. In the case of geography, for example, the western Great Plains became a very important part of national politics and the development of the West beginning with the 1803 Louisiana Purchase. American explorers began traversing the Great Plains with some frequency once the United States purchased much of the region from France at an incredible bargain. Travel through the area meant early efforts at mapping the region, cataloging the flora and fauna, making contact with the Indians, and probing the boundaries that the United States shared with Spain in the Southwest— all of which represented American attempts to determine exactly what it had purchased and whether it had value. Many of these early explorers, beginning with Zebulon Pike in 1807 and continuing with Stephen H. Long in the 1820s, traveled through present-day Colorado while performing reconnaissance throughout the Great Plains and toward the Rocky Mountains. They conducted these explorations with the notion of settlement on their minds, as Americans started to think seriously about expanding the country's domain across the continent, an inkling that eventually became associated with Manifest Destiny. In other words, southeastern Colorado became part of this larger objective to establish American hegemony.

These borderlands, like many others of this period, proved prone to cultural exchange as well as conflict, and records from each of the major explorations offer insight into the complex social, economic, and cultural systems in place. Pike, Long, and John C. Frémont noted these elements, but each considered the place more than the people who lived there, since each thought most about the possibility of potential settlement. By prioritizing development of the Colorado frontier, they focused especially on the region's characteristics, including the flora and fauna, elevation, and climate and emphasizing the human populations in relation to their potential to stand in the way of Americans who wanted to settle the area. Although their reasons varied, none of these early voyagers expressed optimism for American settlement in the western Great Plains. The "great American desert" label that became prominent during the 1820s because of its climate and limitations for agriculture did much to dissuade migrants who had considered a trek into the region, and the nation turned its attention elsewhere.[6]

Colorado became more central to the development of the West in 1858 when William Green Russell's discovery of gold in the mouth of Dry Creek, just outside of present-day Denver, ushered in the Colorado gold rush. The region grabbed headlines because of its seemingly untapped potential, as the possibility to strike it rich invited intense migration. Gold has "unsurpassed power to set people in motion," as nearly 100,000 people crossed the Great Plains during the first year of the rush.[7] Roughly half were discouraged by the surroundings, lack of success, or competition and decided to turn back. The others, devoted to striking it rich, stayed around the foothills of the Rockies, moving from place to place in search of their fortunes. Regardless of the low rate of success, the rush transformed the Great Plains as well as the Rocky Mountains, bringing an incredible number of people to a region that had never felt such population pressure and in the process changing the Great Plains from "a place to get across" to "an essential part of one national vision" that relied on white expansion to the Rockies and beyond.[8]

To that end, town builders like William H. Larimer who understood the temporary nature of the gold rush appreciated the profit potential in catering to miners as well as opening the area for white settlement. With this goal in mind, Larimer led the charge to establish Denver City in 1859, named after Kansas governor James W. Denver, just as Russell returned to his home in Georgia to recruit more help for his planned return the following year.[9] Enough people converged on the Colorado Front Range in the first two years after the discovery that Washington DC designated the Colorado Territory in 1861. William Gilpin acted as the first territorial governor; much of his job revolved around continuing the steady flow of migrants and immigrants by enticing folks willing to rough it on the frontier in hopes of economic success in gold, farming, or ranching.

Only by citing the possibility of the latter two trades could Gilpin hope to revitalize the territory's economy, which had been sagging after the streams of gold dried up. By emphasizing the supposedly limitless possibilities of productive and prosperous long-term settlement alongside the celebration of striking gold, Gilpin and other promoters set in motion impressive migration, developed the railroad, and embarked on the path to statehood. A major component of realizing that growth, instead of just hinting at its possibility, required a similarly impressive influx of agriculturalists who, unlike most of their gold-seeking contemporaries, traveled west with ideas of permanent settlement.[10]

While the two groups differed in many ways, hopeful agricultural-ists who trekked across the plains resembled the gold seekers who had already followed the same trails west. Perhaps most important, members of both groups traveled to Colorado Territory with the same mind-set, one that reflected their firm belief that the West constituted a place for new opportunities, one able to provide a chance at wealth or at least some level of economic success. Miners sought a valuable mineral they could not find in the East, and farmers migrated with the idea that open land and productive soil represented the promise of prosperity. Natural resources were the key—mineral or soil—and each group hoped to extract as much value as it could from the source. Unfortunately, both groups tended to exhaust their respective resources without much consideration for future production. For many eastern and Midwestern farmers, this meant that, in spite of some early calls for soil conservation and responsible stewardship, the predilection for mining the soil in hopes of economic payoff led to rampant erosion and degradation of soils. Many southern and some north-ern farmers had long made a habit of exhausting soil when maximizing production before moving on to another site. The pattern repeated until they ran out of available lands—at which point they turned further west and rambled onto the plains. In this respect, the farmers faced no signif-icant repercussions; the federal government proved willing to abet them in that process and therefore offered its tacit approval.[11]

Boosterism and railroad expansion worked in tandem to make Colorado look more appealing to potential migrants who came to the area in significant numbers by the end of the nineteenth century, and their success afforded Colorado a national, even an international, presence. Demographic evidence shows that new arrivals to Baca and Prowers Counties, which came into existence in 1889 after enough people had moved to the region to warrant their creation, hailed from a considerable variety of places.[12] Many farmers had either immigrated to the United States or were first-generation Americans. Germans and English predominated among that population in what would become Baca County, while German, Russian, Irish, Dane, Swiss, and even Canadians registered in the 1910 census for Prowers County. Mexicans also represented a significant segment.[13] Heavy migration streams entered the state from Missouri, Texas, Kansas, and states in the Midwest and East.[14] The census of 1890 reflects the initial growth in both places, with Baca County home to 1,479 and Prowers to 1,969 residents.[15]

Despite the numbers, however, the prospect of increasing migration to the Colorado plains proved difficult at times and especially during the relatively dry years of the 1870s. Increasingly, promoters developed optimistic rhetoric designed to combat the notion that aridity precluded successful agriculture and to assuage migrants' concerns about their futures. Boosters sent pamphlets to places like Germany and Great Britain, held exhibits at the World's Columbian Exposition, and plastered newspapers in the Midwest and East in hopes of luring potential residents. According to promoters, a bit of elbow grease coupled with faith in Mother Nature would make the Great Plains America's breadbasket. Boosters in the Arkansas valley reassured migrants that river water was both plentiful and available to all comers. Companies like the Arkansas Valley Sugar Beet and Irrigated Land Company and the Colorado Arkansas Valley, Inc. claimed that irrigation solved the problem of aridity and that new arrivals could tap into water provided by the river or one of the newly created canals that crisscrossed the landscape.[16]

Boosters found it fairly easy to convince potential settlers that settling along the river would make their lives easier, but enticing farmers to settle far from available water proved quite challenging. They addressed this problem by changing the narrative. To their credit, as much as it often ended up costing migrants who believed it, promoters across the Great Plains promised that the dry climate would change, and it would change quickly. Many boosters proffered the notion that "rain follows the plow": the lack of moisture, while daunting for initial pioneers, would be remedied by farmers farming. In essence, "the plow, symbol of the American farmer, was to give life to the plains, not just by breaking them, but by producing conditions which would lead to increased rainfall."[17] Few questioned the assertion that sustained agriculture had the power to dramatically shift a region's climate, in part because scientists supported the claim Relying on the scientific climatology theories from men such as Ferdinand V. Hayden and Cyrus Thomas, boosters tried to convince potential migrants that humans, and specifically those involved in agriculture, could effectively become agents of climate change. A similar theory purported that plowing and sustained agriculture helped expand the "rain belt," a theoretical expanse of land that enjoyed increasing precipitation levels. The Syndicate Land and Irrigation Company celebrated the inclusion of what would become Baca County into this growing expanse, thereby identifying it as a legitimate destination for farmers. While most future residents lived too far from the Arkansas or its tributaries to rely on irrigation, the company contended that the growing rain belt should be taken as proof that "successful agriculture without irrigation is an established fact" because the rain would eventually make up for not having irrigation.[18] In effect, they found new ways to justify moving, and their marketing worked.

While most migrants who flooded the region anticipated success, many quickly deduced that the theories about declining aridity had been nothing more than theoretical. Even as the federal government tried to push settlement, presumptive residents found themselves facing a series of challenges that often seemed insurmountable. The Homestead Act of 1862 offered a plot of 160 acres to any head of household, at least twenty-one years of age, who resided on the acreage and prepared it for cultivation—labeled "improvement" of the plot because the potential landowner took previously undeveloped land and made it profitable. The homesteader then gained title to the land after five years of continuous residence and the payment of a small fee. The policy reflected the long-held notion of yeoman farmers as the embodiment of republican virtue, a testament to the idea that hard work and integrity put one on the path to success. It certainly worked to some degree in southeastern Colorado, as roughly 2,000 claimants capitalized on the policy and moved to the region between 1862 and 1900. Yet, while it demonstrated continued efforts by federal politicians to facilitate white expansion across the nation, the policy proved largely inadequate for homesteaders in arid Colorado. An increase in relinquishment or cancellation of homestead claims "brought realization of the fact that one quarter section of the non-irrigable lands of the high plains was, in many instances, not adequate for the subsistence of a family."[19]

Consider the case of Earle Gillis, who lived in Baca County after his father was "lured by the promise of a free homestead." In 1887 the Gillis family relocated to "a land where the sun bears down relentlessly in summer and blizzards of arctic intensity sometimes rage in winter; where hot winds blow in the daytime, and night winds chill." Their new home, a one-room shack built from sod and covered with dirt–"typical of the homesteader's house in that day"–housed them for "two long years filled with loneliness and disappointment, with hardship and poverty." The lack of water constituted the biggest problem for Gillis and his family, who gathered rainwater or took some from nearby arroyos, then boiled it to clean out the bugs and mosquito larvae, but had little respite for their crops. Their first crop "was practically a failure" and only produced enough to feed the oxen; Gillis's mother raised chickens, and his father hunted jackrabbits for fresh meat.[20]

The second year proved just as disappointing, and even with two growing seasons the family had nothing to sell in town, leaving them with just enough for sustenance. Consequently, the Gillis family looked to alternatives to make money and finally found a way to scratch together a few dollars by selling cattle and buffalo bones to buyers in nearby Lamar who then ground the bones for fertilizer. Gillis remembered the most profitable trip that included the sale of 1,300 pounds for $3.40—quite a haul considering it cost the family nothing to collect the bones and it was their only income. In the end, however, it was not enough to keep them in Colorado. They faced consistently harsh weather, isolation, and uncomfortable living conditions, and they failed to last the five years necessary to earn ownership via the Homestead Act. The Gillis family moved back to Oklahoma after the third year of unsuccessful crops and with no promise of future gains in Baca.[21]

Gillis's memories offer a telling portrayal of the difficulties that many homesteaders faced in trying to establish themselves in a foreboding environment without enough land or capital to make it a profitable venture. The size of one's farm had a hand to play in that as well; Glen Durrell, whose family moved to the Colorado plains in 1908 and stayed until 1916, remembered residents talking about summer fallowing to allow moisture to accumulate, but it was never practiced because "land available to any one settler was too limited" to do anything except plant and hope for a healthy harvest.[22] What Durrell seemed to understand, and what Congress eventually embraced as well, is that farming in the arid western Great Plains was made easier with more than the 160 acres of land that came with a homestead. To its credit, Congress addressed such limitations by offering additional legislation designed to tender more land to homesteaders—primarily because the federal government had more than enough land in the West to give away and it wanted successful settlement. Such policies were meant to amend the original Homestead Act and included the Dry Farming Homestead Act of 1909, which enlarged available acreage from 160 to 320 acres on lands that were not irrigable, and the Stock Raising Homestead Act of 1916, which allowed previous homestead entries to increase to 640 acres in hopes of allowing for grazing and raising forage crops to feed livestock. Additionally, the Timber Culture Act of 1873 sought to exchange a 160-acre tract for the applicant's planting of trees to provide tree cover and eventually fuel to inhabitants. The applicant could claim the land after three years if he/she cultivated 40 acres of trees per 160-acre plot.[23] The Desert Land Act of 1877 represented another example of Congress trying to come to terms with aridity in the plains, with a specific focus on invoking personal responsibility for irrigation. It offered up to 640 acres to any applicant willing to find ways to irrigate land that was considered "desert land," defined as "all lands exclusive of timber lands and mineral lands which will not, without irrigation, produce some agricultural crop."[24] The Newlands Reclamation Act, passed in 1902, promoted irrigation throughout the West by devoting a portion of the money made by selling public land to the research and development of irrigation projects on the state level.[25]

While these various acts and policies certainly helped settlers, the fact remained that the prospects of farming this region successfully were never guaranteed. It is somewhat remarkable that agriculturalists in the area had significant stretches of good years. The best years varied according to the crop, but the southeastern counties played an important role in helping establish Colorado as an agricultural state. Indeed, Colorado led the nation in sugar beet production through much of this period and had contributed to the world wheat supply during the 1910s and 1920s.[26] Because of federal policies and settlers' willingness to stick it out in the region, the southeastern Colorado agricultural economy eventually blossomed once farmers determined ways to produce and become more stable if not prosperous. Yet middling and poor years seem to be more common than not, leaving residents and farmers fighting to find ways to solve the environmental limits that made the prospect of successful farming so inconsistent. Farmers in both Baca and Prowers faced inclement weather, dust storms, and locusts, and they struggled with finding workers to fill their need for labor. These challenges remained throughout the period from American settlement through the 1930s and afterward, but not all farmers struggled in the same ways. Where farming occurred mattered immensely, and this became especially true when it came down to having access to water.

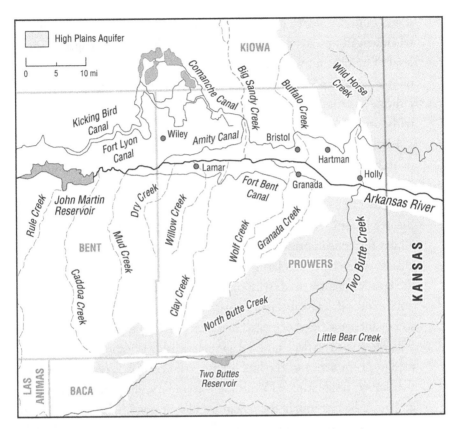

FIG. 3. Most settlement in Prowers County has occurred near the Arkansas River, largely due to the county's reliance on water from it and canals. Map created by Erin Greb.

Water came to represent a panacea to both the government and to settlers, and though it never offered guaranteed prosperity that many irrigation supporters advertised, it still allowed irrigators in Prowers County to capitalize on opportunities to grow cash crops. The proximity to the Arkansas River and its main tributaries marked a crucial difference between Baca and Prowers Counties—the presence of irrigation affected migration rates, settlement patterns, property values, and production regimes. The prospect of farming in Prowers County was much more promising than it was in Baca County, particularly by the end of the nineteenth century when irrigation started to play a starring role. It provided irrigators some level of stability notwithstanding fluctuations in precipitation levels and offered a sense of security to farmers who were coming to terms with farming in a strange environment. It also allowed farmers to try new crops with some reasonable hope of success.

Alfalfa became an important crop for early settlers because it was a highly desirable commodity in a region where so much of the transportation and work had to be done by animals. Local demand meant farmers found quick and easy sales, which alleviated any stress that might have accompanied the need to ship the produce to distant markets. Many farmers also grew alfalfa for consumption by their own animals, which afforded them a chance to diversify their farms and potentially hold enough cattle to meet the family's dietary needs with milk and beef. The big issue was access to water, since alfalfa requires water to grow, and it has a long growing season, so a maintenance level of irrigation is necessary. In Prowers County, where folks often lived on flat land and had some reliable irrigation, alfalfa quickly rose in prominence in the late nineteenth century. It certainly helped that Lamar resident Floyd Wilson pioneered the alfalfa milling industry and built the first mill in the Arkansas valley in 1908, a development that earned him "a place in the economic history of Colorado and the West."[27] After taking off before World War I, the industry continued to hum at a good pace through the 1930s. Indeed, nearly half of all farms in Prowers grew alfalfa in 1934 and 1939, evidence of its place in the crop hierarchy in the irrigated regions.[28]

In 1871 George Swink arrived in southeastern Colorado and began experimenting with crops. Swink hailed from Illinois and made the trip from there to Kit Carson, Colorado, the Kansas Pacific Railway's western terminus. He then walked the last 100 miles to Rocky Ford. Swink opened a store, helped jump-start the region's first irrigation company, and eventually conducted a series of tests to determine what could grow well in the region. He decided to grow squash, watermelon, cantaloupe, alfalfa, and sugar beets. He believed that irrigation allowed him to grow nearly anything he wanted and provided the key to developing the region from "a few rough dwellings" to a valley of settlement and abundant crops achieved by "advanced agricultural methods."[29]

The discovery that sugar beets could thrive in the valley transformed the region by inaugurating the "sugar boom" between 1897 and 1907 and placing Colorado at the top of the list for domestic sugar beet production by 1907. The American Beet Sugar Company and the Holly Sugar company, with factories in Lamar and Holly, respectively, served as prominent examples of the potential for cultivating beets on irrigated lands. Their existence invited local beet growers to initiate contracts with the companies and turn most of their attention toward the cash crop. The sugar boom produced thriving towns and successful farmers throughout Prowers County, giving credence to the rhetoric of boosters and indicating that while "the entire valley was not a garden in 1907 . . . there were a great many fragile oases blossoming along the Arkansas River."[30]

The sugar beet industry flourished because several outside factors converged to support it and the federal government became the industry's biggest booster. Indeed, the industry garnered so much government attention that it earned the nickname "Uncle Sam's Child." The U.S. Department of Agriculture worked with state experiment stations to develop test plots and coordinate research agendas around beet cultivation under irrigation. Additionally, eastern capitalists invested in irrigation projects and railroad expansion, which the federal government helped fund through reclamation projects and land grants. Washington also imposed a series of protective tariffs including the McKinley Tariff and Dingley Tariff to incentivize beet growers and refiners during the 1890s by promising consistently competitive prices for their product. Liberal immigration laws also helped growers find laborers willing to do the backbreaking stoop labor common to the industry. Boosters fueled enthusiasm for the industry and helped entice speculators and investors to consider factory construction along the Arkansas River. Not surprisingly, factories sprouted up in Prowers in both

FIG. 4. The Holly Sugar factory was a six-story brick building in Holly, Prowers County. Photo taken between 1900 and 1910. Denver Public Library, Western History Collection, X-8961.

Holly and Lamar because each growing town sat on the east-west railroad running from Pueblo into Kansas and therefore served as ideal depots for farmers to sell their beets and for refiners to ship the products to market. Irrigation, transportation, investment, and federal assistance came together in the 1890s and made the boom possible throughout the valley.[31]

No state equaled Colorado in beet production, and early factories throughout the Arkansas valley like Rocky Ford and Sugar City pioneered the use of beets on irrigated cropland. The companies developed a practice of contracting with local growers and making sure that they had enough contract farmers to satisfy their needs. The grower promised to plant a certain acreage in beets and to sell his beets directly to that company. The companies then refined the beets and sold the sugar, paying the grower the price that they had decided on. Holly and Lamar both housed factories by the early twentieth century and offered good prices to their growers, creating instant and steady demand.[32] Sugar beets were both water- and labor-intensive, so relatively few Prowers County farmers actually grew them even though companies made growing beets a lucrative business decision. Like alfalfa, the proximity to consumers, in this case the refiners and factories in Lamar, Holly, and elsewhere, meant that farmers had the opportunity to bypass railroad fees to sell their products.[33] The industry experienced a similar postwar boom and bust cycle like the wheat industry, as each cash crop faced gradual decline over the course of the 1920s until the Depression hit. The tried-and-true method of raising tariffs to stabilize prices unfortunately did not work to recover the prices, but the industry started to rebound with federal assistance during the 1930s.[34]

Similar developments fueled the cantaloupe industry's growth into becoming the area's other main cash crop. Though it never found the same footing that beets did in the Arkansas valley, the crop burst onto the scene in the 1890s. Agricultural Experiment Station employee Philo K. Blinn noted the importance of two innovations in expanding the cantaloupe industry around Rocky Ford. First, he claimed that the invention of standard crates meant that every melon grown in the region could be easily placed on the train and sent to market in Denver, St. Louis, and Kansas City. Barrels and thrown-together bushels had previously housed marketable melons, making them tough to ship, prone to bruising, and difficult to tally the total cantaloupes in each shipment. This attempt at homogeneity also bolstered sales by confirming the amount to seller, shipper, and buyer. Second, local cooperative efforts among Colorado farmers gave them leverage to negotiate with the railroad to ensure that it devoted regular cars for cantaloupe sellers—in 1896 the growers supplied 150 cars a day by the end of harvest in August. The use of ice cars on the Santa Fe and other lines helped protect the melons as they traveled to Chicago, Boston, and New York. Unfortunately, farmers could not sustain their cooperation, and fragmentation among growers and growers' associations led to competition. The push for production routinely resulted in a saturated market, leading to the familiar imbalance between supply and demand that shot down price per melon and hurt growers. Moreover, the development of the beet industry enticed many cantaloupe growers to switch cash crops. Thus while cantaloupes represented an important part of the Prowers agricultural economy, their prominence diminished over the 1910s and 1920s.[35]

Even though irrigators faced obstacles, including drought, insects, and mercurial markets, they enjoyed relative stability because their access to water mitigated many of the worst consequences of farming in an arid environment, and their stability helped nearby towns and village prosper. Consequently, the population in Prowers steadily increased from its inception through the 1920s, evincing the stability afforded by irrigation and the impact of successful agriculture on local towns and other industries. Farmers in Baca County never enjoyed this security and were forced to contend with the natural limits of trying to maintain a successful farm in southeastern Colorado without irrigation water. They still found a desirable crop in the development of their winter wheat economy, and through the course of World War I farmers in Baca helped to fuel the war effort.

Baca County had little farming until the 1880s because raising stock dominated the county economy and ranchers owned most of the land. For example, John W. Prowers, one of the most successful ranchers in the area, owned more than 800,000 acres of range for his herd of roughly 10,000 cattle, much of it inside the borders of Baca County.[36] During the 1880s, however, declining meat prices, a presidential decree to remove unauthorized fences from the public domain (a decision that forced cattlemen to find alternative feed options for their cattle), volatile weather, and the push to homestead federal lands combined to take some of the best grazing land from public use and cut into the ranching economy. New homesteaders often strove for a balance between crop and livestock, but these diversified farms never came close to the expansive operations employed by Prowers. Consequently, what had been mostly open prairie

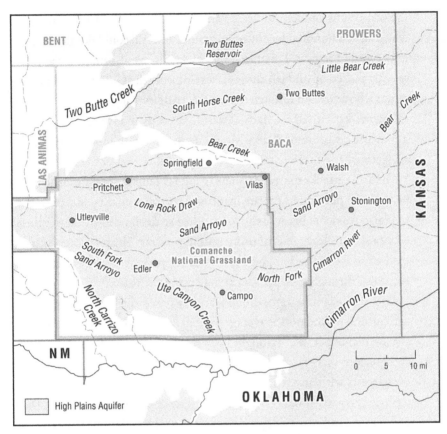

FIG. 5. The Comanche National Grassland and the aquifer represent two legacies of the Dust Bowl, namely, the retirement of submarginal land and the increasing reliance on irrigation that began in Baca County after World War II. Map created by Erin Greb.

slowly became home to small, scattered plots and burgeoning towns, a decided shift in the landscape.[37]

Baca County agriculturalists faced several problems once operations started to intensify in the late 1880s. The vicious regularity of drought and tough winters made any type of farming in Baca County incredibly difficult. Indeed, even some town builders found it impossible to deal with the conditions and left the area. Earle Gillis remembered a few migrants who dug two hundred feet into the ground and still could not tap into the underground reserve. Frustrated and feeling defeated, nineteen claimants quickly picked up and left the area, leaving only the Gillis family and two others who remained through the winter.[38] Additionally, outbreaks of crop-eating pests occurred regularly, devastating crops and frustrating farmers. The worst outbreak of Rocky Mountain locusts in the 1870s "inflicted a staggering $200 million in damage on agriculture west of the Mississippi." Such locust attacks spurred farmers to ask for federal assistance and compelled both public and private institutions to offer charity and assistance to plains residents because the locusts had so decimated the agricultural economy as to leave tens of thousands in poverty. The locusts eventually died off in the 1890s, but plagues of grasshoppers ravaged the plains from the 1880s through the 1940s, peaking during the 1930s.[39]

By a significant margin, though, aridity posed the biggest problem to agricultural production as average rainfall in Baca County hovers around sixteen inches per year. Drought broke out in 1865–72, 1892–95, 1901–4, 1907–8, and 1911–12. [40] The inconsistent weather made it tough to plan on harvests with any reliability. It improved after the turn of the twentieth century when farmers started to appreciate the need to adapt their techniques, and many adopted dryland farming techniques designed to sustain agricultural production in arid conditions. Promoting such methods and convincing farmers to employ them was largely a group effort on the part of experts and practicing farmers. For example, the first meeting of the International Dry Farming Congress met in Denver in 1907 to devise ways to stabilize production in arid environments. In addition, Agricultural Experiment Stations sprouted up across the state, serving as outposts for recent graduates of the State Agricultural College trained in modern agricultural methods designed to ensure success for every farmer, even those living on the Colorado plains.[41] In a stark statement of a harsh reality, the opening line from a Colorado Experiment Station bulletin on dry farming in eastern Colorado from 1910 announced rather ominously that such farming "is a continual fight against relentless, unfavorable conditions" and that even "with the best seeds and methods of tillage there will be some years of total failure and many others of short crops." It suggested that farmers diversify and not focus solely on grain; while wheat should proliferate in the arid conditions, the farmer should think about forage crops like milo maize and kaffir corn to feed his dairy cows, hogs, and poultry. The danger of not diversifying meant that the farmer could not attain self-sufficiency. In addition, the bulletin reminded farmers to plant household gardens of drought-resistant crops like cherries and sweet corn, as well as starches like potatoes and fruits such as melons, to ensure that the family would not starve during even the worst years.[42]

The real boon for dryland farmers, however, came with the adoption of wheat, specifically hardy winter wheat and the turkey red variety that grew well in arid weather. Wheat quickly became the most important cash crop for Baca County farmers and signified that dryland farmers had a chance to prosper despite aridity. Unfortunately, farmers' intense focus on wheat and consequent impact on the plains is a familiar story, concluding in the Great Plow-Up following World War I. As wheat came to fetch considerable market attention during the war and those with enough credit or hope descended on the plains to capitalize on the price, farmers discounted the push for diversified farming akin to what the 1910 bulletin promoted. Indeed, the motivation to acclimate to the harsh environment dissipated when rain fell and farmers could capitalize.

This panned out for some Baca County farmers—at least initially. World War I demand and consecutive years of good rainfall convinced farmers that breaking land for wheat was a smart investment, and with wheat prices over two dollars per bushel in 1917, those not involved in wheat cultivation quickly turned their attention to that crop. Indeed, this represented a broader increase in wheat cultivation across the state, as wheat acreage in Colorado tripled between 1909 and 1919.[43] Such enticement then opened the door for suitcase farming, ownership of farmland by someone who did not make his or her primary home on that expanse but who bought the land as an investment. Relying on tenants or farm laborers and using new technology like the tractor and combine—generally shifting toward mechanized agriculture when possible—absentee landowners had the capital not only to plant their whole acreage to cash crops but to expand their holdings by buying up land that had been relinquished. This trend started with the war and increased in the 1920s.[44]

The price for wheat, sugar beets, and other cash crops did not remain high once the war ended, challenging (and sometimes breaking) farmers who had expanded to capitalize on wartime demand. The general narrative about agriculture in the 1920s is that farmers bought machinery or land on credit, hoped that prices would stay high or stabilize, and then broke more sod to plant more acreage to pay off loans and high interest mortgages. Obviously, the drop in prices left many in considerable financial distress. As they could not pay off debts, many continued to plant as much as possible, breaking more sod in trying to make up the difference.

Economic data from the U.S. agricultural census tell the story about both land use and population in Baca and Prowers from 1910 to 1930. The number of farms in Baca in 1910 was 540, and by 1930 that number sat at over 1,700. Tenants serviced only 16 farms in 1910, but by 1930 the number had skyrocketed to 621 (over a third of farms noted on the 1930 census). The numbers suggest that many of these tenants probably worked on wheat farms, as the number of acres under wheat was merely 453 in 1910 but sat at 87,551 in 1930. Broomcorn production, the second most popular crop for Baca farmers, moved from only 3,805 acres in 1910 to 52,764 in 1930, a spectacular increase in a short time. The rise in mortgage debt demonstrates another part of these developments, namely, the hard path that many expanding farmers took to capitalize on grain prices. Only 12 farms reported any mortgage debt in 1910, a figure that suggests most owners had their finances at least within their control. Conversely, the 1930 number shows the impact of large-scale debt, as only 359 were debt-free. Certainly, the amount and type of debt varied, but when we consider mortgage debt specifically, and then figure it as one instance of debt in addition to the debt accrued buying seed, tractors, and other products necessary for mechanization, we can assume that the average Baca farmers faced significant financial strains by 1930.[45]

Similar development occurred in Prowers County, and although the central agricultural products differed, farmers still accrued debt between 1910 and 1930 as the nature of agricultural shifted to mechanized, profit-maximizing cash crop production. In terms of the number of farms, Prowers expanded from 991 in 1910 to 1,382 in 1930, a smaller rate of growth than the heavy wheat farming endeavors in Baca. Because of the labor-intensive sugar beet industry, however, more tenants worked in Prowers than in Baca: 205 tenants in 1910 and 625 tenants in 1930. Farmers only devoted 23,279 acres to wheat in 1930, though up from the 1910 total of 5,006. This represented a significant difference to Baca land use but similar to county production of oats and rye, all of which paled in comparison to the 30,912 acres devoted to irrigated alfalfa. Sugar beets and other commodities made up the difference. Prowers farmers planted beets on 5,520 acres in 1910 and on 6,810 acres in 1930. The issue of debt further evidences the trend toward expansion in the face of increasingly tenuous economic times. The 1930 census shows that 64 percent of Prowers farmers faced mortgage debt (476 of 744) while that number in 1910 sat at only 199 of 705 farmers, a much lower rate of 28 percent.[46] Again, it is fair to assume that the amount of debt varied by farm, but the fact that so many Prowers farmers faced mounting economic challenges suggests the heavy financial strains they knew even before the Crash in 1929.

Most residents had no idea that the worst years were yet to come, however. It is doubtful that J. L. Farrand could reasonably assume that even his dire predictions would be nothing in comparison to what he would eventually see during the 1930s. Yet he wanted to be part of the effort to stabilize and improve the Baca agricultural economy. Unfortunately, Farrand knew that he needed some significant community support if he wanted to stay in Springfield beyond 1929. He noted the importance of making connections with prominent locals, including members of the Baca County Chamber of Commerce, because he wanted the people to understand his value. Unfortunately, Farrand failed to convince them, and Baca County residents refused to pay for a county agent again until 1934.[47] Times were indeed tough for people in both Baca and Prowers, more so in Baca, but most believed that they could weather the economic storm and get out from under debt once demand returned and prices rebounded. In the meantime, they thought that plowing, planting, and harvesting for markets represented their best chance at recovery.

In that respect, southeastern Colorado maintained a perspective during the 1920s that closely resembled that of their predecessors. Certainly, the modern world slowly infringed on life in southeast Colorado as it did in other western locales through developments like the railroad, the newspaper, and the radio. Yet, for better or for worse, many aspects of life in the region remained very much like they had when explorers first scoured the area. Of course, the key example is that Anglo explorers noted and settlers eventually understood that agricultural settlement in southeast Colorado would prove difficult. Most boosters and town builders dismissed the problems or succeeded in persuading potential farmers to move to the area regardless of environmental constraints like aridity and isolation from markets. The federal government also played a role in inciting exploration, making room for settlement by removing the Indians and taking land following the Mexican War, and promoting settlement with the series of Homestead Acts. It may not have been a unified effort to settle the western plains as quickly and fully as possible, but while various groups came from different angles, boosters on the plains and in Washington pushed the American empire west.

The migrants constituted the key cog in the machinery of expansion. Certainly, some may have been duped, but if we consider the process by which the federal government, boosters, and railroad companies made land available and promoted settlement, then we need to hold the settlers accountable as well. Put another way, the hardship and distress that most farmers experienced after their arrival did not necessarily grow from some nefarious plot devised by boosters or bureaucrats. Many may not have known exactly what they were getting into, but they came. Whether motivated by the potential benefits of homesteading or out of desperation because they had exhausted their options at home, settlers arrived in significant numbers and continued to make the trip into the region through the 1920s. With hindsight, they may have better understood the daunting task before them. Long's description of the "great American desert" proved quite apt in many respects, but people came anyway. What was at first a trickle became a flood once mineral discoveries helped inform the rest of the country and the world of available lands on the western plains.

Agricultural development in southeast Colorado resembled the simultaneous drive for mineral resource extraction in the mountains and foothills. Both miners and farmers rolled the dice on finding a resource base plentiful enough to satisfy their needs and provide economic stability. They seemed to have little concern that the decks were stacked against them, that the likelihood for success depended on many variables, and that they had no control over many of them. They needed the right location, to either find a lode or locate a viable water source or homestead on fertile soil. They benefited from startup capital, though few seemed to have enough to sustain themselves during periods of drought or dry lodes. They acted as speculators, planning a future on a resource base largely unpredictable, fickle, and easily abused, that often failed to return much of their investment or even meet their needs. The farmers exhausted the soil trying to make good on debts and mortgages. The miners took jobs for mining companies or in nearby towns. Many who had no other recourse simply left the area for good.

To their credit, farmers started to think more intently on how to succeed on the Colorado plains by the late nineteenth century. Irrigation certainly helped. Railroad connections to distant markets did as well. Dryland farming techniques gave farmers at least some hope that sustenance remained possible and that good years might provide profit. In these ways, the southeastern Colorado farmers of 1929 had it a bit easier than their predecessors, and more of them found ways to stay on the land in spite of market fluctuations and drought cycles. But most farmers still thought only of production, only about how to plow up enough land to pay back debt or scratch out some profit. Such a mentality led to the "Great Plow-Up" of the 1920s. By 1929 farmers knew how to dig canals and learned that some strains of wheat or broomcorn grew better in dry conditions while cantaloupe and sugar beet did well with irrigation. Yet neither the farmer nor the government could alter the arid environment or mitigate a climate prone to drought, one likely to tend toward harsh imbalance. In that respect, the biggest lesson that Long, Pike, and other early explorers tried to pass on to their successors, that farming may very well not be a successful endeavor on the Colorado plains, had not taken hold by 1929 when the Great Depression spread across the country and shook the nation to its core.

2

The County Agents Take Root

Writing as "The Unofficial Observer" in the Prowers County *Lamar Daily News*, Joe T. Lawless celebrated the election of Democrats to various posts in the 1932 elections, specifically victories by Franklin D. Roosevelt for president and Edwin C. Johnson for governor. Lawless believed in a direct correlation between politics and the weather, arguing that since the Democrats' victories, "nature has blessed our farming sections with timely and abundant contributions of rain and snow." This moisture, greeted with elation and cause for heightened expectations in the arid region, promised recovery for local farmers and a way out of the economic doldrums that they had been navigating for years. Lawless hoped that the rise in precipitation could persuade "grateful residents" to support the Democratic Party, the group deserving credit for bringing the rain, and see to it that "it will not be such a long time between Democratic administrations in the future."[1]

While the association of weather with political tides may seem a stretch, Lawless's view demonstrates the optimism with which many agriculturalists greeted the Roosevelt administration. Most Americans who voted for FDR hoped that the transition from Herbert Hoover could remedy the Great Depression. Hoover's political life and legacy were, and generally remain, tied to the economic catastrophe. For right or wrong, he personified the Depression to millions of Americans, and voters turned to Roosevelt in

1932 largely because he offered hope at a time when millions of Americans struggled to meet basic needs, for which they blamed Hoover. Yet Hoover campaigned extensively on his desire to revive the agricultural economy and became the first president to consider using the federal government to subsidize agriculturalists. Indeed, he had been a friend to the farmers in many ways. Given the gravity of the Depression and its continued grasp on the country in 1932, however, few residents of the Colorado plains seemed to credit Hoover for his efforts and instead focused on Roosevelt's promised agenda. The *Springfield Democrat Herald*, the primary Baca County paper and ardent supporter of the Democratic Party, extolled Roosevelt's virtues and celebrated his victory as the start of a new chapter in American history, one filled with optimism and prosperity.[2] Lawless embodied the exuberance surrounding the election, and his hope that FDR's promise of increased government intervention, his talk of a "new deal" for the American people, could lead the country out of the crisis. Whether or not Lawless seriously considered the Democrats capable of changing the weather, it is apparent that he searched for some positive omen in a time of widespread desperation, comforted by the thought that change would come—and come quickly.

FDR's election represented a watershed moment in American history, as the New Deal ushered in unprecedented federal involvement in American lives. In terms of agricultural policy, however, early New Deal programs did not break new ground by reaching out to rural constituents or trying to stabilize the rural economy. Indeed, while historians have long derided President Hoover's unwillingness to extend more federal assistance to suffering Americans, Hoover was in fact quite sensitive to the plight of American farmers. He wanted to help farmers help themselves without getting the federal government too heavily involved, so he promoted a federalism that gave control to the states. He did this in part by utilizing the Smith-Lever Act of 1914, which created an infrastructure perfectly tailored to his federalism by introducing a system of state-level Cooperative Extension Service bureaucracies that bridged federal, state, and local efforts to address problems in rural America. The act aligned with Hoover's vision because it afforded government assistance to state programs through federal funding but prioritized employing local experts, most of whom were Extension agents who were working with farmers to address their problems. It proposed a synergy between the local, state, and federal levels to promote modern farming techniques, food conservation, youth programs, and focus on stabilizing the farm economy. The Smith-Lever Act demonstrated one of the more important steps Congress made in the first part of the twentieth century to facilitate federal activism in rural America, and Hoover built on that foundation in pushing more federal influence in the face of the Depression.

For all that Hoover did to capitalize on the policy and apply his federalism to rural problems, he could not help farmers as much as he, or they, had hoped. His inability to turn around the agricultural economy contributed to his reelection defeat, and FDR realized upon his victory that his political fortunes hinged on how his administration could help agrarians. To do this, FDR relied on the Extension Service to identify rural problems and to strengthen the relationship between farmers and the federal government. The New Deal depended on local agents to ensure that residents were abiding by new regulations and satisfying requirements to attain newly created subsidies. Those subsidies proved crucial for farmers to survive the worst parts of the Depression and became critically important in southeastern Colorado during the Dust Bowl years. The New Deal also increased funding for the Extension program, including a provision to pay the agents from federal and state coffers rather than have the individual counties pay the agents' salaries. This lessened the financial burden on impoverished counties and thus made residents more amenable to the agents themselves. That funding helped agents solidify their place in the community, making it easier for them to build relationships with their constituents and establish themselves as interlocutors between the local and the federal.

Since the agents became so much more prominent in local communities because of FDR's policies, many southeastern Colorado residents came to view the agents as a personification of a growing federal presence in the countryside. In southeastern Colorado, this meant that locals sometimes embraced them and at other points dismissed them, as agents represented federal largesse to supporters as much as they reminded critics of bureaucracy. Locals' reception of the agents and of federal intervention more generally changed according to the severity of the Great Depression and the Dust Bowl, such that the worse the situation the more they found reason to embrace the Extension Service and the New Deal. This meant that by the end of the 1930s the two groups worked fairly well together toward the shared goal of keeping farmers farming in southeastern Colorado. The relationship between the two bodies remains contingent on this similarity in vision as farmers and the federal government work together today to keep American agriculture stable. This whole transition—the establishment of the Extension Service in these communities, the changing relationship between locals and the government, and their embrace of federal policy then and since—reflects some of the more profound legacies of the Dust Bowl.

That transition started in 1914 with the passage of the Smith-Lever Act, which reflected a growing sensitivity among politicians to the plight of rural residents who were, in one way or another, losing their place in an increasingly modern (and modernizing) country. Representative A. Frank Lever of South Carolina and Senator Hoke Smith of Georgia introduced the bill to establish a national cooperative extension program and attain federal funding for it. At its core, the Smith-Lever Act represented a concerted effort by the federal government to use seasoned experts to help the farmer manage his or her life, economy, family, and land. Each state that participated in the program received federal funding to offset expenses accrued by program employees, including printing costs, travel expenses, salaries, and other fees from working in the field.[3] The key to the program was the relationship between the expert and the farmer. As described by Lever:

This bill proposes to set up a series of general demonstration teaching throughout the country, and the agent in the field of the Department and the college is to be the mouthpiece through which this information will reach the people—the man and woman and the boy and girl on the farm. You cannot make the farmer change the methods which have been sufficient to earn a livelihood for himself and his family for many years unless you show him, under his own vine and fig tree as it were, that you have a system better than the system which he himself has been following. The plan proposed in this bill undertakes to do that by personal contact, not by writing to a man and saying that this is a better plan than he has or by standing up and talking to him and telling him it is a better plan, but by going onto his farm, under his own soil and climatic conditions, and by demonstrating that you have a method which surpasses his in results.[4]

The bill aimed to place agricultural experts in each locale to work at "diffusing among the people of the United States useful and practical information on subjects relating to agriculture and home economics, and to encourage the application of the same."[5] This differed from the motivation behind the Agricultural Experiment Stations that emerged with the 1887 Hatch Act, since those stations focused almost exclusively on conducting research and much less on working directly with farmers on the county level.[6] Indeed, the emphasis on experts and their role in disseminating valuable information to "the people" reflects the bill's Progressive qualities. The federal government's use of experts trained in applied science to ensure efficient and productive use of natural resources constituted a hallmark of Progressive conservation efforts.[7] The Smith-Lever Act promoted the role of educated men in the U.S. Department of Agriculture (USDA) and at state agricultural colleges in leading farmers toward economic stability through instruction. If these experts had the opportunity to demonstrate proper farming practices to agriculturalists, then their knowledge, in Lever's words, "would work a complete and

FIG. 6. "Three sets of brothers, French, Gausman, Vahles." Jack French and Claude Gausman (far left and third from left, respectively) knew each other in college at Colorado Agricultural College (later renamed Colorado State University) before working in the Extension Service as agents. They were perfect as the kinds of experts that the Smith-Lever Act promoted, and they had productive careers in Extension, even working in Baca and Prowers Counties during the 1930s. University Historic Photograph Collection, Colorado State University, Archives & Special Collections.

absolute revolution in the social, economic, and financial condition of our rural population."[8]

Proponents of the Smith-Lever Act owed much to Liberty Hyde Bailey. He designed the New York Extension Service and created one of the first county agent programs to facilitate Extension policy. He viewed Extension as a combination of "sympathy with nature, a love of country life, and a scientific attitude, expressed by a habit of careful observation and experimentation." In this way, Bailey believed that Extension could "advance the larger cultural ideals of a 'self-sustaining' agriculture and personal happiness," demonstrating his concomitant promotion of economic, social, and even cultural growth.[9] He chaired the 1909 Country Life Commission, a body of advisors tasked with identifying the problems that plagued rural Americans and suggesting solutions. The commission understood that the best way to comprehend the challenges farmers faced was by ascertaining their perspectives. To that end, the commission devised a set of questions designed to get farmers' perspectives on country life and then distributed 550,000 copies of the list to farmers around the country. Members of the commission also traveled throughout the country to afford local and state politicians, professionals, academics, and "country people" the opportunity to raise their concerns as well as offer their insight about rural life, whether that meant sanitation practices, elementary education, or connections to the outside world.[10]

While it identified several problems, the commission's report stated clearly and definitively that a national Extension program would remedy the myriad ills facing rural residents. The commission's purpose was "to develop and maintain on our farms a civilization in full harmony with the best of American ideals." In his introduction to the report Theodore Roosevelt contended that "the men and women on the farms stand for what is fundamentally best and most needed in our American life" and they "will be in the future, as in the past, the stay and strength of the nation in time of war, and its guiding and controlling spirit in time of peace." The commission's submission stated that such harmony required "nothing more or less than the gradual rebuilding of a new agriculture and new rural life" where "the business of agriculture must be made to yield a reasonable return to those who follow it intelligently" and "life on the farm must be made permanently satisfying to intelligent, progressive people."[11]

In a way largely indicative of the Progressive Era, the Country Life Commission's report emphasized rural social and cultural life as key elements of promoting such "progressive people." The commission proved willing to intervene in individuals' personal lives when it felt such action proved necessary, even in terms of social and cultural matters. Indeed, most of their suggestions for how rural families might conduct themselves reflected an urban reformer sensibility that was in many cases dismissive of rural people altogether. For example, the commission discussed the issues of intemperance and the rural church because it believed that "the country life problem is a moral problem" and that the church could play a larger role in reorganizing rural life.[12] As one can imagine, however, many rural people "resented what they perceived as an act of usurpation" and largely considered the commission, and urban reform more broadly, both paternalistic and presumptuous.[13]

Members of the commission saw cooperation between local and federal as key to building strong community bonds among people united in common cause. The commission's report noted that "local initiative" must be "relied on to the fullest extent" to ensure that "federal and even state agencies do not perform what might be done by the people in the communities."[14] At the same time, however, the commission posited that outside motivation in the form of someone akin to the county agent often propelled local efforts. They contended that farmers could always benefit from having an expert to consult. The combination of agent and farmer thus united outside expertise with local knowledge and initiative in hopes that the two groups could come together, identify the primary issues, and develop worthwhile solutions.

The commission's report suggested that resource conservation, and specifically a stronger effort to take care of the soil, would prove absolutely critical to stabilizing the rural economy and population. The report tied the land and people together, noting, "The social condition of any agricultural community is closely related to the available fertility of the soil. 'Poor land, poor people,' and 'rough land, rough people' have long since passed into proverbs." The commission found, "Our farming has been largely exploitational, consisting of mining the virgin fertility. On the better lands this primitive system of land exploitation may last for two generations without results pernicious to society, but on the poorer lands the limit of satisfactory living conditions may be reached in less than one generation."[15] It thus identified one of the underlying justifications to support using trained experts to assist American farmers: experts understood the need to take care of one's land, to protect one's resources, and to safeguard one's future.

The architects of the Smith-Lever Act adopted much from the commission, and with the notable exception of addressing social and cultural elements of rural life, they included much from the report in their explanation for the 1914 policy. They emphasized the role of the agent, cooperation with farmers, and conservation as three ways to stabilize the agricultural economy and protect the landscape. During discussion surrounding the Smith-Lever bill, for instance, its supporters argued that the Extension agents' most important responsibility was to correct abusive practices:

> The soil of this country . . . is the storehouse of all wealth. Every living soul is dependent upon it. The very best thought of this Nation should be directed to its conservation. We hear much of the conservation of our timber, our mineral lands, and our rivers and harbors. The soil is the mother of all, and by who or when has a voice been raised and decisive active stand taken to conserve and perpetuate the soil? . . . The soil—the land—is an inheritance, handed down to man for humanity. It belongs to future generations, and as it passes through our hands, we are as responsible as the man with the talents. Let us do our duty—pass this bill—and receive the plaudit, "Well done, thou good and faithful servants."[16]

While lacking specifics for exactly how Americans could save their soils, this call for Extension to remedy the already-evident and pressing issue of declining soil fertility represented one of the more compelling arguments in favor of government activism. Extension Service supporters knew that conservation was critical for prosperity and posterity, as only concerted protection of soil and water could ensure "that future generations also may have a good living and the general welfare be thereby safeguarded."[17] With that reasoning they tied America's progress as a nation with its ability to protect its resources, an argument that proved successful enough to win the bill's passage.

This early twentieth-century national discussion about the importance of sustaining rural America—its economy, its soil, and its families—proved crucial to how farmers, the state and federal governments, and Extension responded to the crises of the Depression and the Dust Bowl in two ways. First, the commission and the passage of the Smith-Level Act evidenced public trepidation over exploitative agricultural practices and argued how such damaging habits threatened an entire segment of the American population. It demonstrated that many people inside and outside Washington DC considered the correlation between the health of the land and the health of the people who lived and worked on it. Second, the practice of sending experts into the field to advise rural America represented an initial foray into extending federal intervention into the countryside. The Country Life Commission report provided some rhetorical firepower to the argument in favor of rural reform and improvement, but the Smith-Lever Act actually placed experts in the field.

While the Smith-Lever Act created a foundation that could potentially help stabilize the agricultural economy and improve agricultural conservation, it had little impact in southeastern Colorado until the Great Depression. While farmers in that region had been suffering since the early 1920s, the Great Crash and subsequent decline marked a critical moment for the United States and proved enough to push Herbert Hoover's administration into action. In some respects, residents in Baca and Prowers seemed almost disinterested in the national response to the Depression or to its causes. As longtime Baca County resident Ike Osteen noted, "Few people in Baca County even knew what the stock market was, and I don't know of a single person in Baca County who killed themselves because of what happened on Wall Street."[18] Residents instead largely turned their attention inward. For example, both the *Lamar Daily News* and its Baca County counterpart, the *Springfield Democrat Herald*, actually covered a local bank robbery more attentively than they did national news during the fall of 1929. Five men walked into the Lamar First National Bank in May 1928, killed the bank president, his son, and a bank teller before abducting another teller and stealing more than $500,000. Authorities either caught those attempting to escape or killed the robbers as they resisted arrest; their trial in the fall of 1929 invited intense local media coverage.[19] The crime, a rare violent episode in the region, might suggest that the criminals felt a level of desperation because of the Depression and chose to rob the bank as a result. The coverage of the crime and trial also demonstrates the level to which these local papers focused on the local rather than the national.

This attention to local issues might explain why residents seemed disconnected from developments in Washington DC during the Hoover administration. One might postulate that Colorado farmers would have been more invested in Hoover and his administration because he had campaigned on the need, and his desire, to help rural America. Certainly, Hoover addressed the farm economy during his time as secretary of commerce, and over the course of the 1920s he strove to make it more stable. Hoover preferred an "associational" plan, designed to combine efforts between locals and federal experts, utilize scientific research to determine the best crops for each soil type, promote stock raising, and help farmers build personal credit. Hoover's approach attempted "to fit farmers and agriculture to an advanced capitalist economy without either resorting to formal intervention to support prices or bowing to interest-group demands." The Agricultural Marketing Act of 1929, the "first attempt by a presidential administration to develop a comprehensive farm policy," is the best demonstration of that approach and the most prominent agricultural policy of his administration.[20]

Hoover determined that farmers faced their biggest issue in trying to market their products, which he believed reflected too many farmers working to sell their goods independently rather than as part of a larger group that could maximize profit. Large, national cooperative bodies could effectively stabilize prices by regulating the flow of products and thus negotiate the price point; this allowed farmers to help themselves with only minimal federal intervention. A federal farm board would then supervise the marketing strategy and verify that the associations in wheat or cotton or other products judged the domestic and international markets correctly, ensuring that farmers garnered the best prices for their goods. That act combined personal responsibility with minimal federal involvement (education and coordination of the federal farm board) to mitigate the problems caused by fluctuating markets.[21]

Evidence from Baca indicates that Hoover was onto something and that farmers suffered from increasingly volatile markets. Indeed, Ike Osteen remembered a surplus wheat crop in 1931 even though 1930 was a "spotty year for moisture." Osteen noted with dismay that such surplus simply drove down the price so that what had been 68 cents/bushel in 1930 became 25 cents/bushel by 1931.[22] The crop production for the county actually peaked in 1930 at an estimated value of more than $3 million, far surpassing its previous high of roughly $2.5 million set in 1920. Even in 1931, Baca County farmers produced an enormous amount of crop value; they simply could not sell it off for what they believed its proper worth. That was key to the entire situation—estimated or assessed value had nothing much to do with what farmers could get for their products.[23]

What is somewhat remarkable is that Hoover had been trying to rectify the supply/demand imbalance since the mid-1920s, when he backed parts of the McNary-Haugen bill in 1924. That legislation likewise relied on a federal farm board system but focused on the issues of export and tariff as ways to ensure solid prices for American products. It centered on the "two-price system," a federally designed cooperative marketing scheme to set up a domestic price as well as a lower international price so that American produce could compete in the world market and sellers could also make money at home. It included an equalization fee that was effectively a tax that farmers paid to offset the loss accrued by selling their surplus on the world market for lower prices. The tax would be much less than what the farmer would receive from higher domestic prices, so the farmer would ultimately benefit. The bill eventually passed both the House and Senate in 1927 only to be vetoed by President Calvin Coolidge. Most McNary-Haugenites rightly believed that Hoover pushed Coolidge toward the veto because Hoover did not agree with or support the idea of the equalization fee. Everything else about the McNary-Haugen approach seemed to align well with Hoover's views, and therefore he passed his Agricultural Marketing Act in 1929, a bill he later called "the McNary-Haugen bill stripped of the equalization fee."[24]

While it is evident that the Hoover administration came into office with the firm conviction that depression in the agricultural economy required remedy and had a sense of how it wanted to proceed, it largely failed farmers. Hoover could not convince his rural constituency that he had done much in their favor, in part because he did not utilize county agents to build relationships with locals or drum up enthusiasm for administration policies. Residents of southeastern Colorado seemed isolated from policy discussions, so they approached the Hoover administration with little fanfare. The issue of farm boards or equalization fees, while certainly important in the abstract and obviously in some circles around the country, had little to offer farmers who had been facing economic challenges for several years. Hoover seemed disconnected and callous, which contributed to his election loss in 1932.

To suggest that he might have won had he done more to utilize the Smith-Lever Act and reached out more directly to farmers is perhaps a bit revisionist, but he certainly would have fared better than he did in his battle against Roosevelt. Hoover's unwillingness to employ a more confident rhetoric left Americans unsure of his convictions. More to the point, his inability to capitalize on building a stronger relationship between farmers and Extension may have cost him some support as well. Most observers in southeastern Colorado had realized by 1929 that their economic outlook remained gloomy and that they had few solutions for their continuing problems. Some started looking to Washington. Indeed, newspaper coverage of the Depression and what options farmers had often centered on whether the federal government, and more specifically the Extension Service, should have a larger role in southeastern Colorado. Most seemed to support the agents and Extension, though some trepidation existed, which demonstrated that many residents resisted outside "help" despite their continued economic distress. Yet most hesitation abated as the economy worsened, and criticism of federal involvement in local and even regional affairs had almost completely disappeared by Roosevelt's election in 1932.

A submission to the *Springfield Democrat Herald* from 1929 initially spurred a public conversation about Extension and summarized the key points against employing the county agent. The author chastised Colorado Agricultural College for educating students with books rather than empirical knowledge and for then ushering them to places like Baca County where residents had to foot the bill. Moreover, the "county-agent racketeers" only worried about irrigation and "wet farming," and because most Baca farms lacked irrigation, the agent was effectively unnecessary. The author asserted that Baca County farmers were as adept at their craft as farmers anywhere and, therefore, they had no need to support the agent because he offered nothing of value. The agent's salary was perhaps the key issue, as the author cited the "outrageous" taxes already levied against locals and the importance of having more money at their disposal.[25]

The newspaper included a rebuttal to the submission criticizing the program that defended the new agent and tried to drum up support for the Extension Service. Marlon D. Lasley applauded the county commissioners who invited an agent to work in Baca and predicted that this would be a tremendous boon to local farmers. Considering the Extension Service in Baca County was still in its infancy in 1929, Lasley argued for patience: "If our farmers will just give him a chance to help them, I am sure they will find that he will be right on the job. I expect to call on him for advice on several things I have in mind, regarding my orchard and grapevines, and both hogs and sows."[26] Many other submissions, editorials, and columns similarly supported Extension. Joe T. Lawless, notable in The Unofficial

Observer, which was a semiregular column in the *Lamar Daily News*, criticized the Prowers County commissioners for temporarily dispensing with their agent. Lawless believed that agent Frank Lamb "was unusually efficient and always willing to work overtime. His advice and assistance enabled the farmers to improve the quality and increase the output of all crops, thus aiding them to earn many times over the small fractions of taxes they paid for the benefit of his counsel and aid. The County agent's office is an investment instead of a liability."[27]

The agents entered the conversation about their relative worth and how they served the community at various points during the public debate. They did this by writing editorials, penning regular sections in the back pages like the County Agent's Column and Country Correspondence, and public meetings to discuss the program and agents' roles. Local agents generally wrote such columns to explain some aspect of pending legislation or the proper ways to fill out credit applications or which crop variations were doing especially well at that particular time—all demonstrations of their worth to the community. Agents continued writing these columns well into the 1940s, but their efforts to publicize their impact seemed more pressing during the first years of the Depression when they were still trying to gain a foothold in the community. In one of the more straightforward examples of an agent defending his position, Frank Lamb calculated what he considered to be his relative worth to explain that his salary was indeed worthy of locals' sacrifice. Lamb argued that his $2,150 annual salary was incredibly inexpensive considering his ability to educate farmers, help them become better producers, and assist them in marketing their goods. He deemed those services to be worth nearly $25,000 annually, a cost ten times his salary.[28] Again, he deferred to money to show his worth, demonstrating that he fully realized that most locals who resisted the program did so because of the cost to the county. If Lamb could find some way to convince them that he had saved them money, then they might be more likely to support him.

Roosevelt's election and the development of New Deal agricultural programs incited growing support for the Extension Service, and as their jobs became more complicated, agents' relative worth became more apparent. The flood of federal legislation inaugurated during the First Hundred Days promised tumult for American farmers as they came to grips with new programs, regulations, and eventually the opportunity to qualify for federal subsidies. Many hoped the agents could help them navigate these changes. For example, the editor of the *Lamar Daily News* argued in favor of reinstating the Prowers County agent to serve area farmers. The editor claimed to disagree with the initial decision to remove the agent and believed that an agent could now undoubtedly help local farmers and could do so at a minimal cost to the county government. He contended that an agent familiar with the local terrain, practices, and products and someone friendly with local farmers could advocate for county farmers and protect their interests. Rather than have the federal government supply an "emergency administrator" for government programs, which would occur if the county refused to appoint someone from Extension, the editor pushed for a permanent agent who represented Prowers residents. Only then could farmers ensure that they qualified for federal subsidies based on drought or satisfied policy requirements in terms of the corn and hog program, for example.[29] This portrayed locals' growing appreciation for the county agent's primary role, namely, his post as interlocutor between locals and the vastly expanded (and still expanding) New Deal state.

Even letters from detractors present widespread acceptance that the New Deal represented something quite different than the Hoover years or anything that came before. In a letter to the *Lamar Daily News*, Mrs. C. L. Nickelson expressed her belief that the agent served as nothing more than a cheerleader for local farmers. Given the financial strain that his salary would exact, she argued that he was not worth the expense. She claimed that while the initial phase of New Deal policy was going into effect, the federal government had already proven capable of reaching agriculturalists under Hoover's administration without an agent's help. She ceded that the situation was dire enough to warrant federal intervention, and dryland farmers specifically deserved assistance because of drought and dust, but she worried that the agent would require unnecessary spending on the part of county residents. Mrs. Nickelson argued, "The facts alone should determine as to whether or not dryland farmers should receive federal benefits; a county agent should not be expected to 'doctor' the facts." By suggesting that an agent merely served as a "press agent of the community," she simultaneously discredited Extension and accused it of discounting dryland farmers in favor of bigger, irrigated cash crop farms. While her perspective largely aligned with earlier criticisms of Extension agents, it is important to note that Mrs. Nickelson was sensitive to the changing context when she offered her perspective in 1933. She understood that Roosevelt's election brought an increasingly active federal, state, and county government. She welcomed such intervention to some degree, but she believed that federal employees and programs could filter through the area without help from the county agent. She contended that the federal government had already proven "quite capable of getting to the farmers in counties where there was no county agent" and that consequently locals could access government services directly.[30]

The *Lamar Daily News* included George B. Long's rebuttal to Mrs. Nickelson. Long felt that the sheer number of New Deal programs, to say nothing of their complexity, necessitated some kind of translator for local farmers to communicate with federal bureaucrats. Long wrote that "our Federal Government was developing to contact the individual farmer" and "the wheat farmer and the corn and hog farmer has been brought under the protecting arm of our Uncle Sam." Because of the drought, Long continued, the county would enjoy even greater federal largesse and consequently needed a contact person—the local agent. The government moved toward price control over "every commodity of importance produced" in Prowers County, and the USDA recognized Extension as the "one channel thru which authentic information goes to Washington." Because he fully expected government intervention, Long reiterated his support for some kind of mediator to work on behalf of local farmers and refuted Mrs. Nickelson's view that leadership from Washington was enough.[31]

Long's argument identified a key rationale to support the program: only the agent could promise financial stability by supplying federal subsidies, and in that way Extension became vital to farmers during the New Deal.

In theory, the agents' approach to their posts changed very little with the onset of the New Deal as they adhered to the same fundamental goals outlined in the legislation that gave birth to the program. In practice, and per agency records, however, the onus for protecting locals and ensuring that they obtained what they considered their share of New Deal agricultural subsidies meant that the agents became much busier in 1933. The agents kept meticulous records documenting their experiences, utilizing both a Narrative Summary and a Statistical Summary to explain how they executed their assignments. As the titles imply, the agents included information about their day-to-day activities in the field, submitting quantitative evidence in the Statistical Summary to demonstrate their involvement in the community. The bulk of the agents' Narrative Summary delved into personal experiences with farmers and community leaders as well as their perspectives on local problems, potential solutions, and interpretations of local conditions as more qualitative analysis. Additionally, the agents sometimes offered other documentation to evidence their impact, including pamphlets, newspaper clippings, posters, and other notifications about agent-led activities.

By scribing such detailed accounting of their exploits, agents seemed conscious of justifying their position to the community and to their superiors at the state office, as either of the two had the power to replace the agent or conclude the program. For instance, agents tracked their time in the field by noting the number of visits to communities in the respective counties. They also recorded the numbers of letters and articles written, bulletins distributed, and meetings/training sessions held for interested locals. The numbers for each of these endeavors stayed relatively consistent from 1929 to 1933 when activity rose, but judging by agents' time spent with locals, it is fair to conclude that they represented an important resource for local farmers throughout that period. Indeed, as had been their goal from the outset, the agents tried to keep farmers on their land and the local economy stable. In effect, they found themselves responsible for determining ways that locals could survive, then convincing farmers that they should follow their advice. This was often accomplished by one of two demonstrations, labeled method and result. The method was "given by an Extension worker or other trained leader for the purpose of showing how to carry out a practice" while the result was "conducted by a farmer, home maker, boy, or girl under the direct supervision of the Extension worker, to show locally the value of a recommended practice."[32] For instance, the Baca County agent held demonstrations and conducted visits with local families regarding issues like the family diet, food preservation, furnishing the home, tending to home gardens, and even planning the family's wardrobe.[33]

Yet agents more often handled issues of cooperative marketing, attaining credit, paying mortgage, handling debt, and taking care of general financial concerns. Agents' ledgers and notes demonstrate two key components of the Hoover administration's approach to stabilizing the rural economy: limited direct federal intervention and reliance on local cooperation to spearhead reform. The numbers for Prowers County agents show that the Hoover administration relied heavily on them to advise farmers on how to attain federal assistance. Even then, however, Prowers County agent Frank Lamb conjectured that the limited federal involvement would not suffice to meet the crisis. Lamb noted that the level of desperation in Prowers by 1932 required at least some help from the federal government because local/county/state government failed to provide for all parties. This was especially true when the county was not recognized by the Governor's Relief Committee for additional funding in 1932, leaving county farmers in the lurch. Surprisingly, government assessors who traveled the Colorado plains to determine which counties should be considered drought stricken, and therefore subject to federal assistance, decided that Prowers did not meet the criteria. Rather than rely on federal subsidies, then, Lamb hoped that he could increase the number of farmers taking low-interest federal loans to satisfy their needs. Since most farmers had little capital, they ran into debt each year as they prepared for the season's planting. Thus one of Lamb's principal jobs was to help people coordinate loan applications to take advantage of federal loans. This was not a great consolation prize considering the federal government offered some direct assistance to people in drought-stricken counties, but Lamb helped some 188 farmers garner a total of $30,852 in 1932 alone.[34]

Agents' efforts during the early Great Depression years also exemplified Hoover's philosophy by looking to foment relationships with local advisory boards, prominent individuals, and county organizations. Indeed, agents tried to facilitate community relations to help those most in need to attain assistance and to help themselves gain the community's trust. Lamb mentioned the Southeast Colorado Livestock Association and the Prowers County Farmers for their assistance to him in reaching county residents. Additionally, he thanked the Prowers County Farm Bureau and the Arkansas Valley Economic Conference for their help.[35] Such emphasis on local boards and clubs demonstrates the ways that Lamb and other agents immersed themselves in their communities and how they eventually became fixtures in rural life. In Lamb's case, he succeeded where Baca County agent J. L. Farrand had failed; Farrand never made enough powerful friends who could sway the public toward accepting him, but Lamb did, and they rewarded him with steady employment. The Service relied on such networking to survive the 1920s, as farm bureaus and similar groups proved necessary for agents to build bases of support when the program was new in most locales.[36]

Moreover, agents had to generate personal connections to ensure their livelihood, since a fickle constituency could always decide that the program had outlived its utility or if faced with economic constraints could choose not to pay the agent's salary (as Prowers County residents did with Lamb in 1933). By opening his doors to visitors, making phone calls, holding regular meetings, and visiting farmers' residences, the agent tried to curry support with locals to improve the likelihood that they listened to his message. Essentially, the Extension Service spent much of its time between late 1929 and early 1933 focusing on instruction and especially cost-saving measures, whether trying to reach out to farmers about their techniques, or farmers' wives about home economics, or their children about organizations like the 4-H.[37] This diversity in focus represented the key elements of the Smith-Lever Act at work, as such activity satisfied the dual goals of reaching entire families and promoting education.

Yet the Hoover administration did not utilize the Extension Service as well as it could have in building support or in helping rural Americans. Agents like Frank Lamb understood that the Extension agents represented a powerful resource, one that did little for anyone if allowed to lie dormant. He hoped that a little federal support would allow him and his colleagues to make a difference.[38] The New Deal afforded them that opportunity by tasking agents with fixing the problems that rural Americans faced in the 1930s, and with the flood of New Deal legislation designed to help farmers, agents found themselves in unchartered territory. This proved particularly true with the advent of the Agricultural Adjustment Administration (AAA), a product of the Agricultural Adjustment Act that Congress passed during FDR's First Hundred Days. The AAA attempted to reconcile the disproportionate relationship between supply and demand that many New Dealers believed caused the Depression. New Dealers promoted the AAA as a means to stabilize prices for rural commodities by limiting production, and thus limiting supply, to increase demand. This initiative represented another tie to the Hoover administration, as the AAA was based in part on the ideas of McNary-Haugenism and the Agricultural Marketing Act that outlined the importance of stabilizing prices by regulating supply. The AAA promised direct federal intervention in coordinating the removal of surplus goods until the price improved. It offered subsidies to farmers who voluntarily decreased their production rather than relying on farmers to freely combine and stabilize prices by cooperatively holding back surplus, all with minimal government involvement and supervision, as had Hoover's policies.[39] In practice, this amounted to programs across the Great Plains that tried to stabilize prices by paying farmers to not put wheat or hogs or other products on the market. The AAA used farmers' incomes from 1909 and 1914 as a benchmark for what might be considered an appropriate income for farmers, and then determined how much the subsidies should be to aid farmers to reach that level.[40]

The AAA had two significant impacts in southeastern Colorado. First, it fundamentally changed how agents worked in their communities. The AAA and its subsidy programs made for an active agent, as "he became an administrator rather than a teacher" and "a promoter rather than an educator" to ensure that farmers understood the new programs and to maximize their participation. He did so by encouraging them to take advantage of government subsidies or conversely by warning them that penalties for noncompliance could be levied against them.[41] For agents, this also meant that the government looked to them to determine whether farmers had in fact satisfied policy requirements and therefore earned federal subsidies. The new role of handling the government's purse strings meant that agents had considerable power in these communities and had more direct influence on counties' agricultural economies. This meant a whole new level of intervention for agents who became intimately involved in building the relationship between farmers and the federal government with the advent of New Deal programs.

The AAA's second major influence in the region was economic, as it had a considerable financial impact through its programs to purchase excess swine, corn, and wheat, three of the most common agricultural products in the region. Once agents informed locals of these new policies, they helped farmers become eligible for subsidies by keeping them in line with federal regulations. Residents of Baca County did quite well. Agent R. E. Frisbie counted 725 farms under the hog reduction program and another 1,040 farmers who agreed to sell cattle or sheep to the federal government, as well as 1,136 farms limiting wheat production. Given that 1,420 farms existed in the county, most farmers complied with some aspect of the AAA, and many participated in multiple programs. They obviously benefited from such involvement, as Frisbie counted $265,000 in federal payments for the cattle and $3,224 for sheep. Additionally, the AAA allocated extra funds to provide for the Extension Service and for agents to hire locals to participate in the process of building compliance in rural areas.[42]

The numbers for Prowers County similarly demonstrate an increase in agent activity in 1933 and early 1934 with the rush to qualify farmers for AAA benefits. Indeed, agents became so busy that the federal government allocated wages for forty locals to work in tandem with the agent to distribute relief funds and get farmers aligned with the reduction programs. Their efforts proved largely successful. Farmers on 1,323 out of 1,473 Prowers farms abided by some form of production reduction, whether of beef, sheep, swine, corn, or wheat. The bulk of the monies supported corn and hog reduction, which garnered roughly $116,000 from the federal government (about $75,000 for the swine and $41,000 for the corn), and the cattle purchase program brought in nearly $187,000 for locals. Agent A. J. Hamman claimed that nearly 90 percent of all corn acreage and 64 percent of all wheat acreage in the county was removed from cultivation in 1934 so that farmers could benefit from the program. Hamman noted that despite some early reticence about selling off produce at government-mandated prices and according to the government's schedule, a series of secret ballot votes at various county meetings showed that locals widely supported the AAA.[43] Prowers farmers warmly embraced the AAA and its benefits, signing up for hog reduction at a higher rate than any other county in the state in 1933.[44]

Hamman appreciated how much agents contributed to this public support for the AAA and felt strongly that continued federal intervention was necessary to support agrarians. Hamman believed that "farmers had been in a declining market since about 1921 or 1922 and Washington had the money to relieve them." He also believed that showing his constituents that the federal government could offer such assistance would pay dividends in earning farmers' participation. He knew how important it had been for agents to clarify policy to local farmers. Hamman remembered many occasions when locals misunderstood federal programs or when non-Extension representatives of state or federal programs failed to adequately explain policy to area farmers. He remembered a specific example when a widow and her family "got the erroneous idea" that they had to sell their two hundred head of cattle to the government. He assured her that she could keep her stock, all in good condition and on some of the best grass in the county, and he watched as the relief came to her face—a common response among those he helped during his years as agent. Prowers residents roundly celebrated his assistance; Hamman mentioned a woman who wrote a poem about him and sent "two nice frying chickens" as well as a man who delivered a leg of mutton to the Hamman household in appreciation of Hamman's help with the sheep buying program.[45]

Locals reacted to other aspects of the New Deal in terms of assessing additional programs as well as thinking about what this newly introduced federal largesse might mean for their economy. For example, the Civilian Conservation Corps (CCC) became one of the more popular New Deal programs in the region, and the announced plan of constructing a camp in Springfield, the county seat in Baca, invited support in local newspapers. Residents appreciated the CCC for its potential employment opportunities as well as for the likelihood that the camp would help local farmers by assisting them in tending to their acreage.[46] Other programs like the Works Progress Administration (WPA), the Public Works Administration (PWA), and the Civil Works Administration (CWA) helped alleviate unemployment by hiring Coloradans for infrastructure projects. The WPA employed nearly 150,000 Coloradans between 1933 and 1942. Funding by the WPA in Baca amounted to $1,064,021 and led to the creation of several landmark sites, including Springfield City Park, the Baca County Courthouse Annex, the Two Buttes Gymnasium, and the Vilas School.[47] In addition, the WPA helped construct airports, roads, post offices, firehouses, bridges, and other buildings across the Colorado plains, many of which still stand.[48] In these ways, the New Deal not only hired up the region's unemployed but also slowly changed the area's landscape.[49]

While many families remained on the relief rolls during much of the 1930s, residents never missed a chance to capitalize on a more activist government that might be willing to do even more to help them survive the down years.[50] Locals' request to establish an Arkansas Valley Authority based on the blueprint provided by the Tennessee Valley Authority represented the most compelling example of such opportunism.[51] The Lamar Chamber of Commerce introduced the idea that the centerpiece of such regional development could be the Caddoa Dam, which could provide irrigation to local farmers by storing Arkansas River water in a reservoir to be used during dry periods. Such irrigation was indispensable, some argued, though "it might be an exaggeration to say that the Arkansas and its tributaries are to the Arkansas valley what the Nile is to Egypt." A dam along the river would ensure production and economic stability, promoting confidence in local farmers who wanted desperately to control some component of their future economic outlook in a time of frustration and desperation.[52] In this climate of federal intervention and largesse, locals like Fred Betz, editor of the *Lamar Daily News*, tried to persuade politicians to subsidize (or outright cover) construction. Betz contended that the Caddoa Dam represented "Colorado's greatest opportunity to share in the expenditure of the public works funds" and promised "an adequate water supply for the irrigated lands of the district, which in turn will mean the future prosperity of the community." He realized that while many locals wanted to construct the dam, it would take "united support of the citizens of this entire region" as well as concessions by the federal government to finance the build.[53] Debates about the likelihood of construction and the various benefits it promised for local residents continued well into the 1930s; however, what had been a rather nebulous and inchoate discussion of what could happen if the dam could be built suddenly became much more distinct and coherent, an issue of when, with the inception of New Deal spending. Indeed, locals understood this switch, and many started to embrace federal intervention.

That seemingly unlimited budget proved counterproductive in some ways. Everything about the early New Deal seemed to revolve around financial subsidies and federal largesse. There is little evidence from the agents' perspectives that many of the thousands who signed up for production reduction thought much about how easing up on their supply may help the land recover from years of intensive agriculture. This is, of course, understandable given the circumstances and the dire projections for the American economy. That a handful of locals had adopted methods to curb soil erosion and protect their farms outside of any federal programs, by planting trees to serve as shelterbelts for their properties or constructing terraces on their land, is notable. These efforts existed on a minute scale during Hoover's administration, but not because the county agents had not advocated conservation. Extension agents had been pushing for local adoption of soil conservation methods including crop rotation, proper tillage, extensive terracing, and reforestation, but soil and water conservation truly became an issue that warranted much attention from southeast Coloradans after 1934. For instance, by R. E. Frisbie's count, the Baca County agents influenced nearly 1,000 out of 1,420 farm families through Extension education in 1934, but most of those affected focused on the potential windfall from federal subsidies instead of their role in protecting the land itself.[54]

Instead, these early New Deal years demonstrated the FDR administration's intense focus on the agricultural economy by looking at supply and demand instead of reforming how farmers farmed to make American agriculture more sustainable. Most Colorado farmers likely supported this prioritization of the more obviously financial side of this and, fortunately for them, they had reliable county agents and a generous federal government to help them weather the worst of the Depression. The Extension Service cultivated important relations with farmers during these trying times, and farmers had become accustomed to working with and relying on county agents by the mid-1930s. This budding relationship became even more valuable once the weather started to work against farmers on the Great Plains. While old-timers had experienced dust storms, not many people could even imagine the severity of the storms that started to afflict the region in the 1930s. While early New Deal programs had some success and farmers came to better understand the new role that an activist government sought, nothing challenged the government or the residents quite like the Dust Bowl. The agents, more firmly embedded in their communities by the 1930s, were there alongside residents providing the kind of support that only they could offer.

3

Dirt

The dust storms that most people now associate with the Dust Bowl began
in earnest on the plains in 1930 and lasted until 1938, devastating every
aspect of rural life in dozens of counties and several states. The worst storms
hit between 1934 and 1936, and they finally forced the rest of America to
notice what was happening. The stories of these storms moving across the
continent are somehow both familiar and unbelievable. One of the first
really threatening storms that became a national event moved some 350
million tons of dirt from the Dakotas across the northern Midwest, eventually dumping 12 million pounds of dust on Chicago, leaving 4 pounds
of dust for each person in the city. The storm moved onto the East Coast
within two days and covered Boston, New York, Washington, and Atlanta
before it moved off into the Atlantic.[1] Thankfully, it is impossible for most
of us to fully appreciate what it must have felt like to live through that
experience.

Many observers in Colorado seemed to anticipate the dust storms, even
if they failed to predict their severity. Colorado Extension soil conservationist T. G. Stewart offered a "Historical Review of the Soil Conservation
Problem in Colorado" in late 1939. "It is probable," he wrote, "that the soil
conservation problem began in Colorado with the plowing of the first acre
of land about 1839."[2] He was new to the field of soil conservation, having
just transferred from his post as an agronomist for the Extension Service.

Over his career he gained extensive knowledge about farming and understood the challenges. The Dust Bowl represented a wholly different beast, however. He noted that the drought's severity, the wind's destruction, and the land's inability to hold any moisture brought the issue of conservation to a head as early as 1935: "In brief, Mother Nature selected this year to tell us we had a real soil conservation problem in the State of Colorado, the results of several decades of misuse of land or use of land without any plan."[3]

Stewart's post as Extension soil conservationist offered him significant credibility once he decided to sound the clarion call for soil conservation. He presided over an expansive state effort to promote soil and water conservation to remedy the Dust Bowl and ensure that nothing like it ever hit the state again. He pushed for responsible land stewardship in hopes that farmers would protect their topsoil—"the most important thing in Colorado and in the world—more valuable than all of the gold plus all of the silver plus all of the oil plus everything else in the world."[4] Although he utilized such dramatic flair to the point of hyperbole, Stewart usually expressed the need to conserve resources more bluntly. He argued that the Dust Bowl forced farmers to deal with soil erosion. "There is no more West to go to—no new land of consequence," he argued. "Therefore, unless those who operate the land begin to consider the soil conservation problem, our grandchildren will have farms of little value to occupy. We are using and wasting more of our soil resources than belongs to this generation."[5]

Drought had ravaged southeastern Colorado for nearly a decade by the time that Stewart intoned the need to conserve resources for future Coloradans in 1939. The storms and the Depression combined to psychologically, emotionally, and even physically beat and batter area residents. Consider that the *Lamar Daily News* noted in 1934 how southeast Coloradans were "nearly driven to distraction for lack of moisture" and "had taken recourse to superstitions, legends and ancient rituals of late to persuade the rain gods to smile upon them."[6] Concerned weather-watchers offered potential solutions, advising residents to shoot a cannonball into the air to strike a rain-filled cloud or fly dynamite to the clouds via a kite or send a formation of airplanes to penetrate the clouds.[7] Farmers often banded together to offer community prayers, and newspapers routinely advised people to request such divine assistance on their own.[8] Their conviction that the clouds held rain and it would fall with some human invention, to say nothing of appealing to a divine power, demonstrates farmers' search for an end to the drought. What made this hope all the more difficult to maintain was the fact that locals looked to every rainstorm or snowfall as marking the drought's conclusion.

In a cruel twist, these storms often caused even more problems. Heavy rains flooded basements or drowned cars.[9] In an almost unbelievable case of irony, a federal investigator sent to the region to investigate the drought's impact found himself caught in a deluge so severe that his automobile got stuck in wet sand and he was nearly swept away by rising water. Such heavy rains tended to wreak their own havoc because the soil remained so packed, so hard, that it could not soak up the moisture. Consequently, the rain behaved like a glass of water being spilled onto a table; it had nowhere to sink, so it simply dispersed across the surface of the land. The effect produced flooding in the region on more than a few occasions.[10]

Droughts and even dust storms had tormented plains residents for generations and would again in the 1950s and 1970s, but never to the level of the 1930s. The causes of the Dust Bowl were many: the willingness to plow anywhere and everywhere, which loosened the grass and exposed the topsoil; unusually high temperatures and scant rainfall that dried out the land itself; regular and devastating winds that picked up the loosened dirt and blew it across the continent; and soil types more susceptible to erosion. The dust killed livestock, decimated crops, and jeopardized area residents' health through increased rates of emphysema, dust pneumonia, and other respiratory afflictions (see fig. 7).[11]

Residents in the area had two choices: stay and struggle to survive the storms or move out and chance finding employment in a new home. Thousands chose the latter option. Indeed, population declined in Colorado and across much of the Great Plains after 1930. The number of residents in

FIG. 7. A dust storm in Baca County. Photo by D. L. Kernodle, 1936. Library of Congress, Prints & Photographs Division, FSA/OWI Collection (LC-USF34-001615-ZE).

Prowers and Baca Counties eroded dramatically. The Prowers population fell from 14,762 in 1930 to 12,304 in 1936, and the Baca population declined from 10,570 to 6,207 over that same period.[12] Many of these folks migrated west to places like Southern California while others moved around the plains or to urban areas like Denver, where they thought they might be better able to take advantage of federal relief programs.[13] The dust and drought compounded already evident problems throughout this part of rural Colorado, especially in Baca County, and served as the final straw for many to leave. Robert T. McMillan was an assistant economist with the Resettlement Administration assigned to assess Baca County to determine the poverty level, the need for relief, and the overall impact caused by drought and depression. He found that nearly half of all dwellings in the county had been abandoned by 1936. He claimed that the county was slowly filling up with ghost towns.[14]

People who stayed in the region faced declining morale and community support. McMillan found that life in Baca was particularly "deplorable" because "inferior housing, low consumption of material goods, meager community life, and general morale" as well as basic diet and hygiene were unsatisfactory.[15] His study concluded that 40 percent of farm families "reported no participation in church, farm organizations, lodges, clubs, or movies." Such inactivity made the decline look even more severe. Fortunately, families increasingly had access to, and eventually came to rely on, federal subsidies to keep them afloat.

The early New Deal years produced several policies designed to provide relief and recovery while the later New Deal policies incorporated a push for reform. Many policies offered production controls or reductions in exchange for funding. Those dealing with reform likewise offered financial assistance but did so in a way to encourage farmers to dramatically reconsider their role in contributing to the storms and altering their relationship with the landscape. They did so in three ways. First, the Roosevelt administration pushed several submarginal land purchasing programs to retire such land and keep farmers from using it. Second, the New Deal included promotion of conservation methods and techniques that could improve land use practices in Baca County. Local soil conservation districts became the most important example of that promotion and represented the most significant way that most farmers embraced conservation. Third, New Dealers initiated a broad and critical assessment of farming in an arid environment in order to better understand how size of farm and stability of the agricultural economy had become linked. Many experts pointed to the Homestead Act and its allotment of 160 acres as the cause of such problems because farmers on the plains needed more land to diversity their crops, leave some fallow, and own enough livestock to broaden their economic base. In effect, the Dust Bowl compelled New Dealers to consider whether small family farms could make it or if only farmers with more land had a chance. Together, these three elements meant that the New Deal response to the Dust Bowl was to critically assess how the federal and state governments could help ensure that such a catastrophe would never happen again. So while the idea of protecting the land was itself not a novel idea, the New Deal worked to conserve soil on a national level and on private lands. New Dealers acted on the theory that the best way to shore up the rural economy was to balance production with conservation, and New Deal programs allowed locals to opt in voluntarily. In these ways, the New Deal marked a new chapter in land use on the Colorado plains.

Locals played a significant role in shaping resource conservation in southeastern Colorado, illustrating that everyday citizens in fact played a part in constructing, employing, and critiquing New Deal policy. Baca County inhabitants greeted this new emphasis on conservation with more enthusiasm than their neighbors to the north, especially when it came to soil conservation. While Prowers County farmers generally accepted the need to protect the soil on at least a theoretical level, they proved less willing to participate in erosion districts, voluntarily terrace their lands or plant shelterbelts, or abide by most of the soil-saving guidelines provided by the county agents and federal administrators. The soil erosion in Baca was much more severe than it was in Prowers for several reasons. The expansive wheat operations in Baca during the 1920s had led to the Great Plow-Up that loosened soil by plowing under the native grasses, and irrigation in Prowers meant that farmers could manage the moisture levels in their soil better than farmers in Baca. Irrigation provided a safety net, and while the drought and depression affected irrigated farmers, dryland farmers faced much more widespread and relentless erosion. The drought forced Prowers farmers to rethink their land use regimens, but the Dust Bowl affected Baca farmers directly. Put simply, the situation in Baca constituted a more dire and immediate problem, so the response from both farmers and state employees proved more urgent and more comprehensive.

The depths of the Depression, compounded by the severity of the storms, left the Colorado plains ravaged and devastated. The people often became desperate and looked to FDR for guidance, beginning in 1932 with his election. His administration tried to lead the push to save residents on the plains in large part by implementing a sort of scatter-shot approach with policies devised to solve what New Dealers believed to have caused the Dust Bowl. Perhaps more than anything else, New Deal agricultural policy acted on the premise that poor land led to poor people and that improper technique jeopardized the soil's health. Thus county agents and federal experts tried to stabilize the agricultural economy by improving land use practices and by identifying land that should not be under production.

This became clear in the report of the Great Plains Drought Area Committee, which FDR commissioned in 1936 to survey the Dust Bowl, determine the main causes, and develop possible solutions. It called for "readjustment and reorganization" to remedy Dust Bowl conditions (see fig. 8).[16] Colorado Agricultural Extension Agency director F. A. Anderson agreed that Colorado farmers needed to think about readjustment and rehabilitation, critically reassessing where farmers farmed and acknowledging that some land should be free from production or needed rehabilitation. He supported the committee's contention that farmers had to reconsider both how and where they farm to better account for environmental constraints.[17] The federal committee and Anderson agreed that farming on submarginal land represented a main cause of the Dust Bowl and a key concern going forward. The various buzzwords, whether readjustment or reorganization, essentially called upon farmers to consider where and what they planted. They believed that some land should not be under production and in some cases needed rehabilitation or even outright retirement, but farmers effectively tried to bring all land under the plow. Consequently, county agents and federal experts hoped to educate farmers on the need to consider the land's health and ability to hold crops before they cultivated it. They tried to introduce farmers to the issue of farming submarginal land, and then they focused more intently on what to do about the presence of such land on the plains.

The label "submarginal land" emerged frequently in correspondence, institutional memoranda, newspaper coverage, and agents' records in the

FIG. 8. FDR's Great Plains Drought Area Committee stopped in Springfield in Baca County during its investigation of the Dust Bowl in the summer of 1936. Photo by Arthur Rothstein. Library of Congress, Prints & Photographs Division, FSA/OWI Collection (LC-DIG-FSA-8b38397).

1930s. Yet multiple definitions of the word existed, and its meaning changed depending on who used it. Fundamentally, agreement existed about the need to protect land that experts deemed "submarginal," but they did not define such land in the same way. As the historian John Opie points out, "The word 'submarginal' remained poorly understood by government agencies, the public, and the affected farmers; it was not measured by soil quality or water quality, but by the more complex capacity of the farmer to sustain himself on his land."[18] The complex rubric to determine submarginality incorporated soil type and quality, potential crops, and relative health of the soil when assessed, but at its core the determination often reflected the land's potential economic productive capacity. L. C. Gray, one of the most prominent New Dealers to consider land rehabilitation and retirement as head of the Division of Land Economics within the Bureau of Agricultural Economics, hinted at one of the underlying problems with the category of submarginality. Gray argued, "Little, if any, of the land in the Great Plains is 'submarginal' in the sense that it is not adapted to agriculture of some type under proper conditions of tenure and size of holdings." In other words, ownership and acreage helped determine what happened to land, such that 160-acre homesteads under intensive cultivation and home to soil exhaustive crops like wheat would more often become submarginal than a 500-acre farm where some land had been left fallow and the farmer diversified his crops. Use determined submarginality as much as the soil itself. Gray believed that the federal response to submarginal land should be to "make possible a change in the type of agriculture" by identifying how farmers worked and reassess agricultural practices that increased the land's vulnerability.[19]

Gray thus identified the tendency for farmers and federal experts to overuse the term "submarginal" to apply to any land subject to or already affected by erosion. As he claimed, the federal government often considered "nuisance" lands to be submarginal. These were "lands peculiarly subject to wind erosion which are therefore the point of origin for great quantities of silt, sand, and dust that injure other lands and are a source of great discomfort to the residents of the region."[20] For example, Gray pointed to supposedly submarginal lands that worked perfectly for grazing; consequently, he argued for a reassessment of land use that better reflected an accounting of productive ways to utilize specific plots of land. This meant a careful assessment of where and what to farm that utilized modern science and land planning. In short, despite what some farmers may have hoped for, farming was not always the answer, and some land lacked the means to support crops or proved better suited for alternate uses.

At least a portion of farmers decided to break such land for crop production despite such problems, and they engendered a cycle of exploitation that loosened topsoil and made such acreage vulnerable to blowing.[21] Much of this land had been broken according to market demand, such that farmers bought and expanded onto lands that they hoped could produce the desired commodity, even when they had no way to determine whether it could.[22] In that way, submarginal could be used to describe any land deemed unable to meet farmers' expectations; yet even expectations often became confused. The issue of unproductive versus productive and poor land versus good land often depended upon a definition of net or gross value, resale potential, acreage, and utility. Echoing Gray's point that submarginality depended in part on use, John D. Black posited in 1945 that submarginal land did not exist; land could be unproductive and there could be "submarginal use of land," but the land itself could be productive depending on the circumstances. In that sense, farmers' practices determined productivity as much as where they farmed.[23]

Despite differences in how people defined submarginal land, most observers like F. A. Anderson, L. C. Gray, and M. L. Wilson argued that stabilizing the agricultural economy meant addressing the misuse of land not fit for intensive cultivation. In effect, their philosophy reflected the notion that farmers' poverty led to land degradation and that denuded land produced economic turmoil. Wilson was especially sensitive to rural poverty. He spent most of his career trying to understand the causes and consequences of rural poverty while working as Montana's first Extension agent, chief economist for the AAA, director of the Division of Subsistence Homesteads, assistant secretary of agriculture, and eventually director of the Extension Service within the USDA.[24] Wilson postulated that "as we look into the future, and think in terms of the future of democracy, of the kind of rural life that our social philosophy sanctions and of the complexities and difficulties involved, low-income farming becomes our Number One agricultural problem." While multiple variables caused or aggravated rural poverty, Wilson argued that land use mattered most and that "rural poverty tends to be concentrated in areas where the natural resources are exhausted."[25] For Wilson and others, submarginal land only represented one part of the problem; farmers needed to adopt new land use strategies and emphasize conservation or else the issue of rural poverty would spin out of control.

Baca County provided an ideal test case for the hypothesis that rural poverty and the misuse of land remained inextricably linked because, unfortunately, both prevailed in the county during the 1930s. Robert McMillan spent time in Baca County studying rural poverty and its causes among farmers, and he identified many of the social problems that had plagued county residents and explained the declining population. He also viewed land use as a cause of rural poverty in Baca. Repeating an iteration of the mantra of "poor land, poor people," McMillan identified the signature cause of most of the problems in Baca County. As he put it, there was a "close relation of marginal families to marginal lands" and "poor families are located on land which will not produce a living."[26] Yet the situation proved complex, because while it became apparent that the New Deal federal subsidies helped farmers mitigate some of their economic problems, McMillan contended that federal handouts were not an adequate response to rural poverty. He posited that "federal subsidies have served to cushion the impact of drought intensity" but that government funding had inflated people's sense of their standards of living by artificially raising their incomes.[27] Furthermore, he argued, the subsidies were not enough to totally alleviate the issue of poverty in an area so devastated by drought and dust. He found that a quick survey of farm families in Baca showed that they were unable to "meet their [financial] obligations even with government assistance."[28] Consequently, farmers misused the land as they tried to "recoup their losses and meet interest and taxes by speculating on cash crops."[29] The dire economic situation bred reckless production because farmers attempted to stay out of debt or pay off old debts by maximizing their production regardless of how that impacted their land's arability.

McMillan and several New Deal agricultural experts posited that tenancy exacerbated the problem, as did the presence of many small-time operators who lived on plots of less than 400 acres. The issue of tenancy has been a largely forgotten piece of Dust Bowl history, even though New Dealers quickly sensed the need to reconcile it to improve land use. McMillan believed that tenants and small operators were prone to poor steward-ship. He found that most tenants had moved to Baca after 1926, but very few of them were likely to spend much time in the community, so they had little chance to contribute economically or socially. He conjectured that their relative stability depended on the length of time spent in the county, which meant that most new tenants had little to no stability in their lives.[30] Unfortunately, only 11 of the 193 operators interviewed for his study actually "gave Colorado as the state of birth," as most came from Kansas, Missouri, Oklahoma, and Texas. McMillan thought that the recent arrivals contributed to local problems because they "have moved excessively since coming to the county; have located on small farms; have retained a tenancy status; and have followed a cash crop system of farming," which led to unsustainable agriculture. McMillan went so far as to call these migrants "problem" families because he found them mostly impoverished, without ties to the local community, and quick to move on, and he argued that longtime residents looked unfavorably on new arrivals.[31]

McMillan posited that proper land use planning represented the best way to remedy this divide by assisting tenants and small owners to sustain themselves during tough economic times. The key to steadying the agricul-tural economy was not necessarily to eliminate tenancy. He believed that the system itself was mostly sound in that it allowed tenants an opportunity to move toward ownership (even if it was not guaranteed to present that chance). But tenancy aggravated land abuse because it did not ensure stable financial relations between owner and tenant nor did it help small farmers, tenants, and part owners overcome indebtedness. Therefore, he believed, tenants often jeopardized the land's health because they sought to maximize production, often to pay off debts or establish capital for eventual investment. He cited one-crop farming as a significant problem for Baca tenant farmers; market demand dictated what they grew, regardless of the land's capacity for producing that crop, because they needed money. According to McMillan, "One-crop farming is the attendant evil of tenancy and small farms. Farmers on small farms are compelled through necessity to raise crops which will produce the largest returns per acre. Also, the landlord is too often interested in collecting the greatest cash return from the land regardless of soil losses."[32]

Albert Cotton, member of the Land Policy Division inside the USDA during the early 1930s, explained it well in his assessment of farm landlord-tenant relations: "From the point of view of the general public, one of the most serious consequences of the widespread prevalence of farm tenancy is the rapid deterioration in tenant-operated farms." He added, "This tendency has already become a serious menace to the nation's soil resources" and "while many farm owners have adopted extremely bad land-use practices, naturally landowners who will receive the resulting benefits will be more likely to cooperate in future programs of soil conservation than tenants who must do the work but will not stay on the farm to receive the benefits."[33]

The intertwining of tenancy and land degradation compelled New Dealers and Extension agents to focus more intently on changing land use to improve the agricultural economy. As McMillan noted, subsidies did not represent a legitimate, long-term, and sustainable solution to rural poverty. In other words, the land itself warranted attention. The federal government devised two strategies to provoke more focus on land: land purchase programs designed to buy up land that experts had deemed submarginal, and conservation programs that enabled farmers to take better care of their land.

The Resettlement Administration (RA) spearheaded one of the first federal efforts to ameliorate the submarginal land problem through a federal land purchasing program. The RA embodied the federal effort to utilize land use planning, a discipline that became more prominent over the course of the 1920s but eventually occupied a central place in New Deal land use policy.[34] The RA represented a decisive shift in how Americans understood land use, a move away from the long-held notion that the West had an unlimited bounty and farmers could farm anywhere they desired and toward the realization that some land was not in fact arable. The central thesis that created the program revolved around rehabilitation; submarginal land could be retired and allowed to regrow native grass without being consistently broken by the plow while the government could relocate desolate farmers who had gone broke trying to eke out a living on degraded land. The RA removed farmers from vulnerable land and relocated them to resettlement camps, effectively government-created sites designed to afford each family a plot of land where they could efficiently produce and therefore sustain themselves. Two such camps opened in Colorado, the San Luis Valley Farms and the Western Slope Farms, and accommodated 200 families, most of them former residents of plains counties hit hard by the Dust Bowl. The entire resettlement project cost nearly $9 million and relocated a total of 760 households on nearly 90,000 acres in the Mountain West.[35] It had a local impact as well, granting eight tenants the opportunity to take loans from the Farm Security Administration, the organization that effectively took over the resettlement aspect of RA policy after 1937.[36]

Despite this assistance, residents offered mixed reviews. The editor of the *Lamar Daily News* celebrated its efforts at promoting conservation, in assisting families to make a living, and providing insightful farming techniques to increase production on marginal lands.[37] He also noted the importance of the $2 million spent by the RA in Colorado to help farmers refinance mortgages or gain low-interest loans to purchase their own property. The editor contended that the program was in fact not trying to push people off their lands. It was designed to help them adapt to the environmental constraints found in arid regions, and those who grew frustrated with the program misunderstood its intent.[38]

Despite this support, however, farmers and residents found room to criticize the resettlement plan generally and the RA specifically. Some believed that it had a limited impact, and when it did benefit farmers, it slanted heavily toward the poorest owners and tenants to alleviate the most extreme cases of poverty. Federal experts decided where to resettle families and when, so farmers often lacked any knowledge of their new land (and thus faced challenges adapting to climatic differences, new soil characteristics, and planting new crops).[39] That became a point of contention among locals, as did the level of bureaucracy that typified the RA's dealings with citizens that left farmers feeling as though they lacked any control over the program. This reaction aligns with how farmers responded to other programs as well, particularly in terms of the sense that government outsiders dictated terms and expected compliance.

The RA furthered the idea that farmers on submarginal land needed attention, and it accelerated the trend of thinking about how to deal with the problem. Other agencies took up the mantle for resolving the submarginal land problem, and the initiation of federal land purchase programs had a much larger impact on Baca County than anything the RA could have accomplished. The Bureau of Agricultural Economics (BAE) effectively took over for the RA in terms of its land buying program after 1937. L. C. Gray, who headed the BAE during the purchase program, made his thoughts about submarginality known and devoted much of his energy to addressing the problem by dealing with the land rather than the occupants. Consequently, the BAE program focused more on retiring submarginal land to rehabilitate it and take it entirely out of production. The BAE effectively assessed county lands to determine if they were worthy of government purchase for retirement or if farmers had a legitimate chance to succeed cultivating it.

The BAE program left an indelible imprint in Baca County. Its proposal for Baca considered 289,200 acres, most of which sat in the southwest corner of the county and had not been productive crop land. Of that total the BAE deemed nearly 200,000 acres fit for federal purchase. The area included "186 occupied farmsteads, 19 rural non-farm residences, 99 unoccupied houses which are not in ruin, and 71 unoccupied houses which are in ruin." The "proportionate number of abandoned houses which are not in ruin is evidence of the fact that abandonment has been somewhat recent." Only 51 percent of the nearly 200,000 acres proposed for purchase remained under operation while roughly 22 percent sat abandoned and 27 percent could be considered "open native pasture" that "should be classed as blown out native pasture, as much of it now lies barren and is as subject to wind erosion as crop land."[40] The federal effort to buy up both unproductive and abandoned land demonstrated its willingness to keep land that farmers had degraded—or land likely susceptible to eventual exhaustion—out of production and let it return to grass (see fig. 9).

A cross-section of the folks living within the proposal's boundaries further indicated why the government chose to execute such an aggressive purchasing program. The report cited 218 operators working within the proposal's boundaries, and most of them struggled to eke out a living. The relatively high number of tenants and small operators explain why the average annual gross income of people living inside the BAE's proposal equaled just over $1,000. The report's findings suggested that the most profitable farms were not only large (over 400 acres) but also raised either livestock or general (meaning mixed crop/livestock) whereas the small, crop-only farms faced the toughest production challenges. Also, not surprisingly, the most successful operators had the longest tenure on their present properties, implying that once the farm became established, then stabilizing the family income could potentially lead to expansion or at least the ability to sustain one's earnings. Tenants and small owners obviously faced considerable obstacles to achieve stability, to say nothing of the chance that many could eventually prosper in this environment and within this framework.[41]

FIG. 9. "Return to the Grass," July 25, 1938. Retirement from production allowed many plots like this to regrow native vegetation even after having been cultivated. Extension Service, University Historic Photograph Collection, Colorado State University Libraries, Archives & Special Collections.

The BAE report effectively reiterated what McMillan had determined but with BAE backing the extensive and expensive purchasing program. Rather than let small-time farmers and tenants struggle while using submarginal land, the government approached the most vulnerable farmers with a proposal to buy their lands or at least buy off their equipment. For tenants on submarginal land, federal purchase of owners' land freed tenants to move on to other endeavors, hopefully on more productive cropland or in other industries. Struggling owners hoped for the same opportunities. The BAE report for southwest Baca County suggested a price of $2.50/acre plus an additional $1/acre for improvements; the total was thus widely appealing among residents who supported the program. Nearly 60 percent of the operators interviewed reacted favorably while only 7 percent looked at the program unfavorably, leaving some 33 percent without strong feelings either way. To cite two examples, A. C. Hoover noted, "This program is the only way to control land," and A. A. Yarborough remarked, "The government should buy every acre." Local tenant John Harper claimed that he would "probably leave if I could sell my equipment," an indication that the program provided tenants with a measure of freedom to determine how to remove themselves from a losing proposition and cyclical debt.[42] Furthermore, Baca County Extension agent Raymond Skitt claimed that by 1938 the BAE had purchased 201,000 acres for $635,000, a significant sum for an impoverished county and an ample demonstration of the federal government's largesse during the New Deal.[43]

In this way, the purchase plan had a dual purpose. First, tenants and small owners could use an avenue like the purchase program to cut ties with a losing proposition. Second, buying up barren land, with exposed fragile topsoil left open to a devastating wind, could help neighboring farmers who had already taken up the task of protecting their own plots from erosion. Constrained by obvious and highly contentious property rights, an individual owner could not assume responsibility for improving a neighboring farm by planting a shelterbelt or allowing for crop rotation and a fallow period. Furthermore, federal purchase of these denuded lands effectively served other county farmers because land that had been exhausted, even abandoned, often blew the worst. That soil then blew onto neighboring farms and compounded problems for those owners trying to control erosion on their own land. The purchase of such land and the effort to stop the blowing provided a real public service. The federal purchase and retirement program, then, benefited the seller and his neighbors and marked one of the more formidable components of New Deal agricultural policy. Indeed, one of the hallmarks of New Deal conservation was the way that New Dealers brought resource use into focus by including private lands rather than simply looking to conserve public resources in parks, forests, and elsewhere.

New Deal policy started to complicate personal property rights by levying fines on owners whose inactivity or disregard for soil erosion threatened their neighbors' productivity. Farmers and New Dealers began appreciating "the incompatibility of human boundaries and forms of mobile nature—water, soil, and organisms—that those boundaries could not contain." In other words, nature, whether weeds or drifting topsoil, would not conform to the "straight edges and right angles" that constituted the grid landscape common on the Great Plains.[44] This applied to abandoned land as well as private property left unattended by negligent owners whose degraded land infringed on neighbors' plots. Such nuisance lands threatened everyone's land, such that even those who tried to address soil erosion on their own property often had to deal with soil blown in from negligent neighbors as well as from abandoned land. For example, the RA report on Baca County showed that individuals owned nearly 900,000 acres while more than 740,000 acres sat outside such "organized units" (abandoned or not privately owned) and were thus "subject to wind erosion and uncontrolled grazing, with the possibility that unless remedial measures are taken, there will be a repetition of the dust storms of previous years, and possibly a continuing growth of this menace."[45] By contemplating action on these acres, some abandoned and some left to pasture, federal officials showed a willingness to extend influence over unclaimed land and address the blowing land problem. Furthermore, with programs like the BAE's purchasing plan and the same exercise under the RA, the federal government could buy out negligent farm owners to protect surrounding acreages. In that way, the government aimed to address the worst land, making it more likely that neighboring farmers might prosper without facing blowing dirt.

In addition to utilizing federal purchase programs and resettlement, the New Deal and Extension Service pushed soil conservation to help private owners protect and reinvigorate their acreage. In the early 1930s the notion of soil conservation was still fairly novel. The Soil Erosion Service (SES), created in 1933, was a temporary organization designed to "serve as a jobs program, not to eliminate soil erosion."[46] The Dust Bowl's severity finally convinced lawmakers, and indeed the public, that the issue of soil erosion was a public problem, what SES head and chief advocate for soil conservation Hugh Hammond Bennett called "a national menace." After convincing Secretary of Agriculture Henry Wallace and Secretary of the Interior Harold Ickes that he should run the SES, Bennett successfully lobbied Congress in 1935 to create the Soil Conservation Service (SCS) as part of the Soil Conservation and Domestic Allotment Act. The SCS had a much larger budget and more personnel; it therefore had many more resources to combat soil erosion in American fields. Bennett used the SCS to implement a national conservation program designed to extend soil conservation across the country.

Soil conservation advocates received federal assistance through the Soil Conservation and Domestic Allotment Act from 1936. That act adopted pieces of the 1933 Agricultural Adjustment Act in that it promoted production reduction as a means to control supply. Yet the act also pushed to conserve the soil, largely by subsidizing farmers who participated in the program. The Agricultural Conservation Program, the name given to the subsidy side of the act, funded farmers directly when they reduced acreage or used SCS-approved practices to curb soil erosion. The program represented a sort of shared responsibility and national response to the soil erosion issue as public monies went directly to farmers for practicing conservation. In effect, the federal government passed two significant pieces of legislation nearly a year apart, both designed to address soil erosion and incentivize agricultural conservation.[47]

While the SCS eventually employed a purchase program, its main goal was to push soil conservation on arable lands in order to keep farmers on the land by promoting good stewardship. A SCS memorandum on the agency's relationship with other government programs announced its direction succinctly: "The basic purpose of the Soil Conservation Service, broadly stated, is to aid in bringing about desirable physical adjustments in land use with a view to bettering the general welfare, conserving natural resources, establishing a permanent, balanced agriculture, and reducing the hazards of floods and siltation." It achieved these goals through "technical and material assistance" and "submarginal land purchase and development."[48] In essence, then, the SCS was principally worried about how to revive abused lands, and it needed the farmers' help to do so because only with their consent and assistance could erosion be managed on private lands. In addition to the educational component, the SCS and the Agricultural Conservation Program offered subsidies to farmers who participated in erosion control programs. That financial incentive helps explain the agency's eventual success, because the New Deal often succeeded only when spurring action through the promise of funding.

Unfortunately for Baca farmers, their soils offered tremendous opportunities to practice soil conservation. John Underwood conducted a soil survey in Baca County in 1944 on behalf of the USDA that covered about 75 percent of the county (more than 1 million acres). Underwood's analysis relied on a 1936 SCS study about the soil's characteristics, including its physical makeup, its composition, its ability to hold moisture, and its susceptibility to erosion. The 1936 study was part of the SCS's effort to conduct similar soil surveys in places across the country to construct a soil map of the nation's lands. Underwood's most surprising conclusion about Baca County soils was that it lacked the ability to maintain moisture. Even though the soil itself was quite fertile, it lacked the means to hold water such that even some soil that could potentially be high in fertility could not be very productive in such an arid climate.[49] Underwood used what is called the "capability classification system" that shows "in a general way, the suitability of soils for most kinds of field crops." It judges the soil's limitations, the risk of damage when it is used, and the potential for rehabilitation. He noted that two categories of soil, class III and class IV, could be farmed with some success, but only class III could truly support high levels of cultivation. Even though it stood as the more arable of the two categories, even class III land "must have intensive erosion-control or management practices for safe and permanent cultivation" because such lands "are highly susceptible to wind erosion." Class IV only offered limited cultivation potential and required intensive management, including extensive terraces and contouring to maximize rainfall retention. Overall, Underwood argued, much of the class IV land should be put back to grass and likely used for a livestock feed-crop economy. Other district land fell into class VI, class VII, or class VIII land and best qualified as range land, with a focus on revegetation or retirement, since much of it had "dropped to low carrying capacity as early as the eighties."[50]

The bleak results from the classification process become even more daunting when one considers the amount of land deemed unfit for cultivation. Class III land, the type most amenable to agricultural production, constituted the least of the four main categories at 235,726 acres. Most land fell into the class IV category that Underwood deemed suitable for limited cultivation with an emphasis on production for livestock rather than market. He tallied a total of 269,150 acres in the two districts that should be immediately restored to native grasses (about 21 percent of the total area), but almost half of that amount had already been put under the plow. Underwood, like many other observers who witnessed the level of erosion in Baca, realized that land use adaptation was necessary for farmers' survival. Underwood recommended a higher percentage of livestock farming, a move away from cash crops—and especially wheat—as well as a reduction in cultivated land, and a change in the land tenure system to promote better stewardship among tenants.[51]

The SCS had the resources to help even though it faced a stiff task convincing Baca farmers that changes to their land use regimens were necessary. The SCS also set out to distinguish between land beyond repair and workable land in need of rehabilitation, and that distinction then became the basis for instituting either a purchase or conservation program for the corresponding acreage. H. H. Finnell, an agronomist and erosion specialist who served as regional conservator of the Southern Great Plains and head of the Region VI Soil Conservation Service during the 1930s and 1940s, noted in 1941 that the SCS had been quite active in purchasing vulnerable lands in Baca County from 1936 to 1941. According to his records, the SCS spent nearly $800,000 to buy 782 tracts of land totaling more than 250,000 acres in Baca County. The s c s worked in tandem with other agencies to procure hundreds of thousands of acres in Baca, and its manic efforts illustrate how much the federal government was willing to spend to gobble up and retire submarginal land. Yet while the purchasing programs represented a considerable federal expenditure, Finnell realized that the land purchase program did not represent a significant victory by itself. He admitted that he had a difficult time gauging the agency's success in Baca because 302 operators who had been working with the s c s moved out of the region. Finnell understood that several factors were at work in pushing people from the region, but the high rate of migration served as a reminder that the s c s had not done much yet to keep people on their land. Indeed, Finnell noted that migration had become so acute in some places that the "small towns of Stonington and Richards have become 'ghost towns.'" In effect, buying out individual proprietors helped salvage the submarginal land and sent the former operators on their way with some financial support. Yet it proved insufficient to tackle the problem of land abuse and did little to redeem the remaining population.[52]

The s c s offered additional resources to southeastern Colorado farmers almost immediately upon its creation by Congress in 1935 to support farmers and keep them farming, albeit under circumstances more amenable to resource conservation and sustainable production. According to Bennett, it was a "research and demonstration agency" that focused on constructing a "research program to determine the best and most economical methods of erosion control." It used demonstration projects to show farmers the benefit of land use planning and various erosion control techniques. It worked on public lands to employ erosion control. It managed several c c c camps to help localities deal with erosion by providing labor and additional machinery. Finally, it worked with State Extension employees to reach local farmers, "to make the facts developed by our program available to farmers everywhere, and to supervise and assist farmers and groups of farmers, wherever possible, in applying erosion-control practices to the land." Bennett's underlying hope, and the s c s's primary goal, was to "make possible a fundamental change, farm by farm, and for agriculture as a whole, from an exploitive type of farming to a conservative type."[53]

Indeed, SCS representatives promoted many methods and techniques that helped farmers address soil erosion. For example, SCS employees (and other advisors including county agents) pushed strip cropping, a technique that meant to provide a natural barrier to wind and water erosion by growing crops parallel to the land's contour to mitigate runoff and blowing. The central idea involved farming at right angles to the natural slope to lessen erosion and protect fertility. Planting cover crops similarly helped arrest erosion because the low-rise crops were usually drought resistant and held the soil in place with their roots. Farmers planted cover between crop rows or as part of a rotation regularly to protect against erosion and replenish soil fertility. Efforts like this included the furrow, where farmers plowed deep enough troughs to catch water, keep it, and diminish the likelihood of run-off (fig. 10). They also constructed terraces to halt erosion; the terrace popped up across the contour to intercept runoff and corral water when possible (fig. 11). They planted shelterbelts to serve as windbreaks and soften the gusts that tore through the plains (fig. 12). Each of these techniques aimed to save both soil and moisture, satisfying the two most important goals that plains farmers had to meet to sustain their livelihoods. In total, the SCS offered a plethora of choices for farmers to utilize.

The SCS presented the Extension Service further opportunity to influence agriculture in the region by helping to foment farmer enthusiasm. In fact, the SCS leaned heavily on Extension to get its word about soil conservation across to local farmers, and county agents had been pushing for reforms since at least the late 1920s. Agents approached local farmers in different ways, yet agents most often remained passive, acting as educators and advisors to folks who sought their assistance. They broached subjects like soil conservation but also invoked agronomy and land use planning. For example, the Extension Service published and distributed a pamphlet in 1935 entitled *Keeping the Farm at Home*. The pamphlet reminded farmers how important proper practices were to protecting one's land, especially given how the primary audience found itself during the Dust Bowl. It emphasized the need to utilize the contour when plowing and planting to take advantage of the soil to retain water if applicable or at least terracing on sloping lands, as "the most economical and effective method of reducing soil blowing on farms in eastern Colorado." The pamphlet also highlighted the importance of crop rotation and the necessity of allowing some acreage to lie fallow to improve the soil's recovery.[54] While only a few pages long, this example of Extension Service literature exemplifies the ways that agents tried to help.

While this federal assistance certainly helped, the most important shift in how farmers adapted their practices in response to the Dust Bowl actually came from action on the state and local levels. The 1937 Colorado Soil Conservation Act constituted the best example of efforts to lead the charge against erosion. Its passage showed that the concern for soil conservation did not reside solely in Washington DC, although the SCS pushed for such legislation so that the state governments could share some of the collective burden for education and funding conservation projects. FDR impressed upon governors that states should initiate legislation to control erosion, and he signed the USDA-sponsored Standard Soil Conservation Districts Law that allowed for local and state representatives to deal with the problem. The legislation reflected the desire to let farmers establish a soil conservation district as a locally led body to control erosion and to stoke support in their communities for federal and state erosion control programs. In effect, the law extended federal assistance to districts to ensure that residents, those most familiar with local conditions and environs and thus best suited to lead the fight against erosion, had support to conserve their resources. The districts thus married local autonomy with federal financial and instructional support, constituting an impressive and powerful weapon against erosion.[55]

Instability in the Colorado agricultural economy and the prevalence of dust storms and drought along the plains finally pushed the state legislature into action in 1937. The act's authors noted that wind and water

erosion affected "approximately six million acres, or one-tenth of the total area of the state." Those losses were "caused largely by improper farm and range practices," specifically the attention to cash crops even when such attention exhausted soil, unwillingness to let land sit fallow or rotate crops, and general malaise toward conservation. They could only be remedied through united federal and state efforts to conserve resources. Only with such legislation empowering local boards could the legislature "insure the health, prosperity and welfare of the State of Colorado and its people."[56] The consequent Soil Conservation Act resembled an Extension Service procedural document from 1935 that emphasized community organization and common sacrifice to combat erosion. The 1935 memorandum stressed the need for county agent and SCS cooperation and that both understood "that little immediate good can be accomplished by working with isolated individuals." Instead, successful and meaningful conservation required "organized associations covering an entire erosion area." Advisors should focus on those areas where land had the potential for profitable agriculture or when submarginal land threatened such areas, meaning to sustain successful farmers and rescue the vulnerable ones. Addressing denuded land only drained resources. The directive outlined the push

FIG. 10 *(top)*. "Soil Erosion Conservation," July 25, 1938. The picture shows furrows designed to hold water, thereby conserving the moisture and preventing runoff. Extension Service, University Historic Photograph Collection, Colorado State University Libraries, Archives & Special Collections.

FIG. 11 *(center)*. "Soil Erosion—Better Crops on Terrace," July 26, 1938. The picture shows crops grown on a gradual terrace system to mitigate erosion. Extension Service, University Historic Photograph Collection, Colorado State University Libraries, Archives & Special Collections.

FIG. 12 *(bottom)*. "Dry Land Shelterbelt," July 7, 1938. This man stands in front of a maturing row of trees designed to reduce wind erosion. Extension Service, University Historic Photograph Collection, Colorado State University Libraries, Archives & Special Collections.

for education as a means to ensure farmers' participation in the program and their willingness to practice erosion control "over a period of years" instead of simply conserving soil as a temporary response to Dust Bowl devastation.[57] The Colorado Soil Conservation Act reflected the sense of cooperation, the Extension Service's centrality to conservation, and the need for local input that Extension had outlined.

As written, the bill appeared comprehensive in addressing the need to concentrate on erosion and also to establish parameters that gave local boards the jurisdiction to conduct their business. A central board in Denver managed the statewide efforts to create districts and presided over any potential legal matters emerging on the local level. If, for instance, neighbors quarreled or litigation arose when individuals tried to opt out of district programs or grew tired of participating, then the local boards could consult the state board. Beyond such conflict mediation, the state board left most of the responsibility for the day-to-day operations to locally elected conservation district boards. Two members of the State Planning Commission, the director of Extension, and the director of the Experiment Station in Fort Collins formed the state's board.[58] The presence of land use planners, Extension personnel, and agronomists shows the prominent place that Extension had in conservation efforts as well as the push to consider local needs. New Deal agricultural programs utilized the same formula, relying on local expertise as well as planners and organizers to encourage farmers' engagement with conservation programs, all run through the Extension county agent. In that manner, the board's composition reflected the broader push to tackle land use problems in response to the Dust Bowl.

The act provided local boards some autonomy by promoting, but not compelling, district formation and by bequeathing locals the power to direct district business. Indeed, any five residents could petition the state board to create a district, and once the board approved, then locals voted on whether they wanted to establish a district. If a basic majority voted in favor, then the district chose a supervisory board; as with the state board, the local representative of the Extension Service had a place in that body.

The local boards included judges to hear and decide on local matters, including appeals, as well as supervisors who oversaw the district operations. These supervisors had an expansive purview but had the financial and organizational support of the state and county governments, which undoubtedly made their responsibilities more manageable. Once farmers sanctioned district formation, then the local district board effectively governed farmers within the district's borders. In sum, individuals looking to practice and promote soil conservation now had an organized, well-funded, and resolute resource to combat erosion in their county, and they had a level of autonomy and control they had never enjoyed when working with the SCS or AAA.[59]

The Colorado legislation empowered the local board to extend its influence over almost any aspect of local agriculture if it determined that something or someone had an adverse impact on erosion control. This meant rather typical responsibilities like assessing local soil conditions and identifying the most vulnerable areas, conducting demonstration projects to show farmers how best to inhibit erosion, and building structures or facilities to arrest erosion. The board could "furnish financial or other aid" to "any owner or occupant of lands within the district in the carrying on of erosion control and water conservation practices within the district." Therefore, it could provide access to "agricultural and engineering machinery and equipment, fertilizer, seeds and seedlings" or anything else that might be of use to local farmers. It could also look to the federal or state government for further assistance, either applying for grants or appealing for loans directly, if the local district faced economic strains.[60] With such support and an appreciation for local concerns, the district then designed a system to care, treat, and improve local lands within the confines of a reasonable budget.

The Colorado Soil Conservation Act offered boards surprising power in that the districts had the ability to implement erosion control techniques on *any* lands within the district, including privately owned parcels. According to the act, the local board had the authority "to acquire, or acquire control of, by purchase, exchange, lease, gift, grant, bequest, devise, or otherwise, any property, real or personal, or rights or interests therein" if that property sat within the district and could be afforded.[61] The freedom to purchase or lease any eroded lands, and the authority to devise a conservation plan and then institute it on that land, gave the board surprising power to wield over residents within the district's borders. Any owner within their jurisdiction effectively found his or her property rights amended by the state law; the board had the right to enter anyone's personal acreage and "do such work as may be necessary in their opinion to prevent the erosion of its soil or damage to other lands within the district." The law required the board to submit a letter in writing to the owner letting him know of their intention, after which point they could do anything they deemed necessary to prevent erosion or repair damage for the next calendar year. The owner had a chance to appeal to the local and state boards, but verdicts favoring the district or that found owners who failed or refused to abide by the district's rules and regulations "shall be deemed guilty of a misdemeanor and upon conviction shall be punished by a fine of not more than One Hundred dollars."[62] The legislation thus allotted considerable power to local boards and provided them with legal rights to proceed as they saw fit in corralling absentee owners, negligent tenants, or resistant landowners who lived within the district.

Farmers seemed more openly and fully to embrace the district idea once it became clear that they would have autonomy and power. Indeed, the decision to create a board in Baca in 1938 garnered significant praise across the region. The *Garden City Daily Telegram* noted that the residents' vote "contains a word of encouragement to western Kansas." A simple majority could establish the district in Colorado, whereas a 75 percent vote was necessary in Kansas, something the paper lamented. The paper implored Kansas farmers to take note.[63] The *Pueblo Star Journal* applauded the vote and claimed that landowners "have drawn up battle lines for a fight against the ravages of wind and erosion in one of the most severely stricken sectors of the nation's dust bowl." The importance lay in their collective decision to fight erosion, a testament to them and the possibility that other farmers may eventually unite, combining "courage and determination" with "all available scientific information" and setting a standard for other agriculturalists.[64] The *Rocky Mountain News* heralded the decision as a step to restore the land, remedying the consequences levied by previous farmers who denuded the land for the sake of "progress."[65]

As much as newspapers supported conservation or legislative permission allowed for it, however, the responsibility for forming districts and promoting conservation still rested with farmers. Baca farmers took up the mantle of conservation and established three districts between 1938 and 1941, with the first district forming in West Baca in 1938 by a vote of 275–142.[66] By 1941, the three districts in Baca County combined to include 1,324,040 acres and organized with more than 700 of the 906 farmers in the county. As author of the state report, A. J. Hamman congratulated Baca farmers for such widespread support.[67] The district formation demonstrates a few key points about the importance of conservation during the period as well as the sometimes fickle relationship that locals had with the expanding state. The districts demonstrated that farmers were willing to unite in common cause, tying their individual fate to their neighbors in a show of solidarity in the fight against erosion. Locals seemed very much aware of their responsibility as well as their opportunity to manage conservation efforts, and they capitalized on the district model.

Their participation suggests the broad appeal that soil conservation via the district model had for Baca farmers. The Western Baca County Soil Erosion District supervisors identified conservation as a key issue for members. The board claimed that "the difficulties in the District are the result of the plowing of lands not adapted to the production of cultivated crop" and "the use of lands for which they are not naturally adapted has been a large factor contributing to the severe wind erosion of soils in the District." The members faced "poverty, heavy indebtedness, enormous relief costs, and other economic and social difficulties, and the picture in many localities is a most discouraging one. Everything seems to be 'gone with the wind.'" The board pushed education, a new tenure system that supported tenants and encouraged tenant conservation, the end of speculative farming, and federal financial assistance when necessary. It hoped for a "complete readjustment of the agriculture of the District" to better account for aridity and farm size, to promote crop diversification and not just cash crops, and to encourage resident operator involvement and cooperation.[68] Board members believed that "the success of the whole program rests upon resident operators. Without their interest, initiative, industry, and cooperation, the technical and financial assistance furnished by the Government will accomplish little."[69]

The district boards often pushed the same agenda as federal programs but did so with local leadership and control. For example, the districts tried to gobble up susceptible lands by buying them with federal and state financing and then either leasing them to folks willing to remain vigilant against erosion or selling them to farmers willing to participate in the district. In that way, the districts received a return on their investment from the renter or buyer and simultaneously ensured that the lands enjoyed proper and conscientious stewardship. For example, the West Baca district leased nearly 40,000 acres in 1938, listing almost half of that to cover crop and additionally removing a portion of that total from production entirely. It then sublet almost 10,000 acres to private operators who had agreed to secure the land against further erosion. In this respect, the districts represented an ideal blend of local and federal action united to stabilize farming by adapting practice and method; this was perhaps the most important indication that the Dust Bowl compelled farmers to change.[70]

The Extension Service played a vital role in working with districts, beginning with the effort to earn their establishment and continuing through the district's formation and management, and thereby expanded agents' role in promoting conservation and allowed them to be more active in the process by working directly with farmers. This became evident in Baca's districts rather quickly. Baca County agent Raymond Skitt typed out the proposal for the West Baca district and submitted that form to the federal coordinator for the Southern Great Plains, Roy Kimmel (see fig. 13). In it he defined his role as cooperating "in the educational field, fostering desirable practices and proper use of land in retirement, restoration and general farm program." In that way, the agent's relationship with locals changed very little, as the agents had been tasked with education about and management of land use reform since the program's inception. What had changed was the newfound opportunity for locals and agents to combat erosion. Skitt outlined the importance of working with federal agencies, the need for federal credit extension to purchase land and machinery, the necessity for the local board to work well with the federal bureaucrats, and the importance of keeping everyone in the district on the same page.[71] The agent helped orchestrate such developments because of his standing on the district board.

The agents' activities on district boards offered another opportunity to immerse themselves in their communities. In Skitt's case, he became familiar enough with the situation in Baca that he presented an economic diagnosis of county problems and outlined the ways that farmers could capitalize on federal and state funding. Most farmers lacked available capital to buy machinery to contour their fields or employ labor to terrace or build shelterbelts, so funding that the district model made available proved crucial for farmers to band together to share the burden of initial expenditures. As he noted, few farmers could expand to the point where they achieved enough crop diversity and maintained enough livestock to allow for self-sufficiency. Most farmers needed low-interest loans or credit extension to start the rehabilitation process. Extension was in a prime position to facilitate lending programs, and Skitt informed his constituents of their availability. Even then, however, Skitt worried that farmers had no collateral: "the ability of operators to offer up tangible security for loans at this time, due to the general wind erosion hazard, is negligible." He hoped that the SCS and Extension Service could work with other federal agencies to negotiate favorable loan terms for district farmers. He also reassured Kimmel that farmers often needed financial assistance to practice soil conservation, a strategy that helped put the onus on the state and federal agencies to offer amicable lending terms to farmers.[72] By doing so, Skitt again showed that conservation required a bit from the local, state, and federal levels, a kind of cooperation that often meant federal financing and local practice.

We can see this in the formation of districts as well as in the ways that agents worked with farmers. Fortunately, most agents kept meticulous notes and tallies about how many people were impacted by conservation education, who proved willing to practice new methods, and how much those new methods helped local agriculture. The number of participants grew over time, and farmers became more willing to adopt conservation by the late 1930s. For example, Leo Oyler served as Baca agent from 1934 to 1938 and witnessed farmers' growing support for conservation. Oyler entered the job brimming with enthusiasm. From December 1935 to November 1936, he made more than four hundred visits to area farms, conducted ten training meetings, published nearly one hundred articles, some of which made their way into local newspapers, wrote more than 1,700 letters, and held seventeen method demonstrations for almost eight hundred people. During that same time twelve local farms had planted trees for reforestation and to act as windbreaks/shelterbelts. Farmers also placed 478,000 acres under "terracing and erosion control."[73] The numbers increased the following year, and Oyler used more specific labels and numbers for exactly what locals did to fight erosion. He claimed that farmers utilized some form of erosion control on 1.2 million acres in the county in 1936, using various methods like strip cropping, growing cover crops, terracing, and planting shelterbelts. These efforts predated the district, an indication that the district helped continue farmers' efforts.[74]

The momentum that Oyler noticed in 1935 and 1936 continued. Raymond Skitt, who took over for Oyler in 1938, counted 1.25 million acres under some form of erosion control, with farmers again planting cover crops and growing crops on the contour, but also retiring some lands to summer fallow, rotating crops, constructing terraces, and building small

FIG. 13. "New Dust Bowl Program for Southwest," June 2. "Secretary of Agriculture Henry Wallace today appointed Roy I. Kimmel, of Amarillo, Texas, to coordinate a broad federal program to rehabilitate the soil conservation, resettlement, and AAA wind erosion program in 100 counties, comprising 140,000 sq. miles, in Texas, Oklahoma, Kansas, Colorado, and New Mexico. Kimmel has been in charge of the RA Rehabilitation Program in the southwest, his appointment is effective for an indefinite term." Shown left to right are Wallace, Kimmel, and Assistant Secretary of Agriculture M. L. Wilson. Photo by Harris & Ewing, Library of Congress, Prints & Photographs Division (LC-DIG-HEC-22798).

dams to control potential floodwaters from further erosion. Per Skitt, nearly 66 percent of the almost 1,800 farms in Baca were practicing some erosion control method, and most utilized multiple techniques.[75]

Other programs also made some headway promoting conservation while Oyler still presided as county agent. For example, the Wind Erosion Control Program, run by the Extension Service, paid unemployed residents to contour plots or build terraces or plant shelterbelts, promoting conservation while helping to alleviate some of the unemployment rampant among townspeople in Baca. The program used federal funding to execute an emergency listing program that did not require much oversight or coordination. Guy Dickerson applauded the program and claimed that Baca farmers were impressed: "Talk to anyone in this county and the answer is about the same, i.e., Best program we ever had. Quicker action with less red tape. More good for the money spent; less cost for overhead. The only way we could have worked our ground." Oyler noted that "the emergency listing program has been the most worthwhile program of any yet submitted and probably with less money spent in administration of any of the programs."[76] In short, federal funding, low overhead, and action translated to a broad base of support; bureaucracy, red tape, and too many experts only slowed progress. If they agreed with the program and the benefits proved tangible, Baca farmers seemed willing to participate.

Notably, similar developments occurred in southern Prowers County, but Prowers farmers seemed less willing to participate. The best indication of this is that Prowers farmers only organized their first soil conservation district in 1943. Yet dryland farmers demonstrated some concern for erosion and conservation, especially in the ways they sought the help of county agents like A. J. Hamman and Jack N. French well before that year. Hamman, a Prowers County agent from 1934 to 1937 and then a soil conservation specialist for the Extension Service, tallied a significant number of letters, bulletins, and demonstration/training meetings in 1936. He claimed that twenty-seven families had planted shelterbelts to protect against wind erosion and that nearly 140,000 acres fell under terracing and listing programs to defend against soil loss.[77] Dryland farmers needed government subsidies to keep them afloat, Hamman noted, but their embrace of conservation measures and federal programs to that end had made some difference. French's summary showed that the number of farmers who employed such tactics increased over the next couple of years. The number of farms using shelterbelts, making improvements for wildlife, growing crops on the contour, and rotating their crops grew from 1937 through 1941, an indication that some Prowers farmers slowly started to practice conservation on their lands. French noted that the federal investment on farms in Prowers County through the Agricultural Conservation Program neared $200,000 in 1938.[78] While this assistance proved incredibly valuable, the lack of a district in Prowers showcases how much more willing Baca farmers were to actively manage conservation in their county.

Districts afforded members remarkable flexibility to influence other farmers' practices, and in so doing they gave residents a chance to compel neighbors to conserve resources. Indeed, members could identify infractions committed by farmers within the district, raise charges against them, and seek damages from the guilty parties. As one can imagine, this unprecedented power that enabled farmers to control what others did on their own property often produced intense debate and even resistance. For example, the district's power came into question when the West Baca Soil Conservation District mandated that no resident within the boundaries could break new sod during 1938. The board of supervisors voted to restrict expansion to inhibit soil erosion; they clearly understood that extensive sod breaking and tearing up the topsoil had contributed to their present plight. They decided that no one should break new land without approval from the Board of Supervisors, no restored land should be broken without such approval, and all land should be treated for erosion. If an owner proved unwilling or unable to follow through with those provisions, then the district could act. One such case of the district seeking redress from a negligent owner exemplifies the push by conservators to protect themselves from neighbors' blowing land.

The challenge to district authority came in late spring 1940. The episode exposed two very different versions of property rights and illustrated the board's authority. The West Baca Board of Supervisors sent a letter to Lauriston Walsh informing her of a potential $100 fine because she had broken new sod, flouting the mandate proffered by the board in 1938 and still applicable in 1940. Mrs. Walsh's response to Jas O. Dougan, secretary-manager-treasurer of the West Baca district, argued that the government must somehow reimburse owners for restricting the ways they chose to use their own property. Dougan replied, explaining the importance of "the future benefit of this land" and his intention to "protect your investment as much as possible since a few years of wind erosion would remove the only top soil you have" and decimate the acreage.[79] Walsh rebuked Dougan's initial request and explained that she had every intention of caring for the land and protecting it; she lived in New York but promised that she would "see to it that the responsibility for its control is definitely placed in the hands of one of your local residents."[80] Walsh added that she had never voted to establish the district and therefore deserved some chance to defend herself from the board's arbitrary decision constricting her rights as a property owner.[81] Dougan's final letter defended local farmers and their efforts to combat erosion by indicting Walsh for being an outsider: "Of course you realize the fact that the Board of Supervisors of this District is composed of land owners who have *remained* in this county for many years and who are interested in making *their home here*. They *are* those who have faith in Baca County and because of this faith they are determined not to permit stabilization work that has been accomplished to be destroyed."[82]

The exchange represented bigger issues at play during the early stages of the New Deal conservation state. Dougan's emphasis on Walsh as an outsider illustrated a fundamental tension between locals and outsiders that existed in many rural locales. Many longtime county residents held absentee and large owners responsible for not maintaining their farms. In fact, ample evidence of animosity between locals and outsiders emerges in agent records as well as local newspapers, and that animus may have caused the district to act.[83] Yet the theme of trying to control private land through a public mechanism like the district was in fact central to the eventual vote in favor of the district and presumably at the heart of why Dougan chastised Walsh. Indeed, one of the more notable regional meetings on drought and the Dust Bowl, held in Dalhart, Texas, in November 1936, gave plains farmers the opportunity to discuss both the main problems and potential means to stabilize regional agriculture. The Colorado delegation complained that the biggest issue confronting plains farmers was their inability to control negligent owners' lands. It implored the state legislature and curried support from other delegations to pass a law to declare submarginal land "a public menace." They wanted to employ some "method by which an owner-operator or community can compel the owner of the land to prevent his soil from damaging adjacent land by approved methods of control."[84] The Colorado Soil Conservation Act included similar language, as did the proposal for the West Baca district.

District board members proved more than willing to exert their authority over district members, and they also frequently challenged the federal government. Locals felt free to chastise government employees or blame federal programs for not meeting their expectations. For example, and unfortunately for H. H. Bennett and the SCS, local farmers hesitated to give full approval to the SCS and to federal intervention in their affairs more generally. To many observers, the federal effort levied by the SCS and other federal agencies often seemed misguided. To them, federal involvement was the equivalent of outsiders using book knowledge to teach locals with farming; the supposed experts had little to share about conditions or climate or crops that could assist the seasoned farmer. This criticism echoed earlier arguments against the county agents as interloping experts taxing the county budgets without offering anything of substance in return. At the same time, just as locals eventually warmed to the agents, they embraced certain components of the SCS message rather than dismissing the agency altogether. That selective approach proved a prominent characteristic of how Baca farmers approached federal intervention throughout the 1930s.

Letters from district members evince this selectivity, and the whole process of district formation demonstrated farmers' reticence to completely embrace federal or state activity. Put another way, farmers did not appear to accept anything quickly or completely without some reservations, regardless of the program. For example, the original vote to establish a soil conservation district in Baca County lost when only 24 percent of 585 farmers voted in favor. In reviewing the vote, the state coordinator of the SCS, K. W. Chalmers wrote to regional coordinator Kimmel with his reaction to the district's defeat. Chalmers explained that residents had "been somewhat spoiled by Federal grants" and "they do not feel that there will be any change in the Administrative policy with respect to Baca County whether or not they have a district." Moreover, personal and regional animosities carried significant weight. Sectional controversy between Springfield and the rest of the county, personal jealousies between proposed members of the directing committee, accusations directed at county supervisors and Agricultural Conservation Program Committee members, and a general distaste for the "dictatorial policy" instituted and controlled by Washington, combined to convince residents that establishing a conservation district carried too many potential problems.[85] Obviously Baca farmers eventually voted to start the district, a fact that suggests the initial trepidation over federal oversight waned once farmers understood the district's benefits.

A letter written by J. H. Neal, president of the West Baca district, to SCS chief Hugh Hammond Bennett in July 1938 offered one of the more telling examples of farmers' attitudes regarding federal involvement. Neal chastised Bennett for what he considered the SCS's failings and attempted to clarify exactly what SCS employees needed to do to "redeem themselves in the eyes of the people." The crux of the problem, according to Neal, was the agency's inability and unwillingness to tackle the issue of wind erosion, "the greatest menace of all," with necessary urgency. Neal claimed that "the Service" needed to spend more money buying up wild lands and putting people to work on stopping that acreage from blowing onto adjacent private land instead of employing office and technical help or renting an office. The

SCS spent too much money on personnel and tried to push the costs onto the farmer by charging high rental fees of materials and machinery, which made it too expensive for the average farmer to employ the tactics that the SCS promoted. Neal seemed to accuse Bennett personally, contending that "your Service has gotten hold of the purse strings of Washington and are making an attempt to perpetrate your Department and to dominate all other Agencies of the individual states. . . . Our people feel that your Service is too expensive, not practical, and it is one of the most wasteful of people's money."[86] In a sense, Neal blamed the SCS for an expansive budget that did not funnel money directly to the problem; instead, Neal believed, the SCS paid for a bloated staff and too much office space.

W. O. Brown included similar complaints in a letter he wrote to regional SCS coordinator Al Hurt. Brown criticized the SCS for its unwavering conviction that it should be the final authority on all things agricultural. He claimed that the "SCS wants to run everything, but they're too slow. They don't do things at the right time." He continued: "22 men in the office and 8 men in the field seems to be the method of SCS, and we just don't approve of that way of working. If they would go on the land with tractors and start listening, they could redeem themselves."[87] Further friction developed over a "Memorandum of Understanding" devoted to outlining the cooperation between the district and the agency. Again, the West Baca County representative worried that the SCS had become too powerful and too bureaucratic, sacrificing the county's best interests in the name of authority. "Instead of various agencies cooperating and advising," he offered, "the District is placed in the position of 'advising' with the Service having the final decision."[88] Conversely, Guy Dickerson celebrated Extension's Wind Erosion Control Program because it produced results without "red tape" and federal obstructionism.[89] Farmers wanted as much control over conservation as possible, and many felt most enthusiastic about conservation programs that allowed them such authority.

Evidence of friction between SCS conservationists and local state-employed conservationists further demonstrated criticism of the SCS mission. A letter from district conservationist F. R. Stansbury and Baca County project manager Norman Fuller, both working out of Springfield, to W. R. Watson, area conservationist with scs, detailed some of their concerns about the purchasing program. They expressed their ambivalence about scs protocols. Fuller noted, "I begin to appreciate how a Balkan premier feels when he signs a treaty at Vienna" because, for all the good that scs accomplished, the agency sometimes employed seemingly backward logic. For example, Fuller believed that revegetation on purchased land represented an important step in restoring native grasses, but he questioned "the feasibility of spending $5.00 an acre on $3.00 land in an effort to make it worth $4.00" in resale. The scs needed help to reach private landowners. Despite its tremendous budget and the army of employees across the plains, they needed individual compliance to make much difference. Indeed, Fuller and Stansbury identified one of the signature problems of New Deal policy: citizens needed to cooperate. The scs lacked sufficient power to compel such participation or to help all of those interested. Take the purchasing program, they argued: "There is no further fact that, even if this method [submarginal land purchase] should be successful, it still will offer no feasible solution of the problem for the hundreds of thousands of acres of private land on which private operators may wish to restore the grass cover."[90] These complaints contained a similar theme, namely, that top-down federal authority that did not account for locals or local circumstances had no chance to succeed. This helps explain why the districts became so popular, even if district representatives continued to worry about issues like bureaucracy and red tape.

The local response to the Civilian Conservation Corps camp at Springfield offers another example of the often-tense relationship between residents and the federal government when dealing with erosion.[91] Over its nine-year existence, the ccc employed three million young men in various tasks across the country. The ccc was one of the more popular New Deal programs in Baca County, as camp members quickly became part of the community, even participating in local softball leagues with business owners and other federal employees.[92]

The Springfield camp became an important cog in the machine to fight erosion from its introduction in 1935 through its closure in 1940. The SCS managed the camp and focused workers' efforts on halting soil loss—the camp earned the nickname "Gusts O' Dust" as a result. Camp residents planted shelterbelts, built dams and canals for water diversion, reseeded exhausted soil, and effectively served at the beck and call of the local SCS representatives.[93] Most of the workers hailed from Colorado, with many from either Baca or Prowers Counties. Additionally, counties in southeastern Colorado like Bent and Otero were well represented in the camp rolls. Many of the enrollees had Hispanic surnames like Avila and Martinez, suggesting a heavily Hispanic worker population at the Springfield camp.[94] Local farmers realized that a positive relationship with the camp could provide significant dividends, as the army of eager workers could be put to many tasks to improve area farmlands. The camp offered not only the labor but also, like the SCS, extra plows, tractors, and other technology to use in local fields. A rumor spread that the USDA was considering shutting down the camp, and a flurry of responses ensued, including an unsuccessful plea from Jas. O. Dougan of the Western Baca soil erosion district. Dougan implored Secretary of Agriculture Henry Wallace to reconsider moving the camp because locals so badly needed the help. The camp offered significant assistance to the district once it formed, and Dougan knew that removing the camp would frustrate conservation efforts. Dougan promised Wallace that county residents, while poor and in need of federal help, would "contribute as much as possible" to keep the camp in Baca. Dougan broached the idea of a second camp because he mentioned, somewhat desperately, that the CCC camp was necessary to keep the soil district itself afloat. Instead, the camp closed in 1940 and district members had to adapt.[95]

These criticisms demonstrate that residents found reason and proved willing to call out federal assistance when it failed to meet their demands and that they often adapted to circumstances when they found the need. Indeed, individuals organized, demonstrated initiative in combating the erosion problem, used government help when it became available, and responded to the Dust Bowl. The districts represented the lengths to which citizens traveled to fight erosion. For example, once it formed, the West Baca Soil Conservation District enthusiastically pointed to land they wanted to buy to prevent further blowing and destruction. The supervisors tried to attain a federal grant but had no luck because they formed so early that no government agency had been authorized to provide financial help. Instead, they took loans from local banks to buy up lands and to borrow or lease equipment, and they then formed a specific association within their district, the Sandy Soil Cooperative Erosion Control Association. That body could turn to the Farm Security Administration for loans to continue buying submarginal land, improving it, and then leasing it back to district members. The district members effectively made do until 1940 when districts could turn directly to the FSA and AAA for financial aid; even after that, however, the southeast Baca district continued to borrow from local banks.[96] Eventually the western and southeast Baca districts garnered the promised state and federal support, which enabled them to cover nearly one million acres in Baca. That coverage ensured that farmers had a governing body to turn to for help and an authority capable of checking production and practice to protect from erosion. Under such guidance and with such resources at the ready, it is perhaps not surprising that the fight against erosion gained momentum.

Unfortunately, neither conservation nor the purchase program addressed the third main point that farmers suggested as a necessary change to plains agriculture. As the Western Baca Soil Erosion District summary noted, the Homestead Act left its largest legacy in terms of constructing 160-acre plots as the accepted norm for land division on the plains: "The unwise policy of the Government's rigid subdivision of lands is still felt today, as many of the farms and ranches are too small to support a family. Such subdivision should never have been allowed in the Southern Great Plains."

By so limiting settlers, the comparatively small plot size left homesteaders with few options but to maximize their plot's productivity and try to make as much money as possible in the process. The 160 acres represented what Congress believed a family needed in a humid region, but it did not account for the challenges posed by an arid environment. The small plots lacked enough acreage to diversify their crop. If they wanted to market their commodities, they focused on cash crops, and they lacked the space to take on much livestock to improve self-sufficiency. This meant most homesteaders tried to maximize profit by planting as much of a single cash crop as they possibly could and hoping that a good harvest could provide some capital. The turn to cash crops like wheat represented a step toward soil erosion because the small owners had accrued some debt in buying machinery or other goods and needed to amass enough money to stay solvent, so they plowed up most, if not all, of their acreage.[97]

Expanding operators' lands to support a higher proportion of self-sustaining agriculture proved the most difficult change for New Dealers to implement in response to the Dust Bowl. The notion that only large outfits could prosper on the plains harkens back to explorers like Pike and Long who doubted the wisdom of settling the plains. John Wesley Powell similarly noticed the obstacles that arid regions presented. Writing in 1879, Powell asserted that the homestead method proved inadequate to meet the demands of an arid West because it discouraged community reliance and stood in contrast to his "colony" plan whereby individuals organized to protect pasture land, access water, share range land, and help each other flourish. While perhaps his position that everyone should have 2,560 acres looked unrealistic and proved unattainable for most farmers, Powell's broader criticism of the Homestead Act is noteworthy.[98] His astute assessment of the natural limits posed in the arid region represented a compelling refutation to the Homestead Act that many New Dealers and farmers echoed during the 1930s.

Extension soil conservationist T. G. Stewart noted that farm size represented one of the biggest obstacles to stability in the agricultural economy.

He wrote, "It is not possible, under most conditions within the region, for an individual to earn a satisfactory income on small tracts of land such as 160 or 320 acres." Stewart reiterated what the district report had found and contended that "the small size of farms has brought about a cash crop type of agriculture, and has practically eliminated general farming or livestock farming in much of the region."[99] Such monocrop agriculture degraded the soil and negated the likelihood that farmers could diversify production enough to even sustain their families during hard times. Indeed, John Underwood suggested a dramatic reorientation for land users in Baca County, away from intensive cultivation and toward diversification with a focus on livestock. He argued for a move away from cash-crop farming because most of the small farms around 160 acres had focused solely on wheat and had failed. He knew that the soil was paying the price.[100]

The Great Plains Drought Area Committee's report summed up this anti-Homestead fervor remarkably well when they identified the roots of plains depression and soil erosion.

> The settlers [those who moved into the tall grass prairies of the western Plains] lacked both the knowledge and the incentive necessary to avoid these mistakes [specifically overgrazing and overly intensive cultivation]. They were misled by those who should have been their natural guides. The Federal homestead policy, which kept land allotments low and required that a portion of each should be plowed is now seen to have caused immeasurable harm. The Homestead Act of 1862, limiting an individual holding to 160 acres, was on the western plains a stimulus to over-cultivation, and, for that matter, almost an obligatory vow of poverty.[101]

T. G. Stewart and others identified several potential solutions to augment small holders' properties. He recommended changes in financing to make available lower interest rates for owner operators to establish "an economic unit" (agricultural plots that met a certain level of income per acre to make farming viable for the owner instead of at a loss). Stewart noted that enough land existed in the counties "to supply economical units for each farmer residing there," but the government was unable to offer such units. That led to his second main point, that the influx and continued presence of nonresident operators negated the chance for residents to buy enough land to support themselves. He blamed "suitcase farmers" for not attending to their land and for dismissing their neighbors' best interests. He ceded that no government agency could offer much assistance to force such operators to relinquish title or at least hire local management to protect their land, but he promoted some way of local control (enter the soil conservation district) over whether operators could break new land. He chastised absentee owners for speculation and in the process prioritized the local owner operator whom he thought worthy of more land and more capable of proper stewardship. The transition from one owner to the other would have saved soil and allowed a higher proportion of residents to expand their holdings and prosper on the plains.[102]

As happened in Baca and elsewhere, the soil conservation districts could restrict owners from breaking new land, thus meeting one of Stewart's goals in corralling absentee owners, but even their ability to challenge private property rights never broached the possibility of land transfer. Extending favorable credit rates to tenants and small operators represented the best that the government could do toward that end, and it proved largely successful in offering good rates. Yet the number of people who enjoyed such assistance never amounted to a dramatic transition in landownership rates in Colorado. The Extension Service did its part to facilitate widespread lending. For example, Claude E. Gausman, agent for Baca County, noted in 1940 that he had helped 169 farmers receive credit from federal programs like the Agricultural Conservation Program and then tried to assist them in managing their finances to make good use of the funding. Extension had tried to offer money to farmers whenever it became available under good rates, and the massive distribution of federal funding during the New Deal kept them busy identifying those most in need as well as those most likely to capitalize on the opportunity. Indeed, Gausman seemed to boast when he noted that FSA clients who had qualified for loans were more conscious of conservation because they had more to gain from attending to their land.[103]

Unfortunately for proponents of expanding individual holdings to stabilize plains agriculture, the government had no real leverage to compel farmers to expand and had no desire to take on the financial burden resulting from excessive federal lending programs. For once, it seems, the weather stood on their side. Between out-migration, foreclosure, and abandonment, the combination of drought and depression wrought havoc in both counties by shaking up the population. Folks who survived the decade came out with fantastic prospects, and many expanded their holdings by buying up neighboring lands. Put simply, the Dust Bowl initiated a trend whereby the number of farms declined but the average size of each farm increased. Drought instigated the kind of reallocation of land that those who had bemoaned the limited size of homesteads had pushed for. Despite earlier efforts like the Expanded Homestead Act, the government had no legitimate way to extend individual allotments without becoming directly and problematically involved. The Dust Bowl indirectly facilitated that transition. Many New Dealers lamented the decline in farmers, but the expanded size afforded those farms that remained a better chance of diversifying their crops, employing livestock to balance production, and letting some acreage lie fallow. Fewer, more economically stable farmers meant a better likelihood for conservation and for the slow decline in extensive federal assistance. Put another way, the more stable and economically viable farms, those more likely to survive the worst years of the 1930s, had a better chance to look to the future. Tenants, small operators, and other marginal farmers were more likely to leave. Consequently, the demographic shift could very well have promoted better stewardship of the land and induced a higher proportion of conservation practitioners in the region. The weather, then, initiated the federal response to the Dust Bowl and indirectly advanced the New Deal conservation state.

By 1941 and the onset of World War II, structures and organizations necessary to promote conservation had firm holds in Baca County. There is no way to measure the amount of soil "saved," but the continued support for conservation districts suggests that locals embraced the New Deal conservation state well beyond the end of the 1930s. By the outbreak of war, three districts in Baca covered almost 75 percent of the county, and by the end of the war two additional districts in Prowers County contained many of its dryland farms. The relative autonomy and shared risk/sacrifice/benefits of district organization seemed to appeal to local farmers. Soil conservation districts still exist across the region, standing as evidence of continued support for the blend of local governance and federal support. They are testament to southeast Coloradans' willingness to adjust their land use and adopt conservation techniques.

While the conservation district may have represented the New Deal's biggest win against soil erosion, its efforts to buy up submarginal land also warrant attention. Through various agencies, including the Soil Conservation Service, Bureau of Agricultural Economics, and Resettlement Administration, the federal government spent money and time trying to assess what land should be retired, who needed to be moved off that land and how, as well as what to do with it once the government had paid for it. Certain programs tried to remove tenants and other small operators from submarginal land in part because of the idea that poor land made poor people. Rural poverty would continue unless you moved the most vulnerable people off the most vulnerable lands. These programs, some of which involved lending money to tenants, helping relocate them, or just buying out marginal owners, removed hundreds of thousands of acres from production. This land slowly returned to native grass and exemplified the push to retire denuded land. The purchase program remained an important weapon in the government's arsenal to promote rehabilitation until the early 1960s, when the government deemed such retired land part of the Comanche National Grassland in 1960. The federally protected grasslands cover more than 440,000 acres in southeast Colorado, nearly 250,000 of which lie in Baca County. Remarkably, many of those acres had been some of the worst eroded parts of the county and most in need of retirement.[104]

Although difficult to consider for many New Dealers, the third part of the Roosevelt administration's approach to an unstable agricultural economy dealt with promoting large operators. Many New Dealers came to believe that every farmer should strive for a large, diversified farm to help him or her withstand such depressions and droughts that were undoubtedly going to recur. The small holders, tenants, and agricultural laborers posed the most serious problems for the rural economy; those with the most land fared well. Yet for all its influence, the New Deal could not really instigate the push to larger and fewer farms. Proponents of such a shift found help from an unusual source when the Dust Bowl initiated a dramatic demographic change. Between out-migration, foreclosure, and abandonment, the combination of depression and drought wrought havoc in both counties by shaking up the population and accomplishing what New Dealers could not: compel tens of thousands of marginal farmers to leave the Great Plains. It facilitated a shift away from the 160-acre family farm, and as much as retiring submarginal land or buying out small owners, it underscored an understanding that homesteading would not work in the region. Only the confluence of the dual crises perpetrated the move away from an agrarian ideal of the family farm.

This is part of the Dust Bowl's mixed legacy on the plains. Those who stayed generally adapted, made do with federal subsidies, and started to conserve their resources. Many who could not withstand the drought left. The folks who survived the worst years looked confidently to the future when rain levels started to normalize again in 1939. Moreover, they stood ready to benefit from the influx of wartime demand that started that same year.

4

Claiming the Arkansas

The dual crises of Dust Bowl and Great Depression challenged Colorado farmers by exploiting weaknesses in their land use regimes. The dryland farmers in Baca realized how a sustained drought exposed the damage they had done in tearing up the topsoil and leaving it to blow away. They gradually adopted resource conservation during the New Deal as the chief way to mitigate dust storms and soil erosion and, therefore, to protect their livelihoods. This turn to conservation reflected a dramatic transition and showed their willingness to adapt in the face of such adversity. No such drastic adaptation or adoption of soil conservation occurred in Prowers County. Soil did not matter as much to Prowers farmers. Prowers did not have the same rate of soil exhaustion, did not experience the Great Plow-Up to the same degree as in Baca, and never attained the "Dust Bowl disaster" status that Baca garnered. Most Prowers County farmers cared about water; more specifically, the free-flowing waters of the Arkansas River and its tributaries that had provided food and drink to humans since at least 850 AD remained a focal point for farmers for decades. When disaster struck during the 1930s, most Prowers farmers did what they had always done. They looked to the river to solve their problems. Their adaptation in the face of the economic and environmental catastrophe was to figure out how to capitalize on their access to water.

Unfortunately, the river never seems to satiate its users, and appropriators constantly search for ways to augment their access and stabilize their supply. Few settlers enjoyed the kind of irrigated Eden that boosters promised when they advertised for the "Valley of Content," and current residents continue to long for such a place. Still, Prowers County residents have always drawn as much water from the river as possible to help them achieve their goals. They maintained hope that the river could meet new challenges posed by depression and drought. They debated how to utilize the water better and how to provide more water to users. They sought to ensure a steady and abundant supply, one capable of providing enough water to stabilize and bolster their economic prospects. Locals agreed that a dam and reservoir system along the Arkansas River represented their best chance at maximizing their appropriation. As many observers correctly concluded, the New Deal provided a window of opportunity, one left just enough ajar to allow southeastern Coloradans to capitalize on federal largesse to stabilize and expand their access to irrigation. They got a break; the 1930s represented the peak period for dam building in America, evidenced by the creation of the Tennessee Valley Authority, development along the Columbia River, dam construction in California, and completion of the gargantuan Hoover Dam. Farmers capitalized on this momentum and successfully lobbied the federal government to build the John Martin Dam and accompanying John Martin Reservoir in hopes that the dam project might protect them from future climate and economic crises.

The fight to garner federal funding and manpower to construct the dam demonstrates how Baca dryland farmers and Prowers irrigators conceived of land use during the 1930s. First, the Dust Bowl compelled all farmers to search for ways to mediate the effects caused by aridity. Dryland farmers looked to soil conservation while irrigated farmers focused on expanding their access to river water. It is testament to the drought's severity that farmers practicing opposite agricultural techniques and growing quite different crops had to adapt. Both groups realized that something had to change if they wanted to continue farming in southeastern Colorado. Second, federal support and local impetus combined to change the landscape. In 1933, members of the Lamar Chamber of Commerce organized a committee to develop a plan to dam the river and construct a reservoir capable of holding water until folks downriver needed it for irrigation. After five years and multiple attempts by local associations to persuade federal officials and politicians to embrace their plan, the federal government decided to fund the construction and provide the workers to build the dam. Indeed, as it had so often during the heyday of the New Deal, the government footed the bill and therefore made dam construction, or soil conservation, possible.

Despite these similarities, however, the differences in economic stability and land use choices between dry and wet farmers illustrate that Prowers County farmers often faced fewer challenges than their neighbors in Baca. Irrigation meant that farmers could experiment with more diverse crops, settle on prominent cash crops, and more easily survive periods of drought than dryland farmers. In Prowers, this meant fewer "lows" and more "highs" than in Baca, even though the Dust Bowl and Great Depression affected farmers in both counties. Certainly Prowers farmers faced lost income and land values, and while the population decline paled in comparison with that of Baca, Prowers still lost 16 percent of its population between 1930 and 1940.[1] Yet the relative stasis of knowing you had at least some hope of water and therefore a chance at producing crops, rather than the Baca homestead pattern of plowing and then praying for rain, meant that irrigated farmers had always enjoyed a sort of safety net that dryland farmers lacked.

The Arkansas River never provided the kind of panacea for which farmers and boosters had prayed. Nonetheless it became the focal point for irrigators searching for ways to protect themselves from drought and depression. The history of irrigation in the Arkansas valley and the Roosevelt administration's willingness to spend money allowed farmers to focus on water as their response to drought, thereby precluding the need to consider changing most of their land use routines. "Wet farmers" embraced water conservation, and evidence suggests that they thought about the ways that irrigated water impacted soil health, but they lacked a sense of urgency about retiring land or the diversification that Baca farmers exuded. As long as they had sufficient water, through additional access and more responsible use, then things would normalize and drought could be bested. There was no need for a dramatic reappraisal of land use practices because irrigation represented a viable and attainable solution to their problems. They still dealt with aridity—during especially tough years even some of the small canals and tributaries dried up—and irrigation had its own adverse effects on soil fertility. But the search for water shows how Prowers farmers responded to the Dust Bowl crisis. The arguments they made in favor of a federally funded dam, the way they learned how to appeal to a growing bureaucracy, and how they defended their interests demonstrate that irrigated farmers adapted in their own ways to the changing political and geographical landscapes in the 1930s. They capitalized on the opportunity to use federal willingness to keep farmers on the Colorado plains to their own advantage and came out of the Dust Bowl much better prepared to handle future droughts. Indeed, Prowers residents' faith in the river never truly wavered, and the 1930s and 1940s exposed how that conviction grew stronger during the Dust Bowl years.

Irrigation had been part of farming life along the Arkansas since humans first developed the ability to tap into flowing water for agriculture. Early peoples during the Pueblo I period (AD 850–900) enjoyed permanent settlement in the Arkansas River valley wherever they could effectively divert mountain or river water to their fields. They used ditches and canals to funnel the water and trapped water in small reservoirs to save it until later diversion to their crops and villages. The Spanish and Mexicans utilized similar techniques to establish villages along the Arkansas River near present-day Pueblo as early as 1787. The Hispano ditches, called *acequias,* were community irrigation systems designed to satisfy a larger and more sedentary population of people than earlier efforts. Anglo settlers like the Bent brothers (proprietors of Bent's Old Fort) similarly manipulated river water to irrigate cropland, as they did in using river water to irrigate nearly forty acres of land during the early 1830s.[2] The size and scope of irrigation efforts increased dramatically in the 1840s and early 1850s when Latinos settled much of southern Colorado and used the river to feed their cropland. The discovery of gold offered the real impetus for this expansion, as Latino settlers realized that selling their products to miners, local businesses, and other new arrivals could provide a significant boon to their finances. The chance for such profit proved enticing enough to compel the farmers to create a more sophisticated and extensive irrigation system so that they could expand their acreage. In effect, they believed that expanding their access would produce more wealth for their community, and that theme emerged in later Anglo American thinking about regional development as well.[3]

In many ways, the American version of irrigation dwarfed Latino efforts even though each arose from the same desire to manage the river. Technological and engineering advances, population pressures, market demands, and corporate structures made it possible for Coloradans to significantly ramp up their ability to tap Arkansas water for agriculture. Early community-based development efforts, such as mutual stockholding companies, fueled the drive to build canals and reservoirs. These companies sprouted up throughout the valley once white settlers came to realize the obstacles to productive agriculture presented by a semiarid environment. They also quickly realized that corporate-led development could do things that no individual or small group could accomplish; they turned to irrigation companies and relied on them to augment their access. The Amity Project and the Fort Lyon Ditch Company are two interesting examples of such conglomeration in Prowers County. The Amity Project shows some of the problems associated with early irrigation efforts, while the Fort Lyon situation demonstrates some early successes at capitalizing on the river's bounty.

The Amity Project, a colonization program started by the Salvation Army in Chicago, aimed to use irrigated farmland as an outlet for urbanites plagued by poverty and despair. With inducement from the Santa Fe Railroad, which tried to entice settlers to lands they had been granted by the federal government, a group of thirty to thirty-five families started Fort Amity in 1898. The Salvation Army provided water shares from multiple companies to the settlers upon their arrival to increase their chances of sustaining the community in the arid environment. The farmers established a small town that attracted additional migrants from Chicago.[4]

Unfortunately, neither the farmers nor their supporters understood the problems that could result from irrigating cropland along the Arkansas. Specifically, the Amity farmers faced a drastic rise in the alkalinity of their produce. The Arkansas River's natural level of salinity explains this rise to some extent, as the river is "one of the most saline rivers in the United States" and especially so as the river moves east toward the Kansas state line. Another explanation for the rise in salinity dealt with the migrants' choice of land and the soil on that land.[5] The soil did not allow for the sufficient permeation of the river water, primarily because the water table was so high that seepage proved almost impossible and water simply sat on the ground's surface. Even heavy rain meant puddles and the high water table meant very little runoff from either precipitation or irrigation, so standing water became common. With no place to go, then, the water sat until evaporation removed it, but a heavy concentration of salt stayed on the ground after evaporation. The high-saline soil became waterlogged and cut off the crops' access to oxygen, resulting in heavy crop failures throughout the colony. Surviving plants were so tainted that the produce was inedible, so the colonists lacked anything for market and even struggled to feed themselves. While the farmers tried to drain the water to alleviate seepage, nothing seemed to offset the damage already done. Left with few options, the migrants effectively deserted Fort Amity in 1910 when the Salvation Army left and sold the land to locals.[6] In this case, irrigation offered the chance for sustained settlement, but obviously just having access to water does not ensure that distribution is sound or that the environment (soil, spillway, access) is amenable to irrigation. For Fort Amity residents expecting water rights to safeguard their livelihoods, these consequences certainly came as a surprise, and they abandoned the site less than fifteen years into the project.

The Fort Lyon Canal Company represented a much more successful community effort to use river water to support agriculture. Started in 1887 when two smaller irrigation companies combined their holdings, the company came to control a considerable irrigation system by the 1930s. Competition for water started as Anglo Americans settled the region in the 1880s, and the competition led to intense development. There were some three hundred miles of ditches dug in southeastern Colorado by 1895, and many of those canals ran through Prowers County.[7] Even though it started comparatively early, the Fort Lyon Company fought for its water against neighboring companies as well as folks upriver who had first access to the free-flowing resource. It also had to fend off companies that could easily amass capital by building ditches and then selling land abutting the ditches for profit. Such competition emerged because Colorado water users abided by the doctrine of prior appropriation as written into the state's constitution from 1876. That doctrine, otherwise known as "first in use, first in right," secured senior water rights to the first individual or group using the water for beneficial purposes (agriculture, industry, and mining). Taking advantage of its senior rights, the Fort Lyon Company quickly established a solid hold over the river. For example, the company used concrete to seal some of its ditches, bought two local reservoirs outright, constructed a concrete diversion dam, and built additional ditches to increase access for additional farmers.[8] The company's efforts seemed to have paid off when, just prior to America's entry into World War I, the company president claimed that the Fort Lyon Company could deliver on its promise of turning the arid region into the idyllic gardens that boosters had once predicted. This optimism for the company's delivery and for farmers' benefits continued until after the war when the agricultural economy bottomed out.

Fort Lyon irrigators also faced problems like those that Amity farmers had wrestled with in terms of salinity. Since most of the Arkansas River water that farmers in Prowers County used was return flow, they were subjected to a high level of salinity. The return flow was water that either moved under the surface through the soil or groundwater that reentered the river as runoff from cropland. The flow picked up natural salt from running over saline and sedimentary materials, and it gained salt from the erosion of saline-heavy soil into the water flow and from concentrated levels of salt left after evaporation and evapotranspiration. Since a significant percentage of river water was consumed by the time it reached Prowers fields, even those fortunate enough to have access to irrigation contended with the unforeseen consequences of having that access. Most of the complications resulted from improper irrigation methods, as farmers often used too much water or used water when unnecessary, all of which passed excessive water through the soil near the root and leached the salt from the soil.[9] Indeed, even the portion of river up from Prowers County, and therefore with less salt concentration, had such a heavy dose of minerals that the Atchison, Topeka, and Santa Fe Railway stopped using river water in their boilers because heavy mineral deposits left by the water degraded the system.[10]

Fort Lyon farmers also struggled with increased sedimentation of the river bottom that challenged diversion implements and muddied up the water. Sedimentation became a problem in part because irrigation effectively eroded farmlands. The river eroded the banks as well, chiseling away at the dirt and depositing it on the bottom while slowly leveling off the soil. Heavy sedimentation complicated irrigation by clogging up transference points, leaving less available room in canals and ditches for running water, and overgrowing dams. For example, a diversion dam that the Rocky Ford Canal Company built in 1923 was completely covered by silt and mud twenty years later. Measurements taken at the Fort Lyon headworks showed that the river bed had climbed some seven feet between 1910 and 1944. Companies constantly had to remove the sediment to unclog their waterworks, and early on the process involved horse-drawn scrapers and eventually tractors, and "continuously battling siltation drained stock-holders' purses and required constant vigilance on the part of the super-intendents." Again, just as the issue of salinity emerged early, the problem of sedimentation hampered irrigators associated with Fort Lyon as well as later users across the Arkansas valley.[11]

That did not stop the proliferation of similar companies across the region. The first canal, the Rocky Ford, gained support from local farm-ers who dug out the future conveyance and extended it so that by 1890 the canal ran sixteen miles from the Arkansas south, providing coveted water to the company's shareholders. In Prowers County, the Amity and Lamar canals both distributed water to farmers in the 1880s and received additional water rights by priority of appropriation decrees that gave them more senior rights. These irrigation systems combined to enable Prowers County farmers to irrigate nearly 2,000 acres by 1889. While Prowers had a relatively small number of canals, the fact that local farmers already started to rely on irrigation before 1900 shows the depth of their dependence on Arkansas waters.[12] Once the early companies like Rocky Ford or the Arkansas Valley Sugar Beet and Irrigating Land Company, which took over the Amity canal, established themselves, they sought to grow their rights. By 1930, for instance, the Amity and Lamar canals funneled enough water through Prowers to satisfy farmers on more than 93,000 acres.[13]

Irrigation was not a cure-all; irrigators still faced the same kind of market fluctuations and tempestuous demand that other farmers faced. Moreover, irrigated farmers had no real control over the river's flow and therefore no say on how much water they could eventually access. Inconsistent precipitation still threatened to derail irrigators even if dryland farmers proved more susceptible to variations. Consider that most of the Arkansas River's water comes from mountain snowmelt that makes its way down to the lower valley over the course of the spring and early summer. Another portion comes from the river's numerous tributaries. The combination, however fruitful in good years, was not enough in down years or during drought. Furthermore, the doctrine of prior appropriation meant that those with the senior access rights felt more secure getting regular water, but the junior irrigators who established their access after 1875 had to wait in line. When water became scarce during periods of drought, those with junior rights often found themselves with little or no access.[14]

Even given these challenges, farmers have always chosen inconsistent access to water over no access at all. This proved true in both Prowers and Baca. Early Baca efforts failed to pan out, and very few Baca farmers had a reliable source until farmers started to tap the Ogallala Aquifer in earnest after World War II. Prior to the aquifer, only a few lucky individuals had access to any water, primarily through deep wells that proved difficult to reach for most settlers. For instance, J. A. Stinson of the Herring-Stinson Cattle and Sheep Company set up windmills to pump water from an artesian well that he located on his property. Stinson enjoyed the benefit of owning land on top of a subterranean water source and had the capital necessary to both drill for access and set up the conveyance to bring the water to the surface. In that way, he garnered credit for helping establish the county after 1900 because he offered those without water enough from his well to satisfy some of their daily needs.[15] The use of the underground repositories of water became more widespread when technological advancement made the process easier and more efficient. For example, electric pumps could be used to bring the water up and divert it for irrigation, making it fairly easy for any farmer to access the resource if it existed under his or her lands. Indeed, as well construction matured and pumping equipment became more efficient, groundwater exploitation exploded across the state. Colorado farmers and ranchers used at least 1,100 pumping plants by 1935. More than a third of that number had been installed in 1934 as the increasingly frequent dust storms continued to take their toll.[16] That dramatic rise in pumping plants is another demonstration of how drought spurred farmers to think about access to water.

Yet the presence of groundwater near the surface, therefore at depths more easily mined by irrigators, was mostly determined by geography. The likely locations to find water-bearing gravels were around streams and canals

FIG. 14. "A little water for a thirsty land. Drought committee inspects artesian well irrigation project. Baca County, Colorado, July–August 1936." Photo by Arthur Rothstein, Library of Congress, Prints & Photographs Division, FSA/OWI Collection (LC-DIG-FSA-8b28195).

that had sustained seepage losses, from percolation losses where irrigation water had already been applied, and in old channels that no longer held water. Even if one found oneself on top of such a deposit, however, it was generally not a long term, surefire solution to the problem of aridity. As associate irrigation investigator W. E. Code noted in 1935, "The fallacy that ground water is inexhaustible is believed by many of the uninformed. The opposite, however, is too well proved by the alarming rate at which the water table has receded in a number of districts in the Southwest."[17]

Code's assessment is telling in that not only did relatively few farmers have a chance to take advantage of groundwater, even those who did would eventually need an alternative.

The most successful effort for Baca farmers to access water in a dependable way was to construct dams, something the Two Buttes Irrigation and Reservoir Company did along Two Buttes Creek in 1909.[18] The company had access to some 22,000 acres of land that had been set aside and maintained by the federal government under the guidelines established by the Carey Act of 1894. The Carey Act allocated up to one million acres of arid federal land to any state willing to transform it into irrigated farmland. It acknowledged that the Homestead Act from 1862 could not work in an arid environment because the lack of water negated the settlers' ability to produce. States that agreed to manage the irrigable lands could then seek out individuals or companies that could develop that land for irrigation. It remained the state's responsibility to oversee these projects to guarantee that the developers eventually transferred the water rights to settlers. The settlers who bought or leased the land paid the state to offset construction costs and supply a bit of revenue. This arrangement transferred responsibility for supporting settlers from the federal to the state governments and still ensured that the West became settled.[19]

The Two Buttes Reservoir held water that the dam diverted from Two Buttes Creek, a tributary of the Arkansas, and serviced shareholders in both Baca and Prowers. The construction crews completed the system in November 1910 at a cost of nearly $700,000, despite engineers initially estimating that the cost would be half of that total. The dam, an earth-filled structure with a concrete core, fed into a canal system that ran twenty-three miles across both northern Baca and southern Prowers County. The source canal extended just over ten miles, and an additional thirty-three miles of lateral canals stretched across the landscape to reach farmers. Unfortunately, the farmers who paid $35 an acre for land with water rights (shares in the company equated to an acre of irrigation) from the canal system never received as much benefit from having access as they had hoped. The canal system did not reach as much acreage as the first board of directors had promised, only including 3,178 acres out of the proposed 22,000 by 1941, nor did it have the capacity to fully avert drought. Eventually the company sold its rights to the Fish, Game, and Parks Service, leaving farmers reliant on deep irrigation wells.[20] In essence, then, most Baca farmers found themselves relegated to dryland farming and handcuffed in ways that Prowers farmers could not fully understand.

Irrigated farmers had more stable prospects and a firmer economic base on which to stand, whereas dryland farmers had very little say over their production. Regardless of what they grew, farmers needed water to grow the two most important cash crops; they believed that they could weather the Depression once they stabilized production. Reasons to be that confident existed, in part, because the rough postwar period hit Baca harder than it affected Prowers. In fact, once private and public entities built canals, diverted water, and started using irrigation, the Prowers system of agriculture remained more stable than the Baca system. The agricultural census paints a telling picture of these differences. Baca farming changed dramatically from 1910 to 1930: the number of farms more than tripled and the acreage increased fivefold; the number of Prowers farms increased by about one-third and the acreage doubled.[21] Consider the demographic shift. Baca population in 1910 sat at 2,516 but rose to 10,570 by 1930, a jump of nearly 260 percent; Prowers numbers in 1910 reached 9,520 and rose to 14,762 by 1930, an increase of roughly 50 percent. Migration to Prowers happened earlier, and those who arrived between the late 1880s and 1900 generally earned water rights and stayed in the area.[22] Indeed, water rights had become firmly entrenched by 1930 so there was less impulse for potential migrants to move into Prowers than the comparative draw to a place like Baca. Furthermore, Baca population changed frequently, primarily because of the weather. Good rain years spurred migration into the county whereas drought years compelled exodus. Thus both the population and the agricultural system were more stable in the irrigated areas of Prowers than in the dryland areas of Baca by 1930, and a dramatic change did not seem necessary. On the eve of the New Deal, then, the system seemed stable and better able to withstand the ebbs and flows of market fluctuations. But farmers still looked for help from the federal government when they hit tough times. For irrigating farmers, such assistance usually came in the form of more water.

Even given the long history of human interaction with water in the region, the New Deal represented a new chapter in the history of water use in the Arkansas valley. The combination of engineering expertise and federal funding emerged under the Roosevelt administration and compelled Coloradans to contemplate how the government could help them deal with their environment. Many southeastern Coloradans understood that the government could do things on a scale and in areas that no individual or private firm could realistically accomplish. They increasingly looked to Washington to help them control water to provide security, a goal that became more important after depression and drought hit. Their search for new ways to utilize the water and insistence on local control of the water, to increase the amount and determine use, represented the two most prominent ways that residents reacted to the Dust Bowl and Great Depression.

In some cases, increasing access meant rather humble projects that the government funded and provided labor to finish. For example, the Works Progress Administration (WPA) and Civil Works Administration (CWA) cooperated on a project along Horse Creek near the town of Holly in Prowers County. The agencies built a small dam to control the flow of the creek and decrease the likelihood that the creek could overflow or, in the worst-case scenario, outright flood by straightening the channel and damming the creek to slow the water. Construction started in 1933, and there had been some headway when a flood ripped through the town and washed out the site in 1935, causing some $250,000 in damage to the city and decimating the dam. The county commissioner and U.S. representative John Martin tried to get the federal government to repair the damage and rebuild the dam to protect against recurrence. The WPA agreed to finish the assignment once the Army Corps of Engineers signed off on its legitimacy and the WPA could locate local labor to do the construction. All told, the dam cost nearly $104,000 and the federal government funded almost $99,000 of that total, providing a dam, a straightened channel, two spillways, and, most important to residents of Holly, a safeguard against later floods. The Horse Creek example demonstrates the federal government's willingness to construct and finance water-control projects in the West during the New Deal.[23]

Such attempts to control river flow, while important, never gained the level of attention that the Herculean endeavors to build massive concrete dams along the nation's rivers did during the interwar era. Indeed, the New Deal marked the most significant period of dam construction in American history. The rise of land use planning and the growing desire to consider regional planning to improve the nation's economy meant that New Dealers thought about ways to transform entire sections of the country. This led to the push for multipurpose dams designed and built to satisfy many apparent needs, including "erosion control, water supply for rural areas and urban centers, drainage, flood control, generation of electricity, irrigation, recreation, wildlife conservation, and forest development."[24] Support for multipurpose development on a regional scale in attempts to modernize sections of the country proved especially strong; the Tennessee Valley Authority (TVA) and Columbia River development represent the culmination of that push. The TVA sought to reconcile the divide between urban and rural America by providing electricity and irrigation and allowing rural residents to share in the bounty of natural resources from which private companies had been benefiting. In essence, the TVA tried to rehabilitate rural America through federal intervention; in that sense it was very much like most of the New Deal agricultural policies designed to help stabilize the agricultural economy in the Plains and beyond. By controlling flooding, providing cheap hydroelectricity, improving navigation on rivers, and offering irrigation, the New Deal tried to use water development as a panacea for many rural problems.[25] In the TVA case and others, despite the assumption that motivation for the projects came from the federal government, locals pushed for dam construction. No individual or small group could accomplish what the federal government could, so people looked to Washington to help them out.

The Roosevelt administration enacted several water projects, and while the Hoover Dam and Bonneville Dam might be more massive and more impressive than the John Martin Dam, enthusiasm for reclamation and smaller projects designed to control water flow is important and often overlooked. The editor of the *Lamar Daily News* seemed to encapsulate the zeitgeist of early excitement when in 1934 he advised readers about the potential benefit of using the federal government to help with irrigation. In addition to noting the fortuitous timing of wanting the assistance precisely when the government was willing to spend money, he applauded the government for making natural resources more accessible to the common man. He contended that there could be a program like the TVA in Colorado, the Arkansas Valley Authority (AVA), which could similarly remake the region using the natural advantages of a nearby water source.[26] The TVA specifically stood out as "the glimpse of what it is possible to do in this great country of ours by the application of intelligence as against the old policy of grab, graft and greed." The "proper control of natural resources, storage of water, planting of trees and intelligently planned crop program all may be used to convert our section from a haphazard farming district, dependent upon luck and the weather man."[27]

Fortunately for those pushing to build dams, local political pressure, a willing administration, a need to extend employment opportunities, and a reliance on regional planning all combined to facilitate dam construction in the region. Local newspaper editors built momentum for construction in the Arkansas valley, but the Lamar Chamber of Commerce made the first concerted effort to get a conversation about a dam along the Arkansas River started during the summer of 1933. Always sensitive to business development and regional economic stability, the chamber broached the idea of providing additional water to farmers in the Arkansas valley.[28] Chamber members started to mount support for dam construction by reaching out to local and state politicians in search of political backing. Unfortunately for proponents of dam construction, Governor Edwin "Big Ed" Johnson seemed ambiguous about the project and worried that expanding the river's irrigation capacity did not appear a foolproof plan to avert disaster on the plains. Johnson feared that increasing access to water would only give rise to additional farms and more intense production, two of the underlying causes of the agricultural depression and the Dust Bowl. Johnson seemed willing to listen to locals' and regional politicians' calls for developing the river (it was after all in his personal interest politically not to upset his constituents) but only if the construction was done carefully, within budget, and without expanding farmers' production to a dangerous degree.[29]

At its core, the debate about the viability and even necessity of constructing a dam and reservoir system came down to competing visions about how to recover the agricultural economy. On the one hand, people argued that New Deal policies should focus on production controls to limit surpluses and stabilize prices. On the other, some believed the dam promised stability and production to buttress against future drought or depression. In no uncertain terms, irrigation meant production. If irrigators produced more goods, then the demand dropped, prices fell, and the entire series of New Deal programs designed to reconcile supply and demand would be toothless. Conversely, the Dust Bowl ravaged plains agriculture, and irrigators represented the only constituency capable of a quick turnaround in production; therefore, New Dealers identified them as viable candidates for economic recovery. Their resurgence could bolster the regional agricultural economy and allow time for dryland farmers and others to adapt or move on. Expanding irrigation represented keeping people on the land and helping them rebound from the Great Depression, one of the key New Deal goals in agricultural policy. There was no real need to reform irrigated farming, it seemed, and no urgency to do more than tweak appropriators' practices by promoting water and soil conservation.

Proponents on both sides found their respective spokesmen in a publicized confrontation between Representative John Martin and Interior Secretary Harold Ickes. Democrat Martin, who served in the House from 1933 until his death in 1939, fought long and hard to provide irrigation for his constituents in southeastern and south-central Colorado. Commencing a "verbal battle" with Ickes, Martin chastised the federal government for not doing enough to help westerners.[30] Arguing not only for Colorado but for an entire region suffering under drought, Martin claimed that "any policy of national recovery under the public works administration which denies or does not take into account the conservation and application of water is tantamount to a death sentence against development in the Rocky Mountain states." He continued: "Failure to develop western water resources will be a permanent and discriminatory injury to that region." Martin wanted some respite and directly blamed Ickes for obstructing western politicians' motions for reclamation.[31]

Ickes defended his reticence to fund additional dam construction by extolling the benefits that the federal government had already provided. He claimed that the government had already spent $150 million for irrigation, power, and reclamation works in the region. Moreover, Ickes also espoused the common refrain that irrigation meant more production, a dangerous possibility at a time when markets remained unstable and consumers had little money to buy produce. In that sense, Ickes promoted convincing agriculturalists to limit production, and he believed that new lands should be protected from cultivation and that the federal programs designed to rehabilitate submarginal land or resettle struggling farmers should have the opportunity to work toward stabilizing American agriculture. He effectively prioritized a federal response to the dryland farmers instead of addressing irrigators' needs, believing that the former required more urgency and dealt with the crisis in rural America more directly.[32]

Even given these sentiments, however, Ickes did not disavow the prospect of opening additional irrigation projects in the West. He noted that the federal government had an obligation to help westerners attain irrigation because "we induced people to move to these lands long ago with a promise of water." By that logic, federal aid acted as the carrot to entice settlers to move onto lands not perfectly suited for agricultural production, as promises of assistance to farmers struggling in an arid environment undoubtedly compelled some to take the chance. Ickes effectively justified the money already spent in the West as a form of compensation to help those who moved to the region and struggled to survive. At the same time, though, he worried about the potential for people to take advantage of federal largesse for personal, not social, gain. He thought federally funded irrigation might provoke a reversion to a more exploitive past, whereby Americans "would today continue, as in pioneer days, to lay bare our forests; to destroy the public range; to attempt to grow crops on land the stirring of which by the plow only serves to provide dust for eroding winds to carry away."[33] He hoped that intervention could break that cycle with more inspired planning and a more sustainable approach to natural resource use. Ickes believed that expert guidance could correct improper land use and ensure that future generations had access to America's natural advantages. He feared that additional water development projects could further jeopardize that likelihood by prioritizing production instead of conservation.

This snapshot of the two sides exemplified the often-contentious relationship between western politicians and members of the federal machinery and reminded westerners that Washington held the purse and had the final say on expansion. Ickes effectively pumped the brakes on dam construction, never coming close to denying the possibility that a dam could be built along the Arkansas but projecting a cautionary tone over the proceedings. Thus proponents of dam construction understood that they would need to address such concerns about production if they could successfully win government support. Proponents set the tone for the discussion about construction by promising a measured approach to irrigation. Rather than arguing for a dramatic expansion of farming throughout the region, proponents argued that the dam would do more to stabilize the flow and, therefore, help farmers who already had water rights and had already become established in the area. A dam would not mean dramatic migration into the Arkansas valley or significant expansion of the current agricultural systems in place along the river. This rhetoric played well in the valley, Denver, and Washington DC, but locals still needed to convince supporters in Congress that dam construction made sense in the region.

The formation and establishment of the Arkansas Basin Committee (ABC) in late 1933 really moved the conversation. The ABC started as the product of an informal conglomeration of local business leaders initially joined as the Caddoa Reservoir Association who formed to levy pressure on the Mississippi Valley Committee. The Mississippi Valley Committee coordinated projects for the Public Works Administration and therefore had tremendous clout in determining viable federal projects along the Mississippi and its tributaries. The ABC constituted the first organized push for dam construction along the Arkansas in southeast Colorado and formed as a political body devoted to securing federal support for the endeavor. ABC members argued for the dam to make sure that producers could stabilize their enterprise; farmers used irrigation as a safety net and needed a consistent and abundant supply to keep them financially afloat. They refuted the notion that they would immediately expand their holdings or plant more ground, and the argument that water created stability became central to their successful campaign. They also presented a convincing argument that a dam along the Arkansas River represented the best way to help residents recover from the Depression and drought as well as the only way to ensure prosperity once the dual crises subsided.

The ABC faced the difficult task of convincing federal engineers and regional planners that their proposal had merit. N. R. Graham headed the ABC and relied on a board of directors consisting of prominent residents to compile the first formal proposal that they could take to the Mississippi Valley Committee in December 1933. The proposal explained the ABC's interpretation of every facet of the project, making sure to emphasize the areas they found most important: need, cost, and potential benefit for residents. The introduction painted a rather bland picture of the project, describing "an earth dam, 14,000 ft. long and 120 feet high, with necessary spillways and outlet gates . . . to form a reservoir of 680,000 acre-feet capacity."[34] The completed project looks nothing like the modest image that they presented in the first report, but the fact that the ABC appeared to embrace an unassuming, low-cost vision may have been tactical in trying to persuade the Mississippi Valley Committee, and thus the administration, that the proposal made financial sense. Indeed, the whole proposal demonstrated the flexibility with which locals were willing to approach construction as long as it ensured construction.

The ABC's attempt to downplay the likelihood that the dam would allow for expanding agriculture represented their single attempt to assuage doubters about their intentions. They assured the Mississippi Valley Committee that they wanted the dam for multiple uses and that bringing in more farmers to utilize the river contradicted their vision of development. Their proposal listed a hierarchy of benefits produced by the project as unemployment relief for laborers working on construction, then "complete protection against all floods," and finally "conserved water available for use by existing irrigation systems." The phrase "existing irrigation systems" connoted no dramatic expansion; the ABC presented the dam as a stabilizing force for farmers who sought a little extra security during periods of limited flow rather than a gateway for more production and more farms.[35] In teasing out the details for flood control and broader efforts for river management, the submission celebrated the possibility to improve navigation along the lower Arkansas and the "conservation for uniformity of flow." In other words, the project could lessen the amount of soil erosion occurring along the river's banks by using a dam to regulate flow. It could also help conserve the peak flows for human use by capturing the water that would otherwise pass through the valley (82 percent of the water moving into Kansas escaped capture). The dam and reservoir could tame the river while providing more water to be used in the irrigation systems. The big dam could pacify the river; the dam and reservoir could reduce the likelihood of floods along the river as well as its tributaries and maximize the river's irrigation potential by corralling the heavy flow.[36]

The issue of creating a steady flow to prevent floods and keep more water in the state proved important enough to compel the Colorado Extension Service to research the river's strength as it moved from Pueblo to the Kansas state line. The research and subsequent report suggested that farmers had a legitimate gripe about inconsistent flow and its adverse effect on locals. Extension employees studied the river's flow over a fifteen-year period, calculating acre-feet in the river as well as many of the larger canals in the Arkansas valley. Extension researchers found incredible yearly variation in acre-feet by the time the river hit Lamar. For example, 523,000 acre-feet flowed through the town in 1923, while only 24,000 acre-feet by the time the river passed through Lamar in 1931 (see fig. 8). Such inconsistency and dramatic fluctuation left farmers with no way to plan on how much water they might have, which obviously affected their production.[37]

Not surprisingly, the numbers for water flow through prominent canals across the Arkansas valley demonstrate similar unpredictability. The Fort Lyon Canal peaked with 248,471 acre-feet diverted to fields in 1929 and bottomed out at 100,675 acre-feet in 1934.[38] Similarly, the Amity Canal varied from a high point in 1924 at 131,040 acre-feet (or 3.85 per acre of irrigated cropland) and a nadir of 31,738 acre-feet (.84 per acre) in 1934. There does not seem to be a direct correlation between rain on the Colorado plains and available acre-feet; 1937 was the driest year in Lamar during the fifteen-year stretch, but appropriators accessed 73,520 acre-feet, more than double the allotment from 1934.[39] Indeed, Extension research illustrates the inconsistent amounts, varying dates that water started and stopped flowing, and the disparate number of days the water flowed. Irrigators faced a guessing game to figure out what they could rely on even when they had supposedly reliable access because of their water rights.[40] Dam proponents contended that the reservoir could balance these numbers and give appropriators more control over the water flow, which could protect them from volatile weather or markets. That way, Prowers farmers could maximize their access and plan their strategy appropriately—the ABC thus dismissed the possibility for expansion in their proposal and instead emphasized the need to stabilize rather than augment production by managing the river.

The group tried to break down the dam's potential economic impact as another tactic to curry favor with federal reviewers. Fully understanding that federal largesse had its limits—even under the New Deal umbrella—the proposal set out the financial costs and benefits to demonstrate its positive impact on the regional economy. In addition to the numerous indirect benefits that dam construction could engender, like increased property values and "social development," proponents identified a direct way that construction made financial sense. Proponents used estimates from the Army Corps of Engineers study on the impact of floods in the region to calculate the amount of money saved from preventing destructive floods at $130,000 annually. The dam could prevent damage to urban and agricultural property as well as to roads, railroads, and bridges, plus the destruction of irrigation and diversion materials (e.g., canals and headgates). In other words, the dam's ability to regulate river flow had an immediate economic benefit for all valley residents, not just irrigators.

Advocates also utilized Corps data to figure the construction cost and outline a corresponding program of work that emphasized the impact such employment might have on the region. The ABC found that it would cost just under $8 million to construct the dam and fund maintenance and operation costs. The construction, at approximately $6,744,700, represented the bulk of the costs while the group calculated rights-of-way costs, engineering, and overhead as well as interest charges and an annual finance cost, to run just over $1 million. Recapitulating the cost versus the benefit, the committee figured that the annual cost at $340,258 and annual direct benefit ("Indirect benefits not included") at $473,000 for the first twenty years until the dam was paid off; they hoped that the difference of nearly $133,000 per year might offer additional reasons to support construction.[41] The proposal also tried to downplay the cost by identifying the "acute eco-

nomic conditions." The construction could provide "considerable employment in the area, and greatly reduce the need for nonproductive relief
expenditures." The A B C acknowledged that irrigation could also help, as
area financial stress was "due largely to a complete failure of the crops resulting from an unusually severe drouth." "By providing additional irrigation
water with which the inhabitants may cope with such drouths ," the dam
and reservoir "will not only ameliorate the existing conditions of distress
but will also, to a considerable extent, guard against their recurrence." That,
then, reflected their central argument. Certainly, they conceded, the dam
would involve government investment. Such assistance would advance the
local economy immediately. It also promised long-term financial benefit
by providing jobs and stabilizing local agriculture. The A B C presented a
persuasive argument in highlighting such benefits, especially given the
economic circumstances in play during the mid-1930s.

The A B C also stressed the dam's ability to reconcile differences between
Colorado and Kansas over the issue of access to river water. The two states
had been battling over water rights since the late nineteenth century, and a
case between them was pending in the Supreme Court to determine how
to allocate Arkansas River water equitably when the A B C put its proposal
together. After decades of fighting over access, the states had finally started
to come to terms with how to share the water, and the A B C believed that
the dam represented "the machinery for carrying out the distribution of
the waters, which has been agreed to by the two states."[42] While the proposal lacked specifics on what that allocation might mean in practice, the
authors included a long list of people who endorsed the project, including Colorado's state engineer, the Kansas chief engineer, the president of
the Mississippi River Commission, Colorado's governor, Edwin Johnson
(who warmed to the project as outlined by the A B C), and a number of
Colorado and Kansas water user/ditch associations.[43] Members of both
state houses and residents on both sides of the border further affirmed the
dam's potential positive impact on state relations, and the group hoped that
such potential harmony provided another reason for federal intervention.

Indeed, supporters contended that the dam would solve several grave issues for area residents, ranging from economic depression to devastation from drought to interstate politics. While the Mississippi Valley Committee received the proposal and embraced parts of the plan, that group had reservations about the federal government's role in funding and building the proposed system. The arguments in favor of construction proved enticing, especially the economic justification for flood control and the assurances that more irrigation did not mean more production. A problem emerged over the ratio of federal to local funding for the project. Colorado and Kansas legislators, both state and federal representatives and senators as well as governors, pushed for federal adoption of the plan with near-complete funding. Mississippi Valley Committee members believed that the project was viable from both an engineering and economic standpoint but that the local people should bear the brunt of the costs. That resembled the closest approximation to shared risk and responsibility for the dam, since the federal government helped with expertise, materials, and labor. By putting locals on the hook for some of the construction costs as well as the maintenance and operational costs, plus paying for rights-of-way and land to make room for the dam, the Mississippi Valley Committee suggested shared risk. Local organizers bristled at the prospect of having to fund the dam, and unity fractured over how to proceed, specifically in terms of how to persuade the federal government to acquiesce with additional money. Ickes sided with the Mississippi Valley Committee and refused to make the federal government the sole funding source for construction, so he removed the Bureau of Reclamation from the list of federal agencies that could build the dam.[44]

Just when it seemed hopeless for dam supporters, Congressman John Martin, one of the project's most strident advocates, received word from the Army Corps of Engineers that it might be willing to build the dam. The Corps offered one caveat: locals needed to manage the purchase of rights-of-way and deal with land issues to make room for the dam and reservoir. Otherwise, Corps representatives assured Martin that they could pick up the project because the Bureau of Reclamation had passed on it. The Corps signed off on the project to the National Emergency Council, which then approved the dam in the Emergency Relief Program for flood control. Martin secured a place for the dam in the Omnibus Flood Control Act of 1936. Such inclusion became crucial because it offered congressional support for the Army Corps of Engineers to manage dam construction in any case that had viable flood control concerns. Indeed, the bill allowed Congress to extend considerable funding for such flood control projects, demonstrating legislators' conviction to control the financial and human costs wrought by floods. Despite such funding, however, the Corps still wanted local financial assistance to deal with two important issues in addition to purchasing the rights-of-way for the project: first, locals needed to pay for moving several miles of Atchison, Topeka and Santa Fe Railroad track out of the proposed dam construction area; second, supporters needed to build levees to protect the hospital at Fort Lyon. These requests, though inexpensive compared with what Ickes had wanted, still threatened to upset the local economy. Consequently, while supporters found themselves closer than ever to seeing the dam become reality, a significant gap remained between their vision of federal help and what the government deemed its role in construction.[45]

That divide narrowed due to Representative Martin and Senator Alva B. Adams, who continuously pestered federal officials to increase their share of the funding. Their "active lobbying" led to House and Senate adoption of a report approving full federal financing for all flood control projects. That approval came in June 1938, five years after the Arkansas Basin Committee submitted its first proposal and two years after the dam gained coverage in the Omnibus Flood Control Act.[46] When the Army Corps of Engineers gained appropriation for construction in 1939, residents in towns throughout the Arkansas valley celebrated the coming of federal construction. John Martin died before construction started in 1940, and the project that had begun with the name Caddoa Dam and Reservoir became the John Martin Dam and Reservoir.[47]

FIG. 15. John Martin Dam and Reservoir on the Arkansas River in Bent County in 1979. U.S. Army Corps of Engineers Digital Visual Library. Wikimedia Commons.

The path toward construction was rife with obstacles, however, and advocates faced several tense moments and fought many battles over construction. One of the more revealing episodes dealt with the debate over establishing government control over regional water through the Arkansas Valley Authority. As the name implied, the AVA represented a similar form of regional development exhibited by the TVA. Clyde Ellis, U.S. representative from Arkansas, presented the act to establish the AVA to Congress in 1941. Ellis ruffled western politicians' feathers by presenting a blueprint for creating a federally managed body to control regional water. Not surprisingly, westerners wanted to control the river and thought that the AVA represented a federal power grab, one destined to wrest water rights and the chance for prosperity away from folks in the Arkansas valley. Moreover, Ellis presented the AVA as one of a long list of potential federal projects to develop the nation's rivers. Somewhat predictably, lines were quickly drawn by promoters and detractors over issues of federal intervention, partisan support for the New Deal, and, irrigation rights.

The debates about the AVA reveal how Coloradans approached federal intervention once the worst of the Depression had passed and rain had returned to the plains. For his part, Ellis hoped to utilize the TVA blueprint for nearly a dozen rivers across the nation, with a focus in this case on the Arkansas. The AVA was set to replicate the TVA's basic functions: controlling flooding, providing electricity, making the river more navigable, halting soil erosion, and providing irrigation for farmers. Per Ellis, the TVA had proven so successful in transforming the region's economic prognosis that such federal regional planning promised similar results for other watersheds. He believed that every state deserved protection from drought and depression through federal intervention. A staunch Democrat, Ellis seemed to favor utilizing the New Deal's wide-ranging power to alleviate hardship. It also appeared that he had considerable faith in the expanding federal government to manage the project without compromising the citizens' best interests. In his view, only the federal government had the expertise and resources to supervise regional development, and the New Deal promised to make life easier for everyone in the Arkansas valley.[48]

Ellis did not find many friends among Coloradans with this push for federal control, although "Big Ed" Johnson was one of his most vocal supporters and a powerful ally in pushing for the AVA. Johnson served as both governor and senator over the course of the 1930s, 1940s, and 1950s, and while he ran as a Democrat, he never toed the party line for the sake of partisanship. He placed a premium on state-level reform rather than relying on the federal government to initiate and fund extensive programs. He reduced taxes, initiated a statewide highway construction program, tried to balance the budget, and worked for civil service reform. He felt torn between his party and his largely Republican constituency from the western slope, and his critics often accused him of vacillating over important issues.[49] As a result, he frequently demonstrated a willingness to act as a maverick, making his philosophy difficult to characterize. His efforts to garner support for the AVA represented one of the more notable examples of Johnson's quixotic nature. Even as a staunch advocate for the state whenever it competed with the federal government, Johnson wanted Colorado to join the AVA and thus share responsibility for managing the river's flow with a federally controlled body.

Yet, rather than accept Ellis's plan as written, Johnson introduced a slightly different bill to the Senate for its consideration. Johnson viewed the bill as inherently positive but needing some tweaks to make it palatable for Coloradans. Consequently much of the bill that he sent to the Senate echoed the bill that Ellis sent to the House. Like Ellis, Johnson identified the supposed benefits of federal management, including the possibility for rural electrification, power, and erosion control. Johnson's resistance to allowing the federal government to dictate irrigation constituted the key difference. He contended that each state should have control over the water within its borders. As the Arkansas River west of the 100th meridian was not navigable, the federal government had no legitimate reason to leverage control over the river's flow. As Johnson argued, his version of the AVA "shall have no right, authority, or power to make any demand or place any burden upon the Upper Arkansas River Basin, or any part thereof, for the delivery of water for the benefit of the Lower Arkansas River Basin or for any other purpose."[50]

In so articulating the need for Colorado to control its own water future, Johnson tried to quiet most criticism targeting the AVA. Southeast Coloradans had been willing to entertain federal intervention throughout the 1930s even though they often tried to adapt policy to meet their needs. Most Arkansas valley residents, however, dismissed the AVA out of hand. Unfortunately for Johnson, Ellis, and their supporters, Johnson was the only Colorado politician who favored the AVA, and Representative Laurence Lewis, Democrat from Denver, was indifferent to the proposal.[51] Governor

Ralph Carr became perhaps the most vehement critic of the AVA. Carr scheduled a trip to Washington DC specifically to renounce the AVA and Ellis's intentions to take the state's water, and he also targeted the AVA as the central part of his inaugural speech in early 1941. In both examples, Carr identified the importance of having state control over irrigation and protecting long-held water rights by Coloradans along the Arkansas. His rebuke to Ellis and Johnson relied on political and economic arguments about the Authority's infringement on states' rights. Proponents claimed that the AVA needed to manage flood control and supply hydroelectricity, but Carr reckoned that neither held any reasonable application to Colorado residents. Carr contended that there was no need for flood control along the Arkansas. He continued that the river did not offer viable hydroelectricity options from its headwaters in the Rockies well into Kansas.[52]

Carr thus flayed most of the reasons why the ABC pushed for a multipurpose dam and focused exclusively on the need for states' rights in controlling natural resources. Of course, he did not present his critique as one chastising the ABC or its intentions; rather, he argued that the federal government's creation of a bureaucratic committee, appointed by the president and only truly beholden to him, left too much leeway for abusive power. Carr claimed that establishing the AVA was akin to opening Pandora's box, giving momentum for an already powerful state to grow even more authoritative, continuing down a dangerous path. He fashioned himself a watchdog for all Americans when he announced, "If we are to control the lives, the property, and the rights of American citizens by a new kind of government, then the people should have some voice in determining whether that innovation is to be attempted." Carr considered the AVA a threat to democracy and to vested personal property rights. He intimated that its creation might provoke outright rebellion against an increasingly tyrannical central government.[53]

For Carr, and for nearly every Coloradan who questioned the logic behind imposing a federal body to manage the river, control of the water remained the sticking point. Coloradans from seemingly every corner resisted the Authority, and while they may have cloaked their complaints in various forms, the right to use water sat at the heart of their concerns. They had abided by the proposition "first in time, first in right" to deal with questions of water rights since the state constitution first used the adage to adjudicate conflicts over water. In adopting this legal definition, the state formally recognized seniority as its determinant in deciding access. Consequently, Carr spoke for Colorado users who believed that they had long-held and legally bound access; the prospect of handing over their rights to a government-appointed board left many uneasy and some outright volatile. Moreover, the likelihood that the board would mandate that these users share their water with Kansans downriver only added fuel to the fire.

Coloradans became extraordinarily protective of what they considered to be their water, and the years of drought only aggravated their anxiety and made them more suspicious of ceding control over the river to the federal government. Those who took exception to the proposed AVA over the issue of water rights hailed from throughout the valley and indeed across the state. Some, like Clifford H. Stone, director of the Colorado Water Conservation Board, echoed Carr's concerns that the AVA would have executed too much control over water. Stone said, "No authority of this kind over a river basin where irrigation is practiced can ever be given congressional sanction and authorization without defeating effectually the rights, including the vested rights, of the state to control water for irrigation and other purposes." He believed that there was in fact no constitutional basis for the AVA and worried that if Coloradans supported the AVA, then all other major river basins—and people residing in them—could be subject to the same treatment. He thought that the authority could jeopardize local water rights and exclude local farmers from having an opportunity to influence irrigation development along the river.[54]

Coloradans joined Stone in criticizing what they deemed a federal power grab. Many of these individuals and small companies sent Representative J. Edgar Chenoweth letters that dealt with the water issue. Wilbur B. Foshay, secretary of the Salida Chamber of Commerce, supported Chenoweth's resistance to the bill. Foshay claimed that the Authority would hurt local business because the farmers were the most influential consumers in the valley, and if they continued to suffer from the Depression or if another drought caused further damage, then all business would be waylaid. Chenoweth assured Foshay that he was one of many who disagreed with the bill, writing, "I have had nothing but protests against the Arkansas Valley Authority. It seems that all of the water users on the Arkansas are opposed to the bill."[55] Foshay referenced the head of Denver's Chamber of Commerce to say that if the AVA upset the southeastern Colorado economy, then it would surely disrupt the state economy. But Foshay saved his most sensational indictment of Ellis for his role in extending federal power throughout the region. Foshay compared Ellis's vision of the Authority to Nazi warmongering, because each wanted to divide and conquer to subdue areas "one by one instead of attempting the subjugation of all at the same time." In this case, Colorado represented a significant domino that had yet to fall. Ellis and authority supporters designed a "scheme . . . to attack the [various river] basins one at a time in the hope of preventing united resistance" that would ensue if he proposed the entire plan for the nation's rivers. Because Ellis turned first to the Arkansas as a significant piece of the puzzle, such "subjugation" was an abuse of power and threatened Coloradans' liberty and property rights.[56]

Floyd Wilson's letter to Alva B. Adams offered one of the best expressions of the reticence to sacrifice autonomy or cede individual water rights to a federal body. Wilson, the Lamar businessman who developed alfalfa mills in the area as well as across the Great Plains, suggested that the federal government's ignorance, more than a sheer power grab, represented the biggest problem with the AVA. His letter included a litany of reasons to resist the AVA, including the possibility of federal power running amok as well as adversely affecting both the regional and local economies. Like others, Wilson believed that the AVA was going to "subordinate our ditch companies, water boards, and district engineers" to distant federal controls. The root cause of that disruption was that AVA supporters had no appreciation for farm life in Colorado, especially the bureaucrats who could not grasp the importance of water rights to land values. Wilson argued that "the best men in Washington can't *savvy* why the AVA does not fit into our irrigation picture." Coloradans knew that "water is land, and land is not land without water" and that to understand irrigation is to comprehend "its plan of operation and its definite relation to the price per acre of a farm." Wilson argued that, unfortunately for Ellis and others, irrigation, more specifically the precedence of water rights, was "something one has to live with to know." The AVA threatened stable irrigation, so it jeopardized land values, and its successful passage would cause the suspension of all farm trades in the valley. The Wilson letter exposed an underlying theme of state versus national agendas, where the federal government seemed disconnected from constituent needs. Like most interested observers, Wilson worried that the AVA would eventually take complete control of the river and could potentially sabotage long-held water rights. If the state had control of the water, however, farmers would augment their holdings and not sacrifice their rights. Wilson did not want to jeopardize users' autonomy by allowing federal officials a chance to redirect river flow to people who had not earned their water rights.[57]

Wilson's letter and Governor Carr's speech in Washington both addressed the issue of how jeopardizing water rights would adversely affect property rights. This possibility suggested catastrophe, especially considering the amount of mortgage debt most farmers faced, both generally and specifically during the Great Depression, and the likelihood that most of their money had been tied to their property. Carr contended, "Land without water is worth from ¢50 to $2.50 an acre. Land with any reasonable irrigation right is valued at from $25 to $250 an acre. The water carries the value. The land is merely incidental."[58] While perhaps guilty of hyperbole, Carr evidently believed the potential for interrupting water rights in the valley by inviting federal management would shake the economic foundation of the entire region and leave no farmer or business unaffected. As governor, he could ill afford further deterioration of the agricultural economy when the horrors of the Depression were still fresh in his constituents' minds. He had more clout than nearly anyone involved in the debate, and he used his political connections to leverage pressure on Congress to dismiss the AVA. For instance, Carr organized a conference of western governors to discuss the AVA and similar projects in the West. Carr chaired the meeting, and his hands are easily decipherable in the set of resolutions adopted by the governors in attendance. It was in many ways a reiteration of what he had previously discussed in both Washington and Denver—the issue of federal power run amok, the lack of state control over its own resources, and of course the key point about irrigation. With Carr trumpeting resistance and successfully painting federal efforts in a negative light, the conference resoundingly condemned the AVA and similar federal maneuvers to replicate the TVA in other parts of the country.[59]

In the end, the combination of such political pressure and the widespread disdain for the AVA demonstrated by various groups within the state proved enough to kill the legislation in Congress. Arkansas valley residents widely celebrated, but the editor of the *Lamar Daily News* called readers' attention to the elephant in the room. He noted the contradiction of wanting the federal government to become involved in some issues but not in others. He wondered why it was acceptable for the government to fund and build the dam but not to take management of the facility's production once it was completed. He also pointed to the TVA as a success story that could boost the region's economy.[60] Similarly, Mary Farley condemned Carr and others who selfishly pushed against the bill and jeopardized residents' future. Looking back from 1971, Farley blamed Carr for depriving Colorado of a "key economic resource" and lauded the TVA, which "proved its worth to the area it serves in countless ways which contrast with the plight of the Arkansas Valley."[61] What many then and since have failed to recognize, however, is that the Army Corps of Engineers never pushed the idea for a TVA-like development. They had already determined that they would not build a hydroelectric facility at John Martin Dam, negating one of the main benefits that most observers credited to the TVA. They had also decided that the issue of navigation along the river mattered little in Colorado. Finally, they argued that the proposed AVA would jeopardize progress being made between Kansas and Colorado to settle their ongoing legal dispute over allocating the river's water.[62]

For all the concern over federal involvement, then, most of the critics failed to understand that the Corps never cared about having control. In essence it never really granted their support to the AVA, and consequently it was more than happy to take responsibility for building the dam. Just after the Army Corps of Engineers got the green light to move forward with construction and AVA detractors won their battle in Congress, however, the United States entered the war and the federal government quickly dropped domestic water projects to the bottom of the list of priorities. The Department of Defense removed nonessential personnel from the site and relocated them to the Army Engineers' regional headquarters in Albuquerque in early 1942. Word from Washington to shut down construction entirely came in March 1943; by that point 87 percent of the dam had been erected, leaving only gates and a bridge across the spillway for postwar construction. Yet the incomplete system still had storage for 165,000 acre-feet of water that farmers could (and did) use immediately.[63]

Farmers expressed some trepidation about the adjusted timetable, as the initial plans targeted a completion date in 1943 that was obviously no longer possible with the war on. Conflict over who should supervise and run the completed system, if and when it was finalized, came to the forefront during the hiatus. The key debate concerned whether the Army Engineers should maintain responsibility for the project or if the Bureau of Reclamation should take control. That possibility emerged when Congress considered its annual flood control authorization and contemplated giving authority to the Department of the Interior. That transfer would mean a couple of changes, notably the transition from the Department of Defense to Interior (and thus under Ickes, who never truly warmed to the project). It would also entail a change for local irrigators, as the Bureau would manage and fund its management by charging them for using the water. The Army Engineers had not voiced that possibility in their time overseeing the venture. Consequently, local irrigators like Vera Pointer, secretary of the Caddoa Dam Board, encouraged J. Edgar Chenoweth to fight the transfer. The Army Engineers wanted to see the project through, so they joined Chenoweth in fighting against the transfer and in favor of leaving the dam entirely under the purview of the Corps and its reservoir free to users. That support helped Chenoweth successfully argue to keep the dam a Corps project, ensuring that irrigators could use the new resource at no additional cost. Farmers thus won the debate over initial expenditures and usage costs.[64]

With the project still firmly in the Army Engineers' hands, then, construction continued after the war concluded. With demobilization efforts running full throttle after 1945, the federal government funneled men and material to the site, and by late fall 1948 the reservoir had filled to capacity of 275,000 acre-feet. The dedication ceremony occurred April 1, 1949; Chenoweth organized the festivities, and John Martin's widow released the first batch of irrigation water just in time for spring planting. Later that year Colorado and Kansas finally came to terms on an Arkansas Compact to divvy up the waters, allocating 60 percent of stored water to Colorado users and the other 40 percent to be released down the river to Kansans. After nearly fifty years of battling over the water, with litigation reaching the Supreme Court and tension constant in both state houses, the states decided to cede responsibility for managing the dam and reservoir to the Arkansas River Compact Administration. The Corps advised the ARCA to limit new access and keep the peace while providing water for those who already had established rights. It was effectively local control that respected the Colorado doctrine of prior appropriation and limited new users who would drain the supply. Therefore, the ARCA was the icing on the cake, and it seemed to answer farmers' prayers.[65]

For those fortunate enough to have had water rights prior to the dam's construction, the dam promised steadier and therefore more profitable access to the river water. There was no dramatic upswing in the number of farms utilizing irrigation, just as the ABC had told the Mississippi Valley Committee to get the project approved. The number of farms in Prowers County declined over the course of the 1930s and into the 1940s even as the dam was being built. Indeed, changes from 1935 through 1945 show that the irrigated areas were not immune to the more general trend toward fewer but larger farms. Individual farms became larger on average (from 385 acres to 814 acres), more of the county became classified as farmland (from 565,622 acres to 782,692 acres), and the number of farms dropped (from 1472 to 962).[66] The amount of irrigated land fluctuated a bit over the same span, but there was nothing to suggest that newcomers were taking advantage of irrigation. The number of farms that had irrigation dropped over that period, but whereas the total number of farms declined from 1,469 in 1920 to 1,126 in 1950, the number of irrigated farms dropped from 660 to 647.[67]

These numbers support both the ABC's claim that this would not produce additional users and the fact that irrigated farming represented a more stable and sustainable kind of farming. Though not immune to the Depression and similarly facing problems caused by drought, Prowers farmers maintained their acreages more consistently than their counterparts because they had a relatively stable water source. Consequently, it proved much harder for Prowers farmers to "bottom out" in the ways that their neighbors to the south had. Agent A. J. Hamman noted a difference as early as 1936. Hamman, who started his post in Prowers County in 1934 and understood the depths of the 1930s, found that "the general economic situation has become less acute, with better crops in the irrigated districts" while changes for dryland farmers have "probably been slightly for the worse, in that more farmers have had to have help during 1936 than did during 1935."[68] In another telling example, farmers on irrigated lands in Prowers County "succeeded in producing [a] very good wheat crop" in 1937, the driest year for the county on record, while 80 percent of the dryland farm families in Prowers had left their land for neighboring cities or counties, leaving fourteen schools empty across the county.[69]

Census numbers also suggest that Prowers farmers entered 1940 more economically stable than their Baca counterparts—coming out of depression in better shape just as they had entered the 1930s better able to survive the lean years. Census data for alfalfa and sugar beets demonstrate that by 1934 Prowers farmers had reached a point of stability that they maintained through the war and beyond, despite a drop from 1929 to 1934. For example, Prowers alfalfa production registered at 44,726 acres in 1929, 31,101 in 1934, 28,000 in 1939, and 30,658 in 1954. This was a significant decline of about 25 percent after the first years of the Great Depression, but they reached a point of stasis by the mid-1930s. Sugar beet production illustrates a similar trend: farmers devoted 6,810 acres to beets in 1929, 3,552 in 1934, 3,142 in 1939, and 3,045 in 1954.[70] While these numbers demonstrate a drop in production, the decline in wheat production in Baca over the 1930s (from roughly 88,000 acres in 1929 to 24,000 in 1939) makes the slowed production in Prowers County look comparatively benign.[71]

Demographic change is another indication of Prowers irrigated farmers' comparative stability. The main Prowers County newspaper made no mention of Dust Bowl migrants, unlike the Baca County paper, which routinely kept readers abreast of migrant communities full of former Baca residents living in places like Los Angeles. As Hamman suggested, dryland farmers became much more prone to absolving themselves of their predicament by picking up and leaving. That helps explain the dramatic transition in Baca population from 1920 to 1940 while Prowers enjoyed a more stable and mixed rural/urban population. By 1920 and the postwar wheat boom, Baca had 8,721 residents. That number increased to 10,570 in 1930 before dropping off to 6,207 by 1940, a decline of 41.3 percent. Prowers, meanwhile, had a population of 13,845 in 1920, 14,762 in 1930, and 12,304 in 1940. By 1940, urban residents represented 36.1 percent of the county population, an indication of greater economic diversity, more immediate consumers for agricultural products, and a more stable economic base than that found in Baca County. Again, like the numbers for agricultural production, population statistics suggest that Prowers had an economic and population decline after the start of the Great Depression but had more opportunity to stabilize by the mid-1930s.[72]

Former Extension agent Thomas J. Doherty also noted a sort of psychological benefit, a kind of mental stability that irrigators enjoyed and dry farmers lacked. Doherty polled Baca County farmers about their transition from dryland to irrigated farming during the late 1950s and early 1960s. His assessment of how farmers thought about their prospects for success changed once they adopted irrigation is telling. In his analysis, 92 percent of the respondents, most of whom had access to deep well irrigation, felt either "pretty good or very good" about the change. They commonly noted more stable crop production and higher likelihood for income stability as the most important consequences of their shift. The respondents also reported that they felt more comfortable dealing with credit while having water than without, suggesting that they had more confidence to buy machinery or land or other goods knowing that they could eventually pay back any loans. Doherty found that the biggest deterrent for potential irrigators was the initial investment if the issue of geography and access was dismissed. Once they made that plunge, however, irrigators generally felt more optimistic about their stability and the opportunity for prosperity, and the more successful irrigators proved more likely to adopt Extension recommendations about methods and techniques.[73]

At the same time, however, and as historians have consistently shown, having access does not preclude the need to conserve water or be conscious of use. For all that irrigation provided, it was not fail-safe. It was a technological option designed to mitigate the effects of drought, to provide more options for farmers to choose profitable crops, and, in the end, to ameliorate an arid environment. Many problems existed, including salinity and sedimentation, but perhaps the most frustrating issue for appropriators remained the river's unpredictability, as no one could forecast snowmelt adequately to determine future river flow. As irrigation specialist and onetime Prowers County agent A. J. Hamman claimed, "Don't get in the habit of taking the available water supply for granted."[74] Indeed, even those with water rights found themselves at the mercy of precipitation levels, and the Prowers County agent noted several ditches that ran out of water during the 1930s drought years. The ditch companies that had seniority and preferred access to the river provided for their constituents much better than those with junior rights. Thus some ditches ran without water while others ran at full capacity. Some of the companies with junior rights built reservoirs to counter that fluctuation, but nothing offered as much security as senior rights.[75]

There were also cases of user error, as irrigators often failed to do necessary upkeep on their canals and distribution systems to protect against water loss. Part of this fell on the farmers, many of whom did not know the best ways to divert and distribute the water. Some neglected to seal their ditches and canals to protect against leakage because they did not grasp the need or lacked the money for repairs or had no access to concrete or similar sturdy patch material to do an ample job. Some farmers also watered either too much, too often, or both, wasting part of their allotment and damaging their crops in the process. Many of these individuals lacked much experience with irrigation, so they found themselves not taking full advantage of their good fortune. County agents held seminars and demonstrations to help and announced such meetings in local newspapers to bolster attendance. In some cases, people turned out well to these showings, such as the irrigation seminar that Extension Service employee Floyd Brown taught in January 1939 that some two hundred Prowers County farmers attended.[76] While irrigation specialists, federal employees, and county agents usually offered instruction about the best techniques, and federal workers from agencies like the CCC often provided labor to fix distribution systems, neither specialists nor workers were always available. For example, county agent Jack French noted that a group of farmers requested a demonstration on their ditch to figure out the best way to maximize their water right, but no one was available to help them by offering insight or labor.[77]

As county agents and federal employees came to discover, however, arousing enthusiasm for conservation often lagged behind Baca regardless of experts' availability. County agent Jack French noted in 1940 that the

FIG. 16. "Showing wash off of good soil by faulty irrigation," May 1940. The Extension Service reached out to farmers to promote water conservation by educating them on proper irrigation techniques. Extension Service, University Historic Photograph Collection, Colorado State University Libraries, Archives & Special Collections.

Soil Conservation Service was starting to take hold in the county—five years after the agency's creation and three years after Baca County had established two soil conservation districts.[78] The Webb Soil Erosion District created in 1943 was disbanded in early 1944 when enough farmers from within the district took the governing body to court over charges of faulty elections. The court ruled in favor of the farmers, who claimed the elections were carried out without an adequate number of voters. They argued that 35 percent of the district farmers did not want to establish the board in the first place; the court agreed that the district rules had been violated and had "disorganized" the district.[79]

This apparent reticence to conserve did not become universal. Some agriculturalists never pretended that controlling the river or its tributaries was a definite solution to the issue of drought or insurance for production, so they tried to augment their output by conserving soil and water. A. J. Hamman published and distributed a bulletin that juxtaposed two farms under the same water allowance. The one with contours, an apparent demonstration of conservation, looked much healthier and more productive than its counterpart, which allowed Hamman to suggest that conservation, even with irrigation, still mattered.[80] While the county agents in Prowers found a smaller number of farmers practicing conservation than the agents in Baca, some Prowers irrigators nonetheless practiced crop rotation, planted cover crops, executed summer fallowing, planted shelterbelts, and contoured their fields.[81] The SCS and county agents held regular demonstrations detailing conservation techniques, and it is evident that at least some area farmers found them applicable. Indeed, the Arkansas Valley Soil Conservation District was formed in December 1941 and became "the first demonstration of the use of soil and water conservation practices on a wide scale in a large irrigated area of the Southern Great Plains." The *Lamar Daily News* celebrated its creation with a front-page story that the editors placed beside the announcement that the United States had formally declared war on Japan. The district formed just days before the attack on Pearl Harbor.[82]

Many Prowers County farmers certainly believed that the new dam and reservoir was an answer to their prayers. It is difficult to fault them for wanting to increase their security given the devastation they had experienced during the 1930s and the consistent droughts that plagued the region. Tapping the river had been a part of Arkansas valley life, and as agriculturalists moved from subsistence to market production, their demand for water increased. They turned to technology to augment their access and stabilize their production. The federal government helped them attain their goal of dam construction, and the John Martin Dam and Reservoir dramatically changed the region's landscape.

Indeed, the dam represented a way that farmers changed their approach to resource use. Many residents looked at the dam as another form of adaptation, one made available because of technological advances and federal financing. Irrigation may have allowed these farmers to become complacent about conservation or protecting their soil, but their calls for federal intervention show that they understood that the Dust Bowl and Great Depression threatened their livelihoods. They adapted to the extent that they took advantage of new opportunities to use the federal government to get what they wanted—every constituency in America had done the same and continues to act in its own best interest. Per the editor of the *Lamar Daily News*, the dam represented "Colorado's greatest opportunity to share in the expenditure of the public works funds" of the New Deal, and residents took advantage—to their credit.[83] Certainly, farmers could have done more to conserve the valuable resource, as Extension workers and employees from various federal agencies pointed out, but they made strides in moving toward conservation by the end of the 1930s. Indeed, their efforts at trying to adopt water and soil conservation methods despite promised access to more water suggest that they understood what the water meant to their lives.

The 1930s did not represent a point of divergence from established patterns in Prowers as it did in Baca. Irrigation had been the way of life for many Prowers farmers who enjoyed the all-important water rights that accompanied prior appropriation. The Dust Bowl and the Depression became bad enough to compel irrigators to think about their systems, to consider ways to improve their prospects, and to organize and act collectively to realize their goals. They started to think more seriously about conservation techniques to maximize their irrigation efforts. In effect, they survived the 1930s by trying to get everything out of their water source and looked forward to a completed dam to augment their water allotments. By the onset of World War II, the drought had broken across the region and the river attained healthy levels, offering enough water to producers to make good on promises to supply the war effort.

5

On the Move

Thousands of southeastern Colorado residents simply tried to survive the 1930s until their prospects changed, which occurred at the end of the decade. As Baca County agent Claude E. Gausman noted: "Many rural families have just swept out the last dirt in their homes from the black period 1931–1938 and have produced enough from the farm in 1939–1940 to pay off a portion of those unpardonable debts, thus many of them have just begin [sic] to live again rather than exist."[1] Indeed, those who eventually welcomed the 1940s lived in places that looked far different than they had when FDR entered office in 1933. The landscape had changed dramatically. The Dust Bowl and Great Depression had taken a toll, of course, leaving houses abandoned, schools empty, and fields barren. Yet there were signs of a mended landscape as well. Various New Deal programs had generated the construction of school buildings, post offices, gymnasiums, bridges, and other infrastructure projects. The federal conservation program had purchased some land and rehabilitated other plots, reconfiguring the grid pattern of private ownership that had dominated the plains since the Homestead Act. In addition, construction crews dug a new reservoir to contain water held by a mammoth dam, an installation largely surrounded by fields that showed the return of native grasses.[2]

The social landscape also changed over the course of the 1930s. The ecological and economic crises initiated several demographic shifts that altered the regional population. The population in southeastern Colorado had been relatively steady from the onset of the 1910s to Roosevelt's election in 1932. The late 1920s witnessed the highest population level in Baca County history and represented one of the high points for Prowers County as well. During the Depression, however, the number of permanent residents in both counties slowly started to decline, and that gradual trickle turned into a veritable flood by the end of the 1930s. Some 10,570 residents lived in Baca County in 1930, but 6,207 people resided there in 1940, a decline of more than 40 percent. Prowers County residents faced less severe fluctuation, although they too saw a net drop of roughly 2,000 over that same period, from 14,762 to 12,304, a drop of 16 percent.[3]

That population decline and its consequences were part of a regional transformation engendered by the Depression, Dust Bowl, and the New Deal. The migrants' story has been told by many talented historians and has garnered considerable attention in literature, film, and photography.[4] Unfortunately, this attention misses two important points. First, it is inaccurate to consider this migration as the only demographic change that affected the Great Plains or to assume that every individual or family who moved from the plains during the 1930s and 1940s ended up in California. Dust Bowl–induced migration within the area, especially to cities, represented an important piece of a larger population shift in the area. Even by the onset of war in Europe by 1939, people had started to leave the countryside for the city, many of them to work in growing industrial centers like Denver or to take advantage of urban relief efforts to alleviate poverty and unemployment. In addition, farmers—including many tenants or part owners—simply moved around the countryside in search of superior prospects. Members of the landless population could easily pick up and move without much fear of retribution for breaking contracts. Tenants frequently left one farm for another in another county in hopes of more opportunity or better land. Consequently, much of the rural population was in flux.

The declining population affected the agricultural labor pool and the availability of tenants who would meet local farmers' need for workers. The movement of workers, tenants, and owners influenced land use regimes by challenging New Deal conservation measures and by forcing farmers to find alternate labor sources when World War II started in 1941. The agricultural labor system that had matured by the 1930s in Baca and Prowers Counties, and that relied on the combination of local laborers, tenants, migrant, and immigrant labor, started to dissolve by 1935. Farmers were left scrambling to find a viable workforce by the start of the war when the federal government interceded and provided workers. In a way, the labor disruption offers another perspective of the changes wrought by drought and devastation, a disruption akin to the challenges posed by dwindling water availability and soil erosion. Each element required farmers to adapt to new circumstances; the situation was anything but usual.[5] The lull in demand engendered by drought and depression—the combination led to a dramatic decline in production and therefore less need for workers—ended in 1938 and 1939, then ramped up again with American entry into the war. The population drain left farmers looking for labor outside their families and even their communities and more willing to employ wage labor with the start of the war. This transition is a vital part of how the Dust Bowl and Great Depression changed land use practices in southeastern Colorado. By seeing labor as one element of this change we can better understand the social impact that drought and depression had on area residents and the agricultural economy.

As it looked during the early 1930s, the labor structure consisted of several distinct parts, including a tenant system that emerged particularly after World War I and the wheat boom invited extensive land purchases by outside owners. Absentee owners sometimes hired tenants to work on and manage their holdings while they lived in other towns, often in other states. The two locales reinforced a trend occurring across the state, as the number of farmers who were tenants increased during the 1920s and again from 1930 to 1935 before a steep decline from 1935 to 1940. Consider the numbers. In Baca County, the number of tenants was 621 in 1930, rose to 794 in 1935, and then fell to 398 in 1940. These tenants farmed more than 415,000 acres in 1935 but fewer than 365,000 acres in 1940, suggesting that the tenants who remained in Baca (or who moved in after 1935) found themselves responsible for plots that were larger on average. Prowers County numbers showed the same trend, although to a less dramatic extent, moving from 625 tenants in 1930 to 755 in 1935 and finally 561 in 1940. These farmers tended more than 231,000 acres in 1935 but just under 200,000 in 1940; again 1940 witnessed fewer tenants, but those engaged in tenancy managed more acreage per person.[6]

A few explanations account for the rise and then sudden decline in tenancy in the two counties. First, by the mid-1930s almost everyone had been adversely affected by the drought and depression; many rough years in a row could convince even the hardiest and most resolute residents to pick up and leave. Tenants had a better opportunity to depart, since they had no mortgages, no financial obligations for land, and very little in terms of significant expenditures for machinery—three major capital drains for landowners. Second, early New Deal agricultural programs tried to cut production to stabilize prices. Such programs, including the Agricultural Adjustment Act (AAA) paid owners to limit their production of major commodities like sheep, wheat, cotton, and even sugar beets. If owners took the subsidies and cut production, then there was less need for a tenant to work on land that had been retired; consequently, owners cut tenants to reduce labor costs. In addition, many owners refused to share their AAA checks with tenants and instead kept the money for themselves, paid down debt, or invested in machinery like tractors. Tenants, therefore, had no access to federal assistance, had no guarantee of working a harvest, and had very little leverage vis-à-vis their employers. The Bankhead-Jones Farm Tenant Act of 1937 tried to reconcile some of these issues and make federal help more available to tenants, but by that point many tenants in southeastern Colorado had already moved on to other opportunities.[7]

One of the central problems with the tenant system was that it rarely led to ownership, despite the myth of the agricultural ladder. Proponents of the agricultural ladder argued that tenancy represented one rung on the path to ownership, such that young people who started as tenants could eventually move their way toward ownership through diligence and perseverance. Proponents of the idea supposed that many owners initially started as unpaid laborers on their family farm, gaining skill and experience that allowed them to move up to the next rung, working as hired labor at home or in the community. Such labor allowed the workers to amass some capital and work toward tenancy, the third rung on the ladder that allowed the farmer to make enough money to eventually purchase the land or another plot in the area. The fourth rung, encumbered ownership, allowed the farmer to buy land, farm it, and make a good living until the final step allowed for the farmer's retirement and position as landlord.[8] In some cases, of course, the ladder functioned as its proponents had hoped, and laborers eventually made their way to ownership.

Critics began indicting the ladder idea during the 1930s when it became evident that many tenants had a difficult time ascending the ladder and especially because of the post–World War I depression in agriculture. Critics suggested that more people actually descended than ascended the ladder, meaning that owners more often lost their holdings and ended up working for someone else rather than tenants becoming owners. This not only exposed a weakness in the ladder but it also became a sort of psychological challenge as "the existence of a large number of struggling tenant farmers called into question the cherished American belief that agriculture was the province of contented, independent farm owners and a repository of the nation's civic virtue."[9] FDR appointed the Special Committee on Farm Tenancy in 1937 to determine the causes and consequences of farm tenancy; they found a number of problems with the system that the Depression aggravated and left the entirety of American agriculture vulnerable to collapse. FDR warned, "The American dream of the family-size farm, owned by the family which operates it, has become more and more remote. The agricultural ladder, on which an energetic man might ascend from hired man to tenant to independent owner, is no longer serving its purpose." He continued, "While aggravated by the depression, the tenancy problem is the accumulated result of generations of unthinking exploitation of our agricultural resources, both land and people."[10]

The committee's report and his explanation of it typified the New Deal's response to agricultural problems. The committee identified many problems, notably defective land use, inadequate credit available for tenants, the high rate of debt throughout rural America, and families trying to farm submarginal land or on holdings of inadequate size. Farmers had no guarantee that they could ascend to ownership. The committee found that descent occurred more often and cited "an increasing tendency for the rungs of the ladder to become bars—forcing imprisonment in a fixed social status from which it is increasingly difficult to escape."[11] The committee blamed the tenant system and "sickly rural institutions" that "beget dependency and incapacity to bear the responsibilities of citizenship." It advertised ways to combat the downward spiral and therefore stabilize the ladder. It promoted soil rehabilitation, pushed for the federal government to make more land available to potential owners, stressed the need to extend federal benefits to farm laborers, discouraged land speculation and absentee ownership, emphasized the need to improve contracts between landlords and tenants to better protect the tenants, and supported worker and tenant organizations to defend civil liberties.[12]

Resettlement Administration economist Robert T. McMillan, who spent most of his 1937 report detailing the poverty in Baca generally, also devoted attention to the tenant issue. McMillan argued that deficiencies in the tenant system helped explain the county's ecological and economic degradation since the onset of the Depression. McMillan noted that "nearly two-thirds of all tenants moved to the county after 1926" because they wanted to take advantage of jobs tied to wheat production and simultaneously move away from more impoverished areas. This dramatic influx complicated matters because tenant "stability increased with duration of residence in the region" and instability in tenancy affected the agricultural economy writ large.[13] He found that tenancy was a necessary part of the agricultural economy and that "there is no reason why tenancy should not continue to fulfill its functions in the county agriculture." In other words, tenants still had a place working on owners' land, bringing in crops, and working their way up the ladder. McMillan argued that tenants could achieve a level of economic and social security comparable to owners but without the large-scale indebtedness or risk that owners faced. In his opinion, it made no sense to extend ownership to tenants when many owners in the county failed to survive. The local agricultural economy had been so devastated that even owners who enjoyed considerable government assistance could not meet their financial obligations.[14] Consequently, pushing tenants to ownership did not reflect sound policy to stabilize the rural economy, but the system of tenancy warranted federal attention nonetheless.

Observers hoped that federal attention might also influence how tenants farmed, as some critics indicted tenants for causing much of the problem with soil erosion across the South and West. They remained skeptical of any reforms, whether pushing ownership by extending credit or offering financial subsidies, unless the government also addressed land use. The adage of "poor land, poor people" thus played out in terms of tenancy and soil conservation, whereby the tenant system left tenants in perpetual debt and therefore more likely to misuse the soil. These issues were interconnected, and each needed attention, per Farm Security Administration (FSA) spokesman Flip F. Higbee, who noted that "the problems of impoverished lands and poor people . . . are closely allied with the short tenure system in American agriculture." Higbee continued that the only realistic way to address soil conservation was to invest in improving the plight of tenants and to ensure that they were willing to execute federal programs to protect soil and wind erosion. For Higbee, this meant strengthening the steps on the agricultural ladder and continuing to focus on moving the tenant toward ownership and simultaneously emphasizing the important role that tenants had in conserving water and soil even while they were in another's employ.[15]

Higbee and others contended that farmers who had a financial stake in the land's health proved more conscious of conserving resources, which suggested that tenants or laborers who had no hope of owing land had no incentive to protect it. Tenancy promoted maximum production despite the resource; soil exhaustion often resulted from such negligence. Moreover, critics also indicted landlords for emphasizing production at any cost. While resident farmers often blamed absentee owners for encouraging such activity, especially because they found the "suitcase farmer" guilty of not knowing or caring about the land, their neighbors, or the community, setting the absentee owner up as a scapegoat missed the point. As McMillan found throughout Baca County, the *combination* of tenancy and absentee ownership often resulted in the biggest impact on the environment. He contended that "one-crop farming is the attendant evil of tenancy and small farms. Farmers on small farms are compelled through necessity to raise crops that will produce the largest returns per acre. Also the landlord is too often interested in collecting the greatest cash return from the land regardless of soil losses."[16] The Committee also indicted tenancy for this problem, finding that "the tenant whose occupancy is uncertain at best, and ordinarily does not average more than 2 years, can ill afford to plant the farm to any but cash crops." Furthermore, "the tenant who expects to remain but a short time on a farm has little incentive to conserve and improve the soil; he has equally little incentive to maintain and improve the woodlot, the house, barn, shed, or other structures on the farm." Put succinctly: "Erosion of our soil has its counterpart in erosion of our society," and the tenant system led to the waste of natural and human resources.[17]

FDR's committee agreed that the tenant system's short leases facilitated heavy migration, and they sought ways to ensure longer stays for tenants. Committee members argued that such a dramatic shift in population "not only wears down the fiber of the families themselves; it saps the resources of the entire social order."[18] The report stated that tenancy led to short stays on land and high rates of mobility from farm to farm, county to county, and sometimes state to state. The migration left tenants with no job security and disrupted the family as well as the community. Ralph Swink, Prowers County supervisor of the FSA, estimated that 225 farm tenants in Prowers moved annually, costing each family $57 a year (a total of $12,825 in lost wealth among tenants). These moves, which usually happened during the winter, uprooted the family, pulled children from school, and made the entire family feel discouraged and disinterested, frustrated with their plight. This shift also left the landlord in the lurch. Landlords were often unaware of the tenant's plan and ended up flailing through the winter looking for new employees. Invariably, the landlord found the first available tenant, unconcerned with helping the tenant adjust or weeding through unstable tenants to find the best fit. In essence, Swink argued, landlords promoted "getting the most out of this year's crop, letting the future take care of itself" and treated the tenants in the same way.[19]

Beyond criticizing the system for not leading tenants to ownership, some observers argued that the tenants deserved some blame for their plight. For example, Lewis C. Gray, primary architect of New Deal agricultural conservation programs writ large, as well as economists and social scientists within and outside of the U.S. Department of Agriculture, suspected that many tenants had character flaws. Gray and others argued that some tenants were thriftless, even shiftless, and unstable. They acted like soil-miners, tied to neither the land nor the community, and proved willing to break a contract if to do so was in their best interest. Detractors also claimed that some tenants were dishonest, negligent, and prone to abandon the farm when the crop did not meet expectations, which usually meant that they could not fulfill their financial responsibilities to the owners. The owners then faced multiple challenges to get their enterprise back up and running, a process that cost time and money.[20]

To others, including Secretary of Agriculture Henry A. Wallace, federal reforms could help stabilize the system and resolve many of these issues, most of which were not the fault of the tenants themselves. He contended that tenants could easily become responsible stewards if given a stake in the land's long-term productivity. He believed that longer leases would provide tenants with a stronger sense of responsibility for the land's health. Wallace figured that extending the leases would provide more security to the relationship between owner and tenant and could keep tenants happy, promote soil conservation, and protect owners from broken contracts. He echoed the concern that the *system* promoted land abuse and contended that tenants deserved little blame for soil exhaustion.[21] Indeed, Baca County extension agent Claude Gausman noted in 1940 that Farm Security Administration clients, former tenants who took advantage of tenant purchase programs to buy land that the federal government made available, took remarkably good care of their land. Such new owners proved more conscious of conservation and more willing to protect their land "than the unattached operation," suggesting that ascending the ladder compelled the farmer to be more diligent in conserving resources.[22]

This indictment of the tenants showed that observers had concerns about the system, the landlords, and the tenants, which suggested that only dramatic reforms might stabilize tenancy. As it stood, tenancy offered security to no one and left the land especially susceptible to exhaustion—both potential ramifications countered the basic New Deal agricultural policy. Consequently, FDR's committee constituted one element of a larger conversation about the role of tenancy in American agriculture, the state's responsibility in promoting ownership for tenants, and how to improve the system. If ownership was indeed a goal for most tenants, then the federal government would help them reach it; if that was impossible, then the government could help the tenant leave the countryside. Unfortunately, one problem that plagued the early New Deal effort to right the agricultural economy was the consistent blind eye that it turned to the issue of tenancy. Despite Wallace's apparent concern for the situation, the first incarnation of the AAA only subsidized the owner or landlord and offered nothing directly to the tenant. It also paid owners to retire land, which often negated the need for either tenants or labor, and thus many owners jettisoned those individuals. The subsidy system that proved so vital to helping farmers survive terrible economic years did almost nothing to assist tenants.[23]

The Bankhead-Jones Farm Tenant Act of 1937 was the first congressional legislation designed specifically to help tenants by including them under the umbrella of federal assistance offered by the New Deal. It effectively enacted the recommendations that FDR's Special Committee on Farm Tenancy made in terms of trying to stabilize the agricultural ladder. Congress had briefly considered the tenancy issue a few times between 1933 and 1937, mostly because of John Bankhead and other southern congressmen. The law passed because of the sudden convergence of executive and congressional attention, and it aimed to do three things: "to promote farm home ownership through a system of long-term farm mortgage loans; to rehabilitate distressed farm families (who cannot be aided in purchasing a farm) through short-term loans for livestock, equipment, and supplies; and to provide for the development of a land conservation and utilization program, through the purchase of land submarginal for agriculture and the development of such land into uses for which it is best suited."[24]

It thus expanded the New Deal's efforts to identify and purchase submarginal land and rehabilitate the farm economy through conservation while simultaneously giving the tenant a boost up the agricultural ladder. The act apportioned $10 million the first year, and it allocated $50 million annually by the third year to be managed by the FSA, the successor to the Resettlement Administration and the agency most aligned with tenant concerns. Like most New Deal agricultural policy, FSA representatives and the monies allocated by the act effectively ran through the county agent, who acted as the first resource for interested tenants.[25] Tenants contacted the county agent when they wanted to apply for loans to buy property, and the FSA determined the most qualified for low-interest loans. The program allowed four tenants to buy farms in 1938 in Prowers County and four more in 1939, a small number relative to the tenant population of roughly 775 people, but an indication that the program worked for at least some farmers.[26] Additionally, the FSA offered low-interest loans to tenants to defray costs accrued in maintaining the land, including equipment, seed, and other necessities. The FSA in Baca County used federal money as well, in addition to the low-interest loans for supplies or property, the Baca County office extended direct financial assistance to tenants who wanted to improve their surroundings. For example, the FSA used federal dollars to build or repair over two hundred buildings in Baca in 1939, spending nearly $20,000 and therefore doing the work at no cost to the requesting parties.[27]

According to J. E. Morrison, who eventually served as Colorado Cooperative Extension director from 1952 through 1958, several parts of the FSA-led Bankhead-Jones Tenant Act proved successful across the state. The push for rehabilitating the rural economy through loans and grant-in-aid programs designed to secure livestock, machinery, family necessities, and other goods proved largely successful. The FSA also worked with Extension to promote farm and home economics, as well as quality of life issues like nutrition, attaining a proper shelter, and keeping the family clothed. The tenant purchase program was a boon as well, "probably the most successful part of the Farm Security set up." Even though the program only helped a relatively small proportion of needy farmers—Morrison noted that only a "small percentage of tenants" met the necessary qualifications to become involved in the program—the new owners paid back their loans quickly. Despite these successes, however, Morrison understood that the program existed at a high cost in administration, supervision, and land purchase. He claimed that the resettlement program had "for the most part failed" because the new owners resettled on poor sites or on such small units that they could not sustain themselves. He also questioned the logic of adding more landowners when "if all farmers farmed just ½ as well as the best farmers, agriculture would very soon be swamped in its over-production."[28] His statement effectively identified the crux of the problem with much New Deal agricultural policy, emphasizing the need to keep farmers farming proved counterproductive if the real goal was to balance supply and demand. Morrison thus celebrated the federal effort to stabilize the tenant system but challenged the desire to increase ownership across the state.

Regardless of its sometimes convoluted logic, the New Deal approach to tenancy reflected a concerted effort to address problems with the agricultural ladder. The FSA and the Bankhead-Jones Act were federal attempts to improve tenants' plight, and each had some success in Colorado. But the convergent forces of drought and depression produced considerable migration among tenants. Even while the government tried to restore the agricultural ladder, the trend in southeastern Colorado was away from the tenant system. The declining numbers of tenants, as well as part-owners and managers, left landlords with fewer options to help run their enterprises. The tenants had been an important piece of how especially large owners produced, and their absence disrupted a system that was only starting to mature in southeastern Colorado by the early 1930s. Their departure left a significant dent in the available labor supply for farmers who needed such assistance.

The widespread tenant departure marked one problem for growers looking for workers, and a similar slowing of migration *into* the state during the 1930s made matters worse for Colorado farmers who needed labor. The same factors that allowed for the growth of tenants during the 1920s—expanded production and more focus on cash crops like wheat and sugar beets—led to an increase in farmers using outside labor. Latinos, both migrants from neighboring states and immigrants, made up much of this labor. The push to employ these workers started just before the turn of the twentieth century, when the sugar beet industry exploded and began requiring paid labor during the planting, thinning, and harvesting stages. Corporations like the American Sugar Beet Company and the Holly Sugar company recruited and contracted with outside labor, then made them available to growers who had signed a separate contract with the company to sell their sugar beets to the same refiner. The companies proved remarkably adept at establishing this labor pool, as they had more resources to direct at labor and a vested interest in making sure that their growers had all that they needed to produce.[29] The outside sources became necessary once it was clear that family and local labor could not satisfy the industry's need. Consequently, the refiners looked outside the Arkansas valley and developed labor streams to abet growers. This pattern of attaining workers functioned well until the mid-1930s, but complications from the Depression and drought effectively stalled the migration of seasonal workers into the state.

Sugar companies spearheaded the turn to migrant labor because the various stages of production were each labor-intensive. The first phase involved blocking and thinning the small beet plants once they sprouted from the ground. Immediately upon completing the thinning process, the worker focused on hoeing or weeding to protect the plants from insidious weeds. After a short break lasting a couple of weeks while they waited for the plants to ripen, the workers started harvesting the beets for shipment and processing. Employers often had difficulty finding laborers because the work itself proved so strenuous: "Thinning and harvesting were considered two of the most arduous types of agricultural labor. The tasks required workers to constantly stoop over the rows of plants. In addition, growers exerted constant pressure for speed in both processes, and thinning was done under the hot summer sun, while harvesting took place during the disagreeable weather of late fall."[30] In total, the labor demands meant that most workers spent between 80 and 90 days on sugar beets, and the beet calendar meant that most laborers started and concluded their years in beet fields.[31]

Before the turn of the century and the beet industry's explosion, however, family labor filled most needs in southeast Colorado. Like most nineteenth-century agriculturalists, Colorado farmers utilized family members, neighbors, and locals to fill out their labor needs, only turning to "a few itinerant workers to meet peak season labor demands." Demand increased dramatically at the turn of the century with the beet boom, and local workers proved insufficient to meet production goals. Consequently, beet farmers looked for alternative labor sources. They initially turned to members of various Asian groups, and companies recruited a predominantly Japanese labor force from Denver. Some farmers found that these individuals largely "did not prove satisfactory" because they often ended up starting their own farms once they built enough capital. In an unusual twist, then, their success at moving up the agricultural ladder complicated employers' demands and, in this case, made them competition for farmers who initially hired them. This proved a powerful deterrent to the use of Asian labor, and growers and companies looked elsewhere.[32]

Colorado sugar companies found a solution when they turned to largely Latino migrant labor that filtered between Colorado, New Mexico, Texas, and even Oklahoma. Most of these seasonal employees worked in the San Luis valley or on the West Slope, but Arkansas valley companies increasingly recruited them to work in sugar beet fields. In addition, Mexican laborers became an increasingly important part of these labor streams by the mid-1910s, adding to the number of Spanish-speaking workers available to Colorado farmers. The Mexican revolution started in 1910, and "as civil war raged through Mexico year after year, increasing numbers of Mexicans—a few political refugees, others with a heritage of seasonal migration, and many others mobilized by Carranza's decree in January 1915 liberating them from peonage—fled the war's chaos and its destruction of life and land."[33]

Employers throughout the Southwest embraced these workers with open arms, and immigration continued after the war, despite immigration restrictions, because sugar beet companies worked with cotton growers in Arizona, California, and Texas to earn exemptions for agricultural workers. The Mexican-born population in Colorado increased nearly fivefold between 1917 and 1920, to a total of 11,037, evidence of increased demand among mine owners and sugar beet growers.[34] Although obviously tainted by racism and generalizations, Prowers County agent and eventually head of the Emergency Farm Labor Program during the war, A. J. Hamman contended that Latinos were "essentially an agricultural people." Furthermore, Hamman suggested that farmers were very comfortable with using Latino labor and often preferred them to alternatives. He also claimed that they constituted the "most dependable and generally accepted group of seasonal workers because of their availability and adaptability to a wide variety of hand work."[35] These workers made the sugar boom possible, and Hamman's assessment, though markedly racialized, reflected the thinking that Latino workers were perfect for the work.

This system took some time to mature, and interactions between Anglos and Latinos often became complicated as each side looked out for its best interest. The community of Latino migrants found work in Colorado fields and stayed there in part because they could find a balance between working in one region and living in another, and they moved when and how they wanted to without much interference from their employers. In other words, during the 1910s and early 1920s, many of these migrants had established a "regional community" that allowed them to move from village to field and back again and provided a sense of security away from Anglo pressures.[36] Eventually, though, the Anglo population started to dictate terms for migration and employment, which compromised worker mobility and fractured the regional community. Companies adapted to their changing labor needs by shifting their employment strategies. Initially they employed a "sojourner" strategy, meaning that they did not want any permanent worker settlements and simply wanted them "to remain for

FIG. 17 *(top)*. Mexican beet workers and their housing. Hine Report, Colorado, Sugar Beet Workers, July 1915. Photo by Lewis Wickes Hine, Library of Congress Prints & Photographs Division, (LC-DIG-NCLC-00258).

FIG. 18. "Three adults and six children from seven years to twelve years hard at work on a sugar beet farm near Greeley, Colorado. The father said: 'The children can thin the beets better than grown-ups. We all work fourteen hours a day at times because when the beets is ready they has to be done. About twelve weeks is about all the children can work on thinning and topping. Some of them hoe a little.'" See Hine Report, Colorado Beet Workers, July 1915. Photo by Lewis Wickes Hine, Library of Congress Prints & Photographs Division (LC-DIG-NCLC-00255).

the season and leave after beet topping season ended." After World War I, companies intensified recruitment and tried to entice workers to settle in the region. The companies began to offer opportunities for settlement, offering workers incentives to bring their families to Colorado and stay during the entire year—giving the companies a reliable workforce that they could count on for the following spring.[37] The companies often funded the construction of schools and housing to make the area more appealing to families, facilitating what they hoped would become a steady pool of workers. This then afforded them broader appeal when the companies sent recruiters to find and then ship migrant and immigrant laborers. The employers generally identified hot spots for workers, recruited them, and then paid their transportation costs to get them to the fields.[38]

Those desiring the stability of permanent residence started to move to Colorado in larger numbers during the 1920s and 1930s. Indeed, newspaper coverage in southeastern Colorado supports census data identifying the increasing presence of Latinos in the sugar regions of Prowers County. The U.S. census from 1930 tallied only 48 "Mexicans" in Baca County but 1,436 for Prowers County. Local newspaper coverage evidenced this discrepancy and showed the relevance of Latino communities in Prowers County.[39] The *Lamar Daily News* consistently published articles describing activity in what they deemed the Mexican American part of Lamar. For instance, the paper informed readers of the Mexican Independence Day celebration held at the Mexican Lodge in "Colonia Juarez."[40] The paper later noted a Cinco de Mayo party at the "Mexican colony" and explained that after some considerable debate, the celebrants decided to fly both American and Mexican flags at the function.[41]

These newspaper pronouncements suggested that Latino workers had started to carve out a place for themselves by the early 1930s. None of the notice regarding Mexican American news implied any animosity or racism against the minority, and other than a few articles describing crimes perpetrated by Latino men, the majority of coverage seemed to consider them a productive part of the community. For instance, as Hamman noted, many sugar beet companies enticed these workers by offering housing and schools, and the Lamar newspaper announced school construction in Lamar as well as Wiley.[42] The paper's notification about a community meeting for residents to evaluate the New Deal and discuss how they might tap into federal funding exemplified the way that members of the community had become more settled in the region. The countywide gathering met to discuss employment problems and relief issues that Latinos had faced in garnering federal financial assistance.[43] Such attention to the Latino community illustrated its increasing size and scope in Prowers County before 1935.[44]

These communities often came into being based on some rather unsavory tactics to encourage workers to stay in the area. Companies sometimes employed multiple underhanded methods to force workers' compliance and keep them near the employers by circumventing their mobility. They could stall payments and not remit for two to three years after the work had been completed, thereby forcing the worker to stay in the area to eventually recoup his wages. Refiners also offered free rent and credit at the local grocery store during winter months to entice workers to stay. The company also knew that the worker who stayed would have a hard time skipping out on his grocery bill in the spring if he ever wanted to work for that company again. Finally, the company encouraged workers to build homes on company land. Unfortunately for the settlers, companies often funneled the workers to poor land and withheld the deed until the worker paid the whole mortgage. This started a cycle of indebtedness that held the worker on site until he could pay off the loan—something that often did not happen.[45]

Recruitment methods, nefarious or otherwise, proved so successful in building a deep labor pool that workers eventually started to think about organizing themselves. In effect, the burgeoning community of workers and their families started to think in terms of mutual needs and their rights, and they formed the Federated Beet Workers of Colorado under the American Federation of Labor umbrella in 1935.[46] Their preamble identified their desire to collectively bargain with the sugar beet companies in hopes of better treatment. They sought to unite the native and foreign-born workers in common struggle. The preamble noted the need for unemployment insurance, and it demanded an end to child labor, an improvement in the safety and cleanliness of working conditions, and an expansion of education opportunities for workers and their children.[47]

The rise of urban industrial labor unions and the New Deal's apparent sensitivity to worker organization represented one catalyst for such unionization in Colorado fields. The drought and depression combined to act as another because the dual crises caused economic fallout among growers and refiners. The drought meant that even irrigated farmers needed to monitor production, and the Depression adversely influenced sales. Consequently, neither grower nor refiner had much money. As employers are wont to do, the growers and refiners looked to cut their expenses by reducing labor costs. They used child labor when possible, dropped wages, and failed to do any upkeep on their facilities or employee housing. In effect, worker organization that culminated in unionization in 1935 represented the most obvious response to this mistreatment and indicated the growing proletarianization of agricultural labor in the West.[48]

Refiners' mistreatment of the workers compelled the federal government to intervene in the industry during the New Deal. The Jones-Costigan Sugar Act, penned by longtime Colorado representative Edward Costigan and Texas representative Marvin Jones, addressed both the industry's economic stability and its labor issues. The act included sugar cane and sugar beets as basic agricultural commodities and therefore eligible for inclusion under the management of the Agricultural Adjustment Administration. In addition to offering direct subsidies for production reduction, the federal government tried to regulate and stabilize foreign and domestic prices for sugar—an important boost to the industry when it faced hard times with the drought and depression. In terms of labor, the act stipulated that the secretary of agriculture could intervene to adjudicate labor disputes and create a minimum wage for laborers. Furthermore, the act identified child labor as a significant problem in the sugar beet industry and gave the federal government the power to prohibit child labor from anyone under the age of fourteen.[49]

The act represented the New Deal's attempt to stabilize the beet industry and to protect exploited workers, but it could not preclude the Great Depression's impact on the patterns of migration that fueled the industry. Certainly most observers sensitive to the workers' plight supported the Jones-Costigan Act and federal intervention more broadly for cleaning up the industry and distributing the profits more equitably.[50] The act was in line with the New Deal effort to bring labor to the table and provide workers with a chance to collectively bargain with employers. Labor and capital came to more equitable terms during the New Deal by virtue of federal intervention, and while this became more common among industrial workers, agricultural workers were starting to amass as well and similarly looked for restitution for what they considered decades of mistreatment.[51] The Latino effort in that regard was an important demonstration of both their increased political activity and their comfort working within the system to demand change. Their unity pushed Costigan to consider legislation, and consequently we can consider the sugar beet workers' unionization an important catalyst to gaining Washington's attention for agricultural laborers.

While such unionization illustrated unity among the beet workers, laborers faced many challenges during the late 1930s. For example, dissonance existed between the laborers and members of the surrounding communities. Quite obviously, much of the tension grew between workers and the white population that dominated the economic and political aspects of life in the Arkansas valley. Some animosity grew during the Great Depression because the economic crisis left many white Americans worried about their jobs or bitter about being unemployed. Consequently, debates about the value of employing Latinos, whether from neighboring states or from Mexico, became more frequent, as concerns over Americans' ability to survive the economic catastrophe gave rise to the argument of hiring only white Americans to do any type of labor. White Americans also grew restless because they believed that Mexican Americans and even Mexican Nationals received too much federal welfare and sat firmly on relief rolls, thereby taking potential relief money away from white Americans. Many growers and refiners supported the anti-migrant push and chose to use white labor when possible. Furthermore, the number of farmers able to produce enough to need outside employment diminished during the Dust Bowl, so agricultural jobs for anyone slowly disappeared.

The federal government also played a role in cutting off labor streams into the Arkansas valley by enhancing penalties for illegal immigration and looking at a series of immigration quota bills designed to limit legal immigration to the United States from Mexico. This anti-Mexican hysteria proved enough to start the repatriation movement. Their departure challenged the migration patterns and employment opportunities that had satisfied beet growers; even though production demands had declined over the 1930s, the repatriation saga threatened to cut the Hispanic labor streams off entirely.[52] Indeed, whether they left voluntarily or because of coercion, "whole communities disappeared from Colorado."[53] Certainly, some left by choice. The opportunity to own land in Mexico proved quite a draw when the Mexican government passed an equivalent to the Homestead Act to get more Mexicans to settle in rural areas. The government also encouraged voluntary departure when they offered free-rail transport to returnees. Workers' hours and pay were often the first cuts and the most common tactic that employers used to save money, so these attempts to incite their return might have become more appealing. Instead of waiting around with little hope for surviving the down years, some immigrants with the means to leave did so and never returned.

More commonly, though, American citizens and the federal government pushed for and orchestrated the repatriation of Mexicans and Mexican Americans in the United States. Many repatriates found themselves at odds with important constituencies within the United States and faced significant pressure to leave the country. Some members of the community wanted to allow them to stay, but they faced significant opposition from nativists who wanted to remove immigrants.[54] They also struggled to defend residents from claims made by other repatriation advocates, including members from "all poorer classes regardless of ethnicity" who used an economic argument to justify their demands. Many agricultural and industrial workers argued that immigrants took their jobs by doing work for less money and therefore making themselves more appealing for employers. Evidence suggests that tenants and agricultural laborers who had previously worked in the sugar beet industry but who lost their jobs with the downturn in production had voiced their feelings strongly.[55] In effect, it became easy for whites who struggled to make ends meet to scapegoat immigrant workers for contributing to their problems. Other repatriation advocates also presented an ideological argument by contending that immigrants lacked the means or desire to assimilate to the "American way of life," arguing that since they lived outside of white communities in segregated barrios or *colonias,* attended different schools, and refused to naturalize even when they had been in the country long enough to gain citizenship, they had no desire to assimilate.[56]

While less politically influential than their white neighbors, many Mexican Americans also questioned Mexican nationals' place in the United States and argued in favor of their repatriation. One reason for this animosity was that the backlash against aliens often left residents dealing with collateral damage and therefore nervous about their own standing. Both Colorado residents and migrants to Colorado felt attacked and discriminated against because white Coloradans lumped them in with immigrants as outsiders and undesirables.[57] Self-preservation also influenced the arguments to send immigrants home. Members of Mexican American groups like the American Citizens of Spanish Descent voiced their concerns that immigrants often took jobs and enjoyed federal relief when many Spanish-speaking Americans had been rebuffed in both regards. Indeed, some of the more vocal critics of immigrant aid tried to dissuade relief agencies, and especially the Works Progress Administration, from offering either employment or welfare to anyone unable to prove their citizenship.[58]

These amplified voices calling for repatriation or an alternative to limit the number of immigrants largely succeeded. Between 1930 and 1935, nearly 20,000 Mexicans and many of their American-born children either left Colorado voluntarily or were escorted by state or federal officials to the border, part of the approximately 400,000 total deported early in the Depression.[59] Colorado proponents of repatriation pressured Governor Edwin Johnson to ramp up state-led repatriation, and he responded by proposing to round up all aliens in the state and deport them unless federal officials quickened their pace and expanded their scope to expedite repatriation. In May 1935, with his patience exhausted, Johnson ordered county sheriffs in southern Colorado counties to round up aliens, especially beet workers, to expedite the repatriation process.[60] This action led to the expulsion of 27 Mexicans from Prowers County in early May and seemed to aggravate the situation for Mexicans left in the state and for Mexican Americans as well.[61]

Johnson was not yet done, however, and less than a year later he upped the ante when he decided to close the state's borders. On the morning of April 20, 1936, Johnson sent members of the state's National Guard to the southern border from Utah to Kansas to block migration into Colorado. He ordered the troops to keep an especially well-trained eye on the southern border with New Mexico as well as the southeastern border with Texas and Oklahoma. Johnson gave the Guardsmen explicit instructions on whom they should let in based on the requisites of money and financial responsibility: "If they do not have money for means of support, do not let them pass. . . . Colorado cannot care for indigent from other states and these people become charges of the state after the brief spring labor season ends." Johnson ordered the blockade on April 18 as part of his declaration of martial law along the entire 360 miles of the southern border that included troops patrolling each entry point into the state.[62] This meant that every train, bus, truck, or automobile was stopped and searched and each passenger was forced to prove citizenship and means to survive in the state. Troops stopped 194 cars at the summit of Raton Pass outside of Trinidad in south-central Colorado, and they sent 4 back to New Mexico.[63] Similarly, Guardsmen out of Lamar patrolled the borders in Baca County and stopped several vehicles, eventually seizing two carloads of workers who had wintered in Texas and were en route to Fort Morgan for the spring planting season.[64]

Proponents of such a stiff response to migration certainly applauded Johnson's decision as a step toward regulating the influx of cheap labor into the state. Yet because of heavy criticism he lifted the blockade after less than two weeks. Critics responded quickly, charging the governor for his "law-violating and publicity-seeking" move designed to win votes by hurting the needy.[65] Paul D. Shriver, head of the Colorado Works Progress Administration, argued that Johnson had no power to refuse relief to aliens and migrants, and he reminded the governor that "aliens get hungry, too."[66] One problem for Johnson was that neither he nor the Guardsmen made any distinction between citizens and Mexican nationals, so that they often prohibited American citizens from passing the state line. This mistreatment inspired vocal indictments from various Mexican American groups who, while they may have wanted to protect their jobs from Mexican immigrants, resented the state's willingness to lump all dark-skinned people together.[67] Johnson also seemed to understand that the blockade disrupted agricultural production because it cut off the flow of workers into the state just before planting season started. The consequent labor shortage stirred heavy resistance to the blockade among farmers, and their voices combined with other criticism and caused Johnson to call off the blockade on April 29.[68]

Johnson's blockade further exacerbated an already tight labor market, especially in the beet fields of southeastern and north-central Colorado. The *Lamar Daily News* noted the need for farm labor in Prowers County for use on beet fields in the irrigated district less than a month after the blockade ended. The newspaper cited many local farmers who had been unable to satisfy their labor demand and wanted the city, county, and state to stop providing relief to "employable people" or to families who have "employable members."[69] Certainly climate and crop projections directly influenced the need for such labor, but even as the weather remained dismal between 1936 and 1938, farmers wanted more access to migrant workers. Most of this demand originated from sugar beet farmers in Prowers County, but even broomcorn farmers in Baca County pushed for increased access to workers.[70] The adage of "hope springs eternal" applied for farmers who requested labor and who seemed to approach each planting season with the thought that this harvest would be better, the weather would finally turn, and production could rebound. They stubbornly maintained this outlook despite the continued depression and dust storms, and it convinced them that they needed to have access to workers in case things turned around.

For all of that faith in their future, however, farmers by the end of the 1930s found themselves having to deal with labor issues that they had not anticipated. The Dust Bowl and Great Depression combined to erode what had become a viable labor system for farmers who contracted with workers to meet production goals. The lack of demand and uncooperative weather meant limited production throughout the 1930s; consequently, there had been little need for tenants or hired hands, leaving former employees scrambling to find work for themselves regardless of where that job might present itself. The dual crises thus interrupted the labor regime that had become standard by the end of the 1920s, and growers who started to think about rehiring employees by the late 1930s had few options about how to reestablish the regime or develop a new one. By that point, concerns about having access to a viable labor pool became more pronounced, turning into outright crisis by 1939.

The two major sources of labor during the period would never really be as influential in the regional agricultural economy after the Dust Bowl. The number of tenants never fully recovered to pre-1935 levels. The move away from tenancy in the West became more pronounced after the war, but the writing seemed to be on the wall for tenants who struggled during the 1930s. The arid environment and high price for water rights meant that few tenants could survive on a small plot and without available capital to start and maintain their operations. Despite federal assistance, then, tenancy in southeastern Colorado looked very different after the war than it had a decade earlier. The same can be said for the presence of Latino migrants who had formed a regional community that allowed them to move between the field and their village. The racism and xenophobia that compounded economic concerns and helped facilitate the repatriation movement eventually subsided. It also lessened as soon as farmers again needed inexpensive labor. The circumstances changed to some extent, though, as pressures to curtail immigration during the 1930s led to more federal oversight. Additionally, the seasonal migrant community that moved from village to field slowly became more sedentary. In sum, then, the streams of workers that had been so fluid and abundant coming into the region from both neighboring states and Mexico temporarily dried up after 1935. Farmers again tapped these resources during the war, but the responsibility for contracting with the workers moved from sugar beet company to the federal government, a move that changed the dynamics of employer-employee relations.

Neither Baca nor Prowers residents had a clear picture of how they could deal with the labor shortage, and that uncertainty became a serious issue when rain returned in 1940s and war came in 1941. The workers were largely unavailable, and once the weather turned, there was no ambivalence about the need for a healthy and competent labor pool. The war aggravated the labor problem by instigating a slow trickle of able-bodied workers from the area to wartime industries that sprouted up across the Front Range. This trickle became a flood once the Japanese bombed Pearl Harbor on December 7, 1941, and the United States formally entered the war on the following day. Colorado, like many western states, enjoyed a dramatic increase in federal largesse because of the attack and subsequent declaration of war. A number of men and women either enlisted in the armed forces or were drafted, and nearly 140,000 Coloradans served in one form or another. Additionally, wartime industries in Denver, such as the Denver Arms Plant, the Remington Company, and the Rocky Mountain Arsenal, enjoyed a dramatic increase in federal contracts that accompanied American involvement. Similar examples occurred in other parts of the state, like the Colorado Fuel and Iron Company in Pueblo, which took advantage of its opportunity to create large artillery shells and raked in federal dollars. Additionally, military posts shot up in places like Pueblo, Colorado Springs, and Denver. These operations also required workers, ranging from flight instructors to clerical workers to janitorial staff. The military even bought swaths of open land to practice aviation and artillery, acquiring 800 square miles of mostly open land in Las Animas County and 500,000 acres near La Junta for a practice range. Ranchers were allowed "to use the land for grazing purposes at their own risk."

As was the case across the region, individuals and groups not well represented in the workforce during the Depression, primarily women and minorities, took full advantage of their opportunities and joined white men in the factories. Similar developments occurred across the country, as industrial work invited women and minorities to seize their chance at making money and helping mobilize their country. The war instigated dramatic demographic shifts. Workers moved from the South to the North and West, from the country to the city, and from industry to industry, as employees used the dearth in labor to exercise their mobility in search of their best prospects. Colorado became a popular destination for folks from the plains, the South, and the Southwest. Many of these individuals arrived in the city to find work, then returned to their rural homes after the war or settled in the city outright.[71]

These dramatic population shifts—specifically the departure of workers—hit Coloradan farmers hard and compounded the already evident problem of achieving a stable workforce. Baca County is one example. The county lost nearly 5 percent of its population with the onset of war when 270 men either joined the armed forces or moved for industrial work. Community leaders met to discuss their options at a gathering organized by the county agricultural agent. Their list included shutting down area schools during peak planting and harvest times, decreasing farm acreage to lessen the need for workers, setting a wage scale to keep workers and eliminate both competition and "pirating," and inviting transients to fill the void.[72]

The war amplified that need and led farmers to become more vocal about what they required to produce at the level that the federal government requested and the war effort demanded. As they had before, the federal government helped farmers in ways that few observers could have imagined. The complexion of southeast Colorado society changed as the influx of government-provided outside labor replaced the number of former residents who had chosen to move on. The attack on Pearl Harbor incited a series of changes that left the region dramatically different than it had been in the 1930s, to say nothing of how the region had been during the 1920s.

6

Food for Victory

Rain finally returned to the Colorado plains in 1938. Precipitation levels stabilized in 1940 before a deluge in 1941 brought more rain to the region than it had enjoyed in decades. Baca County farmers started the decade with 16.8 inches of precipitation in 1930, before the numbers bottomed out in 1934 at 8.5 inches, and finally recovered at 14.4 inches—about average for Baca—in 1940. Just in time for war, precipitation maxed out at 29.8 inches in 1941, a level that allowed farmers to temporarily disavow aridity and to produce at remarkable levels. Prowers County farmers endured similar fluctuations, as precipitation levels moved from 16.6 inches in 1930 to 7.8 inches in 1934 and rebounded to 14.1 inches in 1940. Like it had in Baca County, 1941 marked a banner year with 26.1 inches of precipitation as farmers prepared to feed the Allied war effort.[1]

Local newspapers celebrated each rain shower or snow storm with excitement as searing temperatures cooled and rains picked up. For example, the *Lamar Daily News* announced: "Over an inch of rain in one January downpour! Not in years has such a thing occurred. There can no longer be any doubt about it, 1939 is the year!"[2] Prowers residents greeted one and a half feet of snow that hit in February 1939 with enthusiasm—it was the heaviest snow in twenty years—and the local newspaper gave it front-page coverage. There was nothing new in celebrating precipitation like this.[3] But adjacent to the story about the snowfall appeared a story about the best ways farmers could conserve the moisture from the snow to maximize its benefits. The *Springfield Democrat Herald* demonstrated a similar appreciation for life after the Dust Bowl when it reprinted an article explaining the need for farmers to remember the horrors of the 1930s and to continue abiding by federal procedures to stabilize the land and the farm economy. The author reminded readers that even with the return of precipitation, those who lived through the decade should know well enough not to take it for granted or try to institute get-rich-quick schemes to capitalize on improved weather.[4]

Agricultural conservation proponents faced a significant challenge with the onset of war in Europe and Asia because the war seemed to demand production at all costs. The war compelled prognosticators to think about the brewing conflict's impacts on the American economy and specifically its potential benefit for the agricultural segment. Within weeks of Germany's advance into Poland in September 1939, southeastern Colorado newspapers published several disparate interpretations of the conflict's potential influence on America. One article suggested that the war could benefit farmers, as grain futures promised good prices and farmers could maximize production to capitalize on the high prices and heavy demand. The author believed that this promised profits for the resident farmer and not for the speculator; furthermore, farmers could make money from the market while still receiving federal assistance for participating in New Deal programs—the best of both worlds.[5] A more frank, if morbid, response from the U.S. War Department assured locals that "Colorado farmers won't get rich this year on profits from the war in Europe, but they stand a good chance to reap a harvest of extra dollars if the conflict lasts another 12 months or longer."[6]

Part of the reason for optimism was the Allies' dire need for supplies, particularly after Germany overran France and left Britain fighting alone. This heightened demand came while the New Deal was still offering subsidies to farmers for participating in various programs. That local farmers continued to abide by and partake in federal programs is testament to both the New Deal's appeal in the region and to the extent of federal spending that provided such staying power. Indeed, 87 percent of residents polled in Baca County voted in favor of keeping the Agricultural Adjustment Act in 1941, knowing full well that they could continue to make money for conserving their resources and staying actively involved in government programs.[7] Similarly, federal agencies like the Works Progress Administration, the Civilian Conservation Corps, the Soil Conservation Service, the Farm Security Administration, and the National Youth Administration maintained a presence in southeastern Colorado. Extension Service programs spearheaded by the county agent's office continued almost without interruption into the 1940s.

Yet the Japanese attack on Pearl Harbor changed things, and shifting circumstances demanded adaptation. For example, the federal government developed a series of morale-boosting programs, advertisements, and slogans designed to unify the citizenry behind the war effort after the United States declared war on Japan. The key theme underlying these programs in rural America—Food for Victory—recognized the role that food producers played in supplying the military. The federal government organized a system that helped farmers prioritize what materials were most needed, how much acreage was necessary to meet production targets, and why they had to abide by acreage restrictions so as not to oversaturate the market and still reap financial benefits. Certainly, the Pearl Harbor attack sounded the call for unity, sacrifice, and common purpose, and it offered a chance for farmers to free themselves of the Great Depression.

As unusual as it sounds, however, many farmers had problems fueling the war effort and therefore capitalizing on wartime demand because they were unable to harvest their crops. Farmers had yet to reestablish any consistent labor system since the dissolution of tenant and paid labor that had served farmers so well until the mid-1930s. The slow drain of the labor pool that accelerated in 1939 became a veritable flood of workers leaving the countryside for military service or jobs in wartime industries by early 1942. That individual farmers lacked the means to entice workers to stay in the fields when attractive employment alternatives grabbed their attention simply compounded the problem. The federal government interceded in hopes that it could find some way to offset such labor losses and replenish the labor pool that had been dwindling over the previous ten years. The state acted in ways and on a scale that no individual farmer, corporation, town, county, or even state could match, and federal intervention provided necessary agricultural labor. By bringing in Mexican nationals, Jamaicans, American Indians, German prisoners of war, and Japanese and Japanese American internees, the federal government used unprecedented measures to address the labor problem. These groups supplemented local labor that the government also spent time recruiting to lend assistance in the time of heightened need.

The effort to corral workers, create new streams of labor, and fund these activities made the kind of production necessary for war possible. The Emergency Farm Labor Program, managed by the Extension Service from 1943 to 1947, orchestrated the importation, placement, and administration of workers until 1947. Extension's role in helping farmers meet production demands illustrates yet another way that it served its constituents. Agents distinguished where labor was most needed, how to get it there, and in what capacity the workers best met local demand. The agents also remained the biggest organizer of and cheerleader for soil and water conservation, often reminding farmers that wartime demand did not negate the necessity of protecting natural resources. In this way, the Extension Service maintained a presence in the countryside during the war and continued to work as intermediaries between farmers and the federal government. They relied on the relationships they had forged during the 1930s, and the faith that farmers put in them paid off when they turned to Extension agents to fulfill their labor needs.

The attack on Pearl Harbor induced a national sense of emergency, but it took some time for the United States to transition from Arsenal of Democracy to belligerent. Agricultural production represented a core issue; obviously, a military cannot function without provisions. Secretary of Agriculture Claude Wickard inspired farmers with the slogan "Food for Victory" and promised that "food will win the war and write the peace." He reminded farmers of the Depression that followed World War I and convinced them that the same set of circumstances—and economic fallout—could recur unless they took responsibility for managing production and abiding by government directives. Wickard emphasized the importance of focusing on the most needed commodities and explained how blind production, without an understanding of what the military and its Allies needed, would simply disrupt the war effort. In effect, an implicit danger existed because farmers might be tempted to maximize their output, out of both a patriotic and market-driven desire, and if they did then they could easily upset demand, drive down prices, and saturate the market. Wickard hoped that federal directives regarding production goals might negate that possibility and stabilize the relationship between supply and demand; that would prove key to safeguarding production while allowing farmers to capitalize on the moment.[8]

Federal observers like H. H. Finnell worried that farmers might forget the aftermath of World War I and its negative effect on land prices, personal debt, and soil conservation. Finnell, regional director of the Soil Conservation Service (SCS), reminded farmers that high prices and high demand would evaporate once the war ended. Unfortunately, the temporary wartime boon proved enough to entice farmers to expand and mechanize, forcing many into extensive debt. Speculators participated in the mad dash for land only to see their investments dry up with the drop in prices and eventually with the drought. Finnell hoped that farmers remembered the bleak period—and its cause—and that they would approach the postwar period with a better appreciation for sustained prosperity rather than a get-rich-quick mentality.[9]

Finnell and others hoped that farmers would think more intently about conserving natural resources. The push to produce during and after World War I led to the Great Plow-Up and the devastation of plains topsoil, something that conservationists urged farmers to remember. Finnell assured Colorado farmers that conserving soil was necessary "to meet national defense demands" by safeguarding and eventually improving the land's productivity in the time of crisis.[10] Mismanagement of resources or excessive production threatened the soil and therefore jeopardized the war effort. As southeastern Colorado district conservationist R. A. Harris explained, "War time farming means conservation farming." Harris implored farmers to appreciate that since "total war requires total production," they must "build up soil fertility" and "make the best use of every drop of water." Harris echoed Finnell's argument by contending that farmers could ensure their long-term productivity by working with the SCS and other agencies devoted to land management. They both put forward the conviction that postwar prosperity started with conscientious stewardship during the war.[11]

Fortunately, New Deal policies helped prepare farmers to conserve during the war by educating farmers, subsidizing their efforts, and supporting their establishment of county-level soil conservation districts to police local practices. The war emerged when conservation was still ascendant in the region. Consider the *Lamar Daily News,* which announced the attack on Pearl Harbor beside an article announcing the establishment of a soil conservation district in Prowers County: "the first demonstration of the use of soil and water conservation practices on a wide scale in a large irrigated area of the Southern Great Plains."[12] Such districts supplemented local efforts at conservation by offering classes, meetings, and demonstrations on proper conservation methods. The districts also lent equipment to protect fields or repair dams and other irrigation equipment and provided labor to do the work.[13] This Prowers County district joined three others in Baca County, meaning that districts controlled most acreage in the two counties by the onset of war—an important indication that farmers embraced conservation beyond the end of the Dust Bowl era.

Belonging to a district showed a proclivity for conservation, but it did not guarantee that members always possessed restraint. For some farmers, government production limits made no sense. The government identified

FIG. 19. "Get Your Farm in the Fight." Poster created by the U.S. Office for Emergency Management, Office of War Information, Domestic Operations Branch, Bureau of Special Services. Wikimedia Commons.

what it wanted and assessed how much it wanted, but by turning away other products or farmers' surpluses, the U.S. Department of Agriculture (USDA) effectively hurt the farmers. The war promised an opportunity to make up for lost time, to earn money after so many years of small harvests, and to dig out from under heavy debt. The thought that the government could or would impede their financial success left many with a bad taste in their mouths about government intervention, and to some extent this led to a return to traditional antigovernment sentiment that had temporarily lapsed during the New Deal. For example, Baca County farmers approached the county agent with the argument that "Providence" had finally provided rain, but the federal government restricted wheat production and set the sale price, circumventing farmers' chances for profit in two ways.[14] Similarly, local Farm Bureau leaders complained about federal stipulations on both production and prices, noting that 1943 prices for goods were often actually lower than they had been in 1918. Bureau representatives blamed the government for going overboard on such regulations and price controls, which they used to manage the economy to the farmers' detriment.[15]

While the government maintained its influence on pricing and production guidelines, its overall presence in the countryside diminished as priorities changed. For instance, the SCS dissolved its Southern Great Plains branch in 1942 because federal officials believed that the time of crisis in the Dust Bowl region had passed. They argued that improved weather negated the need for a branch principally devoted to the Dust Bowl region, as "favorable seasons" combined with "widespread acceptance of conservation farming methods" to imbue conservationists with confidence that farmers had sufficiently turned a corner.[16] Implicit in this justification and explicit in other cases, however, was the fact that Washington turned most of its attention to the war and focused on international rather than domestic issues. The government often funneled financial and human supplies from agencies not involved in the war effort to agencies tied to it. Thus many agencies became so taxed and short on resources that mundane tasks easily accomplished before Pearl Harbor became more difficult afterward. For example, even when farmers wanted assistance on how to conserve water, the local experts from the SCS sometimes had no time to help. In one case, locals requested a demonstration on how to maximize water from their irrigation ditch, but no one returned their calls. The temporary cessation of construction on John Martin Dam during the war marked another example, and the break meant that locals only enjoyed its benefits after 1948 when workers finally completed the dam, some five years after the projected conclusion.[17]

Although the federal government receded from rural America, Extension Service county agents maintained a constant presence. They continued as intermediaries by executing production regulations, instituting acreage limits to manage how much land farmers devoted to specific crops, and promoting soil and water conservation. In that sense, agents represented a point of consistency between the New Deal and the war. With the agents' help, the government identified the most essential materials for war by early 1942, and fortunately for Colorado farmers the list included many of their principal crops. Wheat, sugar beets, and broomcorn topped the list, and when coupled with vegetables and meat from hogs, cattle, and sheep, southeastern Colorado famers had the opportunity to cash in on government purchasing. The wheat and sugar beet allotments were arguably most important, as wheat became a central part of the soldiers' diet and sugar beets rose in prominence.[18] Agents and Extension employees also spearheaded the federal effort to plan and prepare "for the greatest farm production effort of all time." This meant helping farmers repair their machinery, ensuring that they had access to seed, collecting scrap metal for conversion into military items like tanks and planes, and clarifying a system of financing so that farmers could abide by federal demands. The Extension Service provided experts in plant disease, irrigation, and home gardening to answer questions and lead countywide demonstration meetings once a month in early 1942. Such cooperation proved vital to jump-starting and maintaining the war effort.[19]

The Extension Service also helped the federal government by managing its attempt to provide agricultural labor for American farmers. The dearth of serviceable agricultural labor caused by outmigration and a decline in seasonal labor posed a challenge to wartime production. As Colorado Cooperative Extension Service director F. A. Anderson explained, the labor issue became dire by the summer of 1942 as the already-dwindling numbers of available workers were further devastated by large-scale enlistment and continued migration to urban areas for industrial work. The War Food Administration's demand for increased acreage and food production in crops critical to the war effort brought crisis to Colorado farms. The USDA took responsibility for supplementing labor, but it took some time to win congressional authorization and funds for action. The key moment came when the USDA appointed the Extension Service to address the labor problem, and Extension took over on May 1, 1943. Unfortunately, by that point the spring planting season was already under way, and the Extension Service scrambled to meet demands. Once the Service got its feet on the ground, however, it responded to the labor problem in novel and remarkable ways by reaching out to workers first made available or only made available during the war.[20]

A. J. Hamman, one-time Prowers County Extension agent as well as district soil conservationist during the New Deal years, ended up directing the Colorado Emergency Farm Labor Program from 1943 to 1946. He had experience in the sugar beet industry and had recruited labor for a private company, plus his time as a county agent had taught him how to deal with farmers as well as Extension employees. Hamman only reluctantly agreed to take the position because he realized that the job would be complicated since Colorado "had a much greater and more constant demand for out-of-state-workers than any of the surrounding states." Hamman immediately tried to get "competent people" for his staff and looked to "ex-sugar agriculturalists and persons with that type of experience." He persuaded a trio of men to work as his inner circle, men with whom he had worked in the Arkansas valley and who had maintained ties to various sugar companies in the region. Hamman's decision to use sugar men illustrated the place that sugar held in the state's crop hierarchy, especially during the war. It also showed how their previous experience working with contract farmers and contract labor could prepare them to work with the Extension Service.[21]

Under Hamman's direction Extension turned first to a tried and true labor source, residents and their children, who were well suited—and available—to help on area farms. Starting in 1943, the state superintendent of public instruction allowed for school-aged children to serve as seasonal workers on neighboring farms as needed. Local leaders, ranging from teachers to representatives of church officials to scout leaders, recruited the students and prepared them for work, then made arrangements with local farmers.[22] For its part, the Extension Service distributed pamphlets and held informational meetings as well as empowering community leaders who organized recruitment. A pamphlet from the Service noted that "our farmers' sons and hired men have been called to fight," and so "every available person . . . every able-bodied boy and girl in every city, town, and village" needed to answer the call for production. Without such help, "our fighting men don't get their food and they can't protect themselves against the enemy and fewer of them will come home when the fighting stops."[23] Obviously relying on the sense of common sacrifice and compassion, such calls for action helped bring teenagers into the fields; for example, forty-two boys worked the Prowers County fields in 1943.[24]

This kind of volunteer effort constituted an important tool for farmers and represented one piece of the domestic labor used to meet labor needs. Additional options became available, including some from surprising sources. The USDA helped establish an umbrella organization for all farmworkers called the U.S. Crop Corps, a "national decentralized farm labor program" under USDA and Extension Service jurisdiction. The Crop Corps included the Victory Farm Volunteers, high school students who worked the fields during peak times, and the Women's Land Army, part of the nearly 3.5 million people who labored in American fields during the war.[25] Colorado officials also opened the state penitentiary to allow farmers to access convict labor. If the farmer agreed to cover travel expenses, then the penitentiary supplied guards to supervise the work; roughly 350 convicts made their way into Colorado fields under this arrangement, a demonstration of the lengths to which farmers went to satisfy labor needs. The Extension Service also organized an effort to place conscientious objectors on farms. The state constructed two camps for objectors, and once they had been cleared for work, the Extension Service transported them to area farms and placed them under the county agent's supervision. Conscientious objector labor, like that of the convicts, did not greatly impact southeast Colorado war production. Yet everything helped given farmers' dire need and the Extension Service's lack of alternatives.[26]

Extension employees and federal officials quickly realized that agriculturalists needed still more assistance because local and state options proved inadequate. Consequently, farmers as well as state and federal employees started to mull over their options. They left no stone unturned, and their efforts resulted in a surprising influx of workers. The federal government began by negotiating with other countries to institute guest worker programs and by providing German and Japanese prisoners of war to needy farmers. Consider the sequence of events that produced enough workers for Colorado farmers. The decision to intern Japanese and Japanese Americans, moving them from the West Coast inland to protect sensitive and vulnerable war industries, forced the migration of roughly 120,000 people to internment camps across the Mountain West. American advancement across Africa and eventually Europe meant a constantly increasing pool of Axis prisoners of war that the federal government shipped across the Atlantic to work for the Allied war effort. Additionally, the federal government hammered out deals with Mexico and Jamaica to initiate the recruitment, importation, placement, and payment of eligible workers. Finally, the state pushed for increased domestic migration into areas hard-pressed for laborers.[27]

Coloradans vied particularly hard for government action to bring in workers, and the government responded. The situation proved incredibly complex. Even though the government initiated the guest worker programs with the 1942 Public Law 45 and decided to send prisoners of war into Colorado, each situation required some level of local/county and state complicity in finding adequate accommodations for these groups—to say nothing of pacifying residents, many of whom feared the influx of these outsiders who were ostensibly enemies.[28] Certainly, Jamaican and Mexican labor also required housing and payment in addition to the first step of making contact and then transporting them to their stations. The legislation exempted recruits from military service, and the government covered their transportation and living expenses en route to the fields. Upon their arrival, the workers stayed at housing provided by private agricultural groups who received some federal assistance to defray their costs. The nations agreed that wages should equal the prevailing wage scale and these workers should be free from discrimination by their employers. In effect, then, the worker and employer signed a contract to acknowledge that both sides understood. The dramatic and expansive incorporation of these outside workers caused problems and led to some trepidation among locals. Regardless of any such hesitancy or doubt, however, the federal government's effort and the Extension Service's Emergency Farm Labor Program proved crucial for farmers to produce enough for the war effort.

In some respects, the government-run guest worker program looked very much like the importation of labor by private companies and local organizations. For example, both the East and West Prowers County Farm Labor Associations established contact with Mexican nationals and filed contracts to bring them to work in the beet fields. Large corporations like the Holly Sugar and the Great Western Sugar Corporation often opened their warehouses or constructed camps for workers they contracted and provided to their growers and processors. For example, Prowers County had two such camps that the Holly Sugar and American Crystal Sugar Corporation built to house workers during their off time. Once settled, Mexican nationals filtered out to farms that had established agreements with the corporations. This practice was common during

World War I and the 1920s before becoming less frequent during the Dust Bowl and Great Depression.

The Extension Service effectively took the responsibility of managing labor from these private and public enterprises to centralize labor recruitment, marking a shift away from private firms. An example of this kind of arrangement was the Baca County Civilian Conservation Corps (CCC) camp, which had been abandoned once that program concluded. With the CCC boys gone, the camp offered a perfect place to house outside workers. The government paid for upkeep, repairs, and necessary improvements to ensure that the workers lived in adequate conditions. Extension also offered equipment for a nominal fee, meaning that farmers had a veritable one-stop shop for all things related to labor, particularly for those not aligned with some of the major sugar companies but who wanted access to a reliable labor force. Area farmers could thus stop by these sites, recruit the workers, borrow farm machinery, and even rent cots, tents, and other necessities for housing the workers on their own property—thereby saving them additional transportation costs and lost time in the field.[29]

The government's importation of Jamaican labor has been largely forgotten in the annals of World War II, and the little work that has considered their efforts in the United States has glossed over the workers in the Great Plains.[30] Like most who eventually toiled in Colorado fields, the Jamaican laborers found themselves in unfamiliar surroundings and forced to deal with American farmers and bureaucrats. Moreover, little consistency existed in terms of the number of imported workers—crews changed from season to season and the number of Jamaicans varied per year. This variance reflected a tendency among Colorado farmers to opt in favor of using other groups, primarily because they had previously gained some exposure to the other groups. In addition, Jamaicans quickly earned a bad reputation. Many farmers, and even some county agents, believed that Jamaican workers lacked the effort and expertise that Mexicans demonstrated. For example, Prowers County agent Max Mills claimed that Jamaicans, or "Jakes" as they were called at the time, became the worst of the new workers because they were "not as careful in their work" and "there were many social problems" among the workers, which exacerbated the tenuous relationship between worker and farmer.[31]

The Extension Service reiterated these concerns and noted relatively few Jamaicans found their way into Colorado because of that reputation. Indeed, only four communities used Jamaican workers in 1943 and 1944. According to A. J. Hamman, farmers "only requested and accepted them as a last resort" and "with considerable misgivings."[32] As director of the program, Hamman did relatively little to secure Jamaican workers and had very little patience for them once they arrived. He recalled an instance when nearly an entire train of workers arrived and promptly refused to do any work. Hamman had them put back on the train and returned to Jamaica for not fulfilling their contracts.[33] Some arrived "dissatisfied," which Hamman and others took as a sign that the crew did not want to be in Colorado and that they should be sent home. Hamman believed that such apparent apathy indicated the entire group's predisposition against hard work and cold weather, so he thought that the workers could best be utilized during the warm months in lower altitudes, especially when the farmers warmed them up with "friendliness, flattery, small favors, and fair treatment." Even when this approach proved successful, however, Hamman warned farmers not to rely solely on Jamaicans. Hamman's reaction to the "outsiders" and his unfamiliarity with Jamaicans most certainly influenced his racialized perspective. But his reluctance to contract Jamaicans suggests that many Coloradans felt similarly, and his post as state supervisor gave him the power to determine who made it to the fields.[34] In spite of his misgivings, however, Colorado farmers eventually employed more than 2,000 Jamaicans in 1945.[35]

Because Colorado farmers proved more willing to use bracero labor than Jamaican workers, a considerable number of Mexican nationals found work in the state during the war.[36] Hamman believed that Colorado farmers felt more comfortable hiring Mexican nationals than any other available group. In part, he explained, this was because farmers had become accustomed to

using Spanish-speaking immigrants and migrants before the war began. Many farmers believed that braceros were superior workers, more diligent and knowledgeable than the Jamaicans. Consequently, they searched for alternatives to using Jamaicans, and in many cases the alternative was bracero labor. By the numbers, the peak for Mexican nationals in the state stood at 10,000 in 1944 compared with nearly 2,000 Jamaicans working in 1945.[37] The predilection for Mexican labor proved especially true in the sugar beet industry, which pressured Congress to pass the legislation and used more outside labor than any other constituency in Colorado.[38]

As much as farmers looked favorably at Mexican nationals for their work in the fields, relations between the two groups often included controversy and conflict. While farmers relayed to Hamman that the Mexican nationals were "very satisfactory workers," farmers still had some problems keeping time records, communicating with the workers, and, at least early in the program, getting in tune with federal officials to secure the right number of workers for their needs. Perhaps most important, farmers had trouble completing the stipulations put forward by the legislation that created the guest worker program; this applied to their dealings with both Mexicans and Jamaicans. Farmers had a terrible time finding extra work for the guest workers beyond what they offered in their own fields; this proved problematic because farmers agreed to keep guest workers employed for 75 percent of the season. Consequently, farmers and beet refiners had to coordinate with locals who needed workers.[39]

FIG. 20 *(top)*. "Mexican workers recruited and brought to the Arkansas valley, Colorado, Nebraska, and Minnesota by the FSA to harvest and process sugar beets under contract with the Inter-mountain Agricultural Improvement Association." Library of Congress, Prints & Photographs Division, FSA/OWI Collection (LC-USW33-031868-C).

FIG. 21. Mexican workers display their wages from the train. Library of Congress, Prints & Photographs Division, FSA/OWI Collection (LC-USW33-031869-C).

Conditions in the camps or other temporary settlements to house workers also complicated relations. Braceros usually wanted "able cooks who were Mexicans or who had had experience in Mexican cooking," but this "was a problem that was never completely solved."[40] Jamaicans similarly complained about lacking a Jamaican cook and having to deal with dismal food.[41] Guest workers complained about the pay scale and delinquent payments, both of which became significant problems because beet workers were paid "on a piecework basis" and farmers often had trouble calculating the workers' wages.[42] Finally, racism flared up, although Jamaicans seem to have had a more difficult time in the South than anywhere else.[43]

Hamman noted that while the contractual agreement promised no social or class barriers, the reality on the ground reflected a continuation of such discrimination. He argued that the contract should be changed to reflect that such obstacles existed, because "racial discrimination cannot be overcome by international agreements." The problem was not exclusive to white and Mexican relations, however, as Hamman mentioned significant conflict between "resident Spanish American and alien Mexicans." Such issues resembled the kind of animosity that the two sides lived through during the 1920s and early 1930s when they competed for many of the same jobs. The crux of their problems during the war concerned the benefits that Mexicans accrued by signing their contracts, benefits unavailable to many locals and nearly all migrant laborers who had been in the area for much longer. The Mexicans' relatively high wages, adequate housing, and access to medical care gave them opportunities that most local workers lacked. Residents quickly understood the disparity and complained about their situation. Additionally, friction developed because the two groups shared social situations and in some cases residents resented the Mexicans "crashing Spanish American social gatherings." Some resident and migrant workers even discouraged local farmers from hiring Mexicans, presumably to open better opportunities for themselves while diminishing the likelihood of seasonal conflict. In these cases, as in the examples of white and Mexican national conflict, the Extension Service tried to step in as quickly as possible.[44] They seemed largely successful, as local newspapers rarely mentioned violent outbreaks or vocal criticism of the program from Mexican Americans.

Guest workers often had little recourse when facing racism, discrimination, or broken contracts. This proved especially true for Jamaicans, who had little chance to return home if something went awry. For braceros, however, a trip home was highly possible if the worker so chose. In many cases, workers left farms because of problems with the surrounding community, white or otherwise. They sometimes worked for another farmer in the area, although that meant a breach of contract. Other workers moved from the fields to the nearest urban area to search out the Mexican consul and request a trip home. Of course, workers did such things for reasons beyond discrimination or working conditions. Some simply became homesick or received word of family problems at home. Employers could also have workers sent home for various infractions, meaning that while the contract supposedly made both sides responsible for executing the program, it had little leverage to keep them on the same page.[45]

As much as guest workers offered to Colorado farmers, however, Extension quickly opened additional labor streams. Migrant workers became especially important, and their return to Colorado fields during the war suggested a renewal of earlier patterns. Many migrants came of their own accord while others only arrived in Colorado after heavy government or industry recruitment. For example, the state recruited groups of Navajo Indians from both New Mexico and Arizona to work sugar fields in 1942 and 1943.[46] While enticed by the government, this movement from village to field represented a return to the migratory patterns that the Depression and drought had interrupted. Not surprisingly, the economic and social pressures unleashed during the Depression to keep such workers out of the state abated once farmers needed workers to meet production levels. Consequently, migrant workers from throughout the Southwest proved important during the early war years. Many of them performed stoop labor, demanding physical work in the beet fields, as had other invited labor. Others found work on potato, onion, and vegetable farms as well as during wheat and broomcorn harvests. Workers had opportunities at nonagricultural labor on occasion; in fact, migrant workers were principally responsible for building most of the Amache Internment Camp.[47]

Migrant workers from neighboring states enjoyed the benefit of government support for their endeavors just as did the guest workers. For example, the state government paid to repair and maintain various migrant worker reception centers, where the workers could gather after arriving in the area and wait for work. One such center in Lamar needed considerable attention, and the Extension Service paid almost half of the cost to complete the repairs. Similar migrant centers popped up in small agricultural towns along a line from south Texas to Montana, and Extension agents quite literally shuffled the workers to the fields as soon as they arrived. This, too, marked federal intervention, as Extension maintained responsibility for setting up contracts with local farmers and ensuring that the workers made it to the sites on time. County agents distributed recruiting pamphlets and bulletins to agents in other states so that interested workers could have some information about the situation in Colorado. For example, the Service printed a pamphlet in Spanish and had agents distribute it in the lower Rio Grande valley in Texas. It explained the crops, type of work, and working conditions, and it included a detailed map of Colorado that showed the most direct roads leading to places of high employment. This effort to reach out to the worker proved typical for Extension Service recruitment practices.[48]

At their core, the guest worker and migrant labor developments resembled familiar patterns that the war helped reestablish. But the government's willingness to help coordinate recruitment and fund the enterprise marked a difference from earlier examples. Certainly, the all-hands-on-deck approach that included volunteers, including women and children, further helped farmers during the peak planting/thinning and harvesting seasons. Obviously, American involvement in the war demanded dramatic measures. It also opened the possibility to supplement the war effort in unusual ways with the incorporation of German prisoners of war (POWs) as well as Japanese and Japanese American internees from nearby camps.

The two groups of prisoners shared some similarities with others in the newly amassed patchwork labor force, namely, the presence of a static pay scale, assumed contracts between employers and employees, working away from their temporary residences, and the need to somehow make peace with other workers and surrounding residents. Several differences existed, however, and none was more significant than the fact that the captive labor was considered enemies of the state. The prisoners were subjected to armed guards and lived behind barbed wire under constant vigilance. Made available only because of wartime circumstances, these groups were unlike the other laborers in that no one had much idea of how to work with them once they arrived in Colorado.

The federal government shipped Germans to Colorado once the state and federal government agreed on plans to build three main camps in Trinidad, Greeley, and Colorado Springs. The German POWs' impact on wartime production remains an underappreciated aspect of the home front. The few books that deal with Germans in America do not deal extensively with Colorado, while the few articles on the situation in Colorado tend to focus on prisoners out of Camp Carson rather than those in southern Colorado.[49] Consequently little has been written about the camp in Trinidad or its residents. The best resource is Kurt Landsberger's autobiography about his time spent at Camp Trinidad. Landsberger came to the United States before the war, a Jewish refugee who signed up with the military once the war against Hitler started and who served as a camp interpreter.

Landsberger remembered that most of the residents in and around Trinidad gave Colorado congressman J. Edgar Chenoweth credit for the camp. Chenoweth worked very hard to convince the federal government of the need to open a camp in southern Colorado, and he eventually earned approval in September 1942. Construction started immediately and took about a year to complete; the doors opened to German POWs in 1943. Chenoweth hoped that the camp could act as a financial windfall for his constituents in and around Trinidad. While he likely figured that such a boon would probably earn him reelection, he was nonetheless accurate about the camp's economic impact on the area. The federal government allotted roughly $2 million for camp construction, much of that money filtered to contractors and laborers who then spent money around town while they stayed for the work. Additionally, thirty to forty people worked inside the camp once workers completed the building, and both military personnel and prisoners spent money in local shops and restaurants while stationed at the camp. Federal financing for camp construction helped the southern Colorado economy recover from the depths of the Depression.[50]

The U.S. military started shipping prisoners to the camp in June 1943. Most of the POWs had been fighting in North Africa and surrendered to the Allied forces once they made their move across that continent. The new arrivals settled into their surroundings almost immediately and often took advantage of lax security and an unusually trusting guard. Indeed, Camp Trinidad quickly gained a rather notorious reputation for an extraordinary number of escapees and corruptible camp officials.[51] Yet, as workers, the POWs seemed to perform well once they arrived in the fields, though that may have been because the Extension Service went to great lengths to ensure that no real fraternization existed between the prisoners and the farmers. It distributed information and instructions to farmers describing the healthy respect—and borderline fear—that citizens should have of their employees. The Service reminded farmers that prisoners were in fact enemies of the state and that they remained tremendously dangerous: "What might appear to be innocent conversation and small favors may in reality prove to be acts of treason" as they would manipulate their handlers to escape at any opening. It further reminded readers to remember that the prisoner really did not like them, that he did not like to work for them, and that the farmers should never speak to him unless it related to work. Other stipulations included allowing no women to work in the same fields; the enemy should never see a copy of any official directives or memoranda; and any question about the worker or his work should be funneled to his supervisor in the army. There were certainly strict regulations, but given the near constant army supervision of POWs when they were outside the camp walls and the lack of any public note of transgressions, problems were limited.[52]

Reviews were mixed. Hamman noted that while the populace was "more or less apprehensive of the entire situation from the standpoint of the security of the communities," the army quickly showed its ability to handle the prisoners in the camp as well as in the fields. The army proved so successful, in fact, that Hamman noted, "It finally became difficult in some communities to place any workers other than Prisoners of War because they could be used in large crews and removed as soon as the work was done." This came to fruition only after the Extension, army, and farmers grew more familiar with the process and more adept at filing for and earning certification for workers from the War Manpower Commission. Eventually, the employers became comfortable with the situation, and Hamman claimed that POWs were "the major mobile force of farm workers." He continued, "It is doubtful if the high level of production attained could have been accomplished without the Army and the Prisoners of War."[53]

Yet Hamman also remembered many arrogant POWs who were convinced that they were not only protected by international law while in the States but also that Hitler would eventually recover and the Germans would win the war. Consequently, some of the more confident (and as it turned out delusional) prisoners refused to work much if at all. Additionally, Hamman noted that violence sometimes broke out between prisoners and others despite Extension's best efforts. In one example, he recalled that an American guard assigned to the POWs opened fire on them while they worked in the fields. The guard had been in Europe and had developed an obvious distaste for the Germans. Hamman remembered that "he killed two of them for no reason except his hatred of German soldiers."[54] For the most part, however, locals seemed to embrace the Germans once they had become accustomed to their presence and confident in their security.

Internees from the Amache Internment Camp constituted another component of the new labor force after they arrived in Prowers. FDR signed Executive Order 9066 in February 1942, an act which eventually gave the federal government the power to forcibly relocate 120,000 people of Japanese ancestry from the Pacific Coast to the Mountain West.[55] Many of the internees worked in the fields, helping to fill the void left by locals who migrated to cities or served in the military. This was partly by design, as many of the camps were in rural areas, surrounded by agriculturalists who often needed workers, and many of the internees had farming experience. Indeed, the internment camps represented a distinct opportunity for local farmers to capitalize on this new labor force.[56] Many factors limited worker availability, but internees still played an important part in trying to bring in local harvests, especially sugar beets. These workers were most important from 1942 to early 1944, because after 1944 most of the camp population was evacuated: previously available workers resettled in the Midwest, joined the military, refused to perform menial labor, or worked at the camp farm instead of on private farms. Yet enough workers existed to supplement the work done by other paid labor in the area.

The first phase of the military's execution of Executive Order 9066 included a call for voluntary resettlement, which gave Japanese and Japanese Americans a chance to get ahead of the momentum to remove them from the West Coast. A small contingent of roughly 3,000 resided in Colorado when the war broke out; most of them lived in a limited number of neighborhoods in Denver as well as some outlying towns such as Brighton. In addition, a small population of immigrants and their children (Issei and Nisei, respectively) also lived and farmed in the Arkansas valley. For example, the town of Rocky Ford had a group of Japanese Buddhist residents, and others lived in Prowers County as farmers. Some of them did quite well during the lean Dust Bowl years, which drew some confusion and even ire from their white neighbors who struggled with the elements.[57] A number of voluntary evacuees figured that Colorado would suffice for their resettlement because they knew it sat far enough inland to satisfy the executive order or else they knew people in the state who had previously settled there. Their resettlement, while comparatively small in terms of the total population in the state, effectively raised the hackles of suspicious Coloradans who bristled at the idea of allowing potential saboteurs into the state.[58] Reports of police and laypeople harassing recent arrivals surfaced in places like La Junta, where migrants from the West Coast faced police searches. Governor Ralph Carr assigned the Colorado Highway Patrol to watch the border for migrants, though unlike his predecessor, Ed Johnson, who gave the same kind of order in 1936, Carr hoped to keep the peace and calculate the number of migrants rather than turn them around.[59]

In some ways he did more than assume a passive role in dealing with internment. The federal government called for all western governors to meet and discuss the possibility of moving the suspect population inland after the attack on Pearl Harbor. Wyoming governor Nels Smith offered a typical demonstration of how the governors reacted. He apparently threatened Milton Eisenhower, head of the War Relocation Authority tasked with organizing and executing internment, by announcing, "If you bring Japanese into my state, I promise they will be hanging from every tree."[60] Carr was the only governor who argued in favor of building relocation camps in his state. Carr contended that it was American citizens' civic duty to house the potential subversives, and he felt that they would not threaten domestic tranquility. Additionally, he understood that the evacuees could offer labor for agricultural and industrial efforts in Colorado.[61]

Yet a notable chorus of resistance emerged that tried to compel Carr to change his mind. Residents' fear of competition for good work once the internees arrived explained much of this resistance. For example, members from the Arkansas Valley Cultural and Educational League told Carr that the influx of workers threatened "hundreds of Mexican citizens that work and live" in the valley, and so they organized to defend their interests. Another instance of animosity occurred in Prowers County, where residents formed the Farm Protection Committee to demonstrate against relocation and hiring the internees to do any work in the area.[62] Racism and suspicion certainly explain this reaction as well, as the internees realized once they made their way to Amache. Resistance to the camp became so hostile, in fact, that it largely decided the 1942 U.S. Senate race in the state. Since Governor Carr had vociferously argued in favor of establishing a camp inside the state, he faced a considerable backlash. Indeed, former governor Ed Johnson ran against Carr in 1942 and leveraged Carr's position on the camp against him. Johnson won by a slim margin—just under 4,000 votes out of the 375,000 cast—and observers opined that Carr "would easily have been elected to the Senate had he remained silent on the Japanese American issue." Carr's victory in 1938 had supposedly signaled a return for Republican dominance in the state, but Johnson's electoral win, while he was in fact quite moderate, seemed to have less to do with his political allegiance than with his ability to paint Carr as a "Jap lover."[63]

Despite Johnson's full-throated condemnation of Japanese relocation to the state, construction on the Amache camp began in the summer of 1942 just before the first trainload of 212 internees arrived on August 27. The federal government used the War Relocation Authority to organize camp construction throughout the West, and in some ways the WRA represented a continuation of the New Deal emphasis on public works to lower unemployment. Indeed, the Works Progress Administration (WPA) managed the first stages of camp construction and provided the "personnel, bureaucratic might, and local knowledge essential to executing Executive Order 9066."[64] In effect, the WPA laid the groundwork for internment and then left it to the WRA before the internees started to arrive. The WRA sought out a construction company (a Texas-based company won the contract in a bit of a surprising development) once the Army Corps of Engineers had assessed the site and drawn up preliminary plans for the camp. The federal government then paid to buy 10,500 acres from Prowers County farmers and helped finance some of their relocation costs. While the camp's official name remained the Granada Relocation Center because the site was only about a mile from Granada, Colorado, most internees and employees who worked in the camp referred to it as Amache. Lamar mayor R. L. Christy named the camp Amache after the daughter of Ochinee, the Cheyenne chief killed in the Sand Creek Massacre in 1864. Consequently, the U.S. Postal Service designated the camp Amache to distinguish it from Granada.[65]

Almost immediately, the camp invited criticism and inspired dissent, perhaps because of this dramatic increase in the number of internees and certainly because of residual distrust from locals. Regular editorials in the two main Denver papers, the *Denver Post* and the *Rocky Mountain News*, regularly chastised the WRA. The papers accused the prisoners of laziness and blamed the WRA for allowing them to live the good life despite their enemy status. Critics also condemned the WRA for not paying close enough attention to the prisoners, suggesting that such lax management might allow them to escape or riot.[66] Similar complaints emerged in southeastern Colorado papers, which were much closer to the situation, and in that sense the editors and readers may have had more intense feelings about the camp. The editor of the *Springfield Plainsman Herald*, the Republican paper that took over when the Democratic *Springfield Democrat Herald* went defunct, questioned the camp, its residents, and the WRA for treating them so well. He contended that "Baca county should be put into a concentration camp and the Japs turned loose, it would be much easier for the Americans. They are feeding them better at this camp than Baca county citizens can afford to buy, and the Japs get a salary on top of this elaborate menu and good quarters to live in, and have the buildings all built for them. We wonder if the captured thousands of Americans are getting the same treatment in Japan? If so we are for treating the Japs a little better, and raising their pay."[67]

Yet the converse perspective also surfaced during the months of construction and placement. Ross Thompson wrote several articles describing the camp and interviewing internees to dispel the thought that they had been coddled by the WRA. After visiting the camp, Thompson assured readers that the camp was satisfactory and the WRA had done well to make the camp functional but not extravagant.[68] Supporters of the WRA's mission to protect citizens' rights also voiced their opinions. A week after the Springfield paper's editor vilified the WRA, the paper printed a response entitled "We Must Not Hate," which contained a series of requests for residents to celebrate humanity, respect the internees, and protect liberties. The article argued that part of America's greatness lay in its history and mission: the Bill of Rights, FDR's promise to protect the "four freedoms," and our goals in waging the war against tyranny "are not declarations of hate. There is no mention of race or creed or color. There is no mention of nationality or class. These are pledges for all the nations, all the people of the world."[69] Similar exhortations came from the Lamar paper when it printed announcements from the War Relocation Board (WRB) about the necessity of relocation and the importance of doing the job in "the American way." Everything about dislocation and interment had been done humanely and with the internees' needs in mind: "Even under the stresses and strains of all-out war, Uncle Sam has his feet on the ground and continues to be what he has always been known to be, a very, very human fellow."[70]

Government officials hoped, and many residents eventually realized, that the camp might prove a boon for local business interests. Indeed, the influx of internee labor was not lost on local farmers and businessmen, as many organizations offered proverbial olive branches to camp inmates upon their arrival at Amache. Almost immediately, while the camp was still under construction, the Lamar Chamber of Commerce held a "get-acquainted" dinner for camp representatives to meet Chamber members.[71] There were bumps along the road, however. WRA reports officer Joseph McClelland remembered one of the meetings between the Chamber of Commerce and Japanese leaders—each group on one side of a long table facing the other and airing their grievances—that became rather heated. McClelland claimed that the economic opportunity carried the day and calmed the tension, as one of the businessmen reminded his cohorts that prejudice jeopardized their well-being. McClelland paraphrased the conversation in a later interview: "The Director [James Lindlay, camp director] had said, if you don't want 'em, I won't let 'em in. We'll keep them out there. And it looked like that's the way it was going. And he [the local businessman] said, gentlemen, just think about this. Here we have a city of 7,500 people right to our doorstep, they have money, they need things to buy. Think of the business that we're losing if we just say they can't come in. Well, that sort of turned the tide. They all got to thinking about the business. And from there on, we had no really further difficulty in the city of Lamar."[72] So they effectively became more viable and valuable neighbors when locals perceived them as consumers and merchants. The Holly Lions Club held a party for camp members and used talent from inside the camp to entertain their guests.[73] Even the Lamar Retail Merchants Association got involved as part of the welcoming committee when it allied with the Chamber of Commerce to invite camp residents to shop in Lamar.[74]

Other residents eventually warmed to internees, although it took some time to thaw relations. Mildred Garrison remembered a shift in locals' views of the internees once they had some exposure to their courteous and respectful manner. Garrison's husband, Lloyd, served as superintendent of schools at the camp. She recalled that Prowers residents were cold to all newcomers, regardless of race. She claimed that Lamar merchants initially held back certain coveted items when camp residents visited their stores. In other cases, stores more blatantly refused to serve internees once the storeowner found out that they came from the camp. A small number of stores made their feelings known by posting "No Japs Allowed" signs in their front windows.[75] Shopkeepers eventually became more amenable to camp residents, but only when it was evident that camp residents never caused problems or shoplifted. In effect, internees had to prove themselves to locals, and it took time. Residual racism existed among those who could not move beyond their prejudice, she noted, but most folks in Lamar started to think more highly of camp residents and federal employees the longer they lived in the area.[76]

Indeed, such initial animosity and suspicion seemed to cloud farmers' perspectives on using outside labor during the first year of the war. Before long, however, many farmers realized that their need and the workers' abilities gelled well enough to allow them to hit production goals and enjoy the benefits of wartime demand. Regardless of how they personally felt about the workers, the dearth in agricultural labor made them come to terms with using outside workers. Certainly, no one could have guessed that such a motley array of workers would have found its way into Colorado fields. The federal government and Extension facilitated the considerable undertaking in response to labor needs, and the numbers of workers, their assistance to area farmers, and the impact on production suggest that the labor program worked during the war and set the stage for a major transition in American agriculture following the war.

Regardless of farmers' trepidations about who worked where, the numbers illustrate the Extension Service's effectiveness in supplying labor. It controlled the flow of labor into Colorado from mid-1943 through the harvest of 1947, with the peak years coming in 1944 and 1945. The groups mentioned here—domestic migrants, Braceros, Jamaicans, German prisoners, and Japanese internees—all worked during that period, but German and internee workers only worked heavily until late 1944 and early 1945. Somewhat surprisingly, the number of "local recruits"—though that phrase is never defined in the tally—nearly met the number of imported workers during the period, according to Extension numbers. The numbers for the state are indicative of this trend. The local recruits numbered roughly 43,000 in 1943, 54,000 in 1944, 53,000 in 1945, 55,000 in 1946, and 38,000 in 1947. The number of imported workers followed the same pattern, moving from 48,000 in 1943, 65,000 in 1944, 75,000 in 1945, 63,000 in 1946, and finally 43,000 in 1947. In totals for the two groups, then, the Extension Service provided at least 100,000 for three years and then more than 200,000 in 1945 as the war concluded. Certainly, the number of workers per farm varied, and downtime existed between planting and harvest that did not keep all parties busy or in the same place for the entire year, but the Extension calculations noted the estimated number of farmers served for these years as well. The totals varied from 13,700 in 1943, 16,000 in 1944, 18,000 in 1945, 15,900 in 1946, and 13,000 in 1947.[77]

What is somewhat surprising is that farmers in the two counties actually preferred certain workers and attempted to use their favorites even when others became available. In Baca, for example, most farmers avoided using Japanese internee labor. Baca County agent Claude Gausman held a meeting in 1942 to take the farmers' temperature on the prospect of using internees in their fields. The overwhelming response, despite their "new zest" to contribute to the war effort, was that the only reason to think of utilizing that group for work revolved around controlling wage demands from other workers.[78] Gausman found no one willing to use Japanese internees other than as leverage. Conversely, similar meetings in Prowers County produced consensus that the labor situation required farmers to use every possible means at their disposal, at least during the crisis. A committee of local farmers agreed that internee labor should only be used if workers were heavily supervised and the placement proved temporary; the workers were to be removed from the field as soon as the work had been finished.[79]

There is not much evidence as to why Baca farmers became more willing to use a mix of migrant and imported labor rather than the internees, but they found enough alternatives to satisfy their needs. Their needs revolved around broomcorn and wheat. Broomcorn rose in popularity among Baca farmers soon after a farmer who had previously been successful with the crop on similar land in Kansas succeeded in harvesting some in Baca in 1887. By the 1950s, Baca had ascended the ladder for domestic production and supplied nearly one-third of all broomcorn in the nation. The plant has long fibers at the top that can be chopped, and the fibers, after being separated by hand and left to dry, were used principally in brooms and as packing material. The plant grows well in dryland conditions, especially in the sandy soils of Baca County. It was hardy, cheap to plant, drought resistant, fast growing, and economically viable, especially when railroads gained access to the county in 1926. Moreover, the plant became desirable during the New Deal when conservationists recommended the plant to protect against wind and water erosion. The government paid farmers to use drought-resistant crops to guard against erosion, so broomcorn farmers received financial benefits from market sales as well as federal subsidies. Given the need for brooms in new wartime industries, on ships, and in barracks, as well as the demand for the bristles as lightweight and flexible packing material for shipping overseas, broomcorn boomed. Unfortunately for growers, the crop proved labor-intensive. No mechanical equipment existed because only a few farmers in a relatively small geographical area relied on the crop, which meant there was little perceived need for mechanization. Consequently, hand labor dominated the industry until it started to decline in the 1970s when other materials made better brooms.[80]

Its eventual decline notwithstanding, broomcorn production dominated the labor market in Baca County. Farmers' initial effort to find workers in 1942 focused on local workers and volunteers; officials urged shutting down schools and encouraged farmers to decrease acreage temporarily so that fewer people could still produce effectively. This changed in 1943 when outside labor became part of the labor pool. During the 1943 harvest 1,895 men worked for 170 farmers (thus nearly 20 percent of county farms) even though the Extension Service did not have full control of labor arrangements, and German prisoners, whom farmers identified as the premium choice, were not yet available. Yet with the workers available, 1943 marked a record for broomcorn production, and Roy Haney became the "Broomcorn King of Baca County." Haney devoted 6,000 acres to the crop and hauled in roughly 1,200 tons worth $300,000.[81]

Extension also helped farmers in 1944, leading to another banner year for broomcorn, as more than 2,100 workers assisted Baca farmers on 262 farms. Some of these workers came from Camp Trinidad, which opened its doors to allow 300 POWs to work the fields during harvest in the fall. In addition, the Baca County Labor Association sponsored migrant workers who slept outside Springfield in the former CCC camp. Many of these workers hailed from places like Oklahoma, Arkansas, and Missouri, and they stayed in temporary housing constructed specifically for broomcorn workers.[82] Extension also arranged for 100 Mexican nationals to assist with the harvest. Most of them stayed in and around the small town of

Walsh, generally in broomcorn sheds or temporary worker housing akin to where other migrant workers stayed. The federal government helped offset costs in both cases by providing army guards for the prisoners and managing the contracts to import Mexican workers as well as necessary camp construction and repair.[83]

The following year, with the American war machine at full throttle, similar production demands for broomcorn and wheat necessitated the continued use of outside workers. Nearly 1,100 workers found gainful employment in Baca County, and more farmers looked to Extension for assistance in landing more workers than it had in 1943 or 1944. Again, German prisoners played an important role, primarily because farmers started to search them out when it became apparent that they generally stayed for the entire harvest. Migratory labor from neighboring states sometimes left the area before the harvest started or at least before it was done, meaning that farmers who relied on workers to haul in their crop found themselves scrambling if their workers abandoned them during the season. The prisoners stayed at the CCC camp again, although the West Baca County Farm Labor Association sponsored them in 1945. The combined labor force helped reach federal goals for broomcorn—Baca farmers grew nearly 20,000 tons of it for nearly $4.5 million in 1944 and almost 11,000 tons worth $2.5 million in 1945.[84] There was less need for outside workers in Baca to harvest the broomcorn once demand dropped, but the war had been good to Baca farmers. While most farmers focused on broomcorn, onions, potatoes, and wheat also garnered attention and produced profits, giving nearly every farmer in the county a solid foundation on which to build in the postwar years.[85]

Wartime production led to similar benefits for Prowers farmers, although Prowers farmers who utilized the labor program focused almost exclusively on sugar beets. Despite a drop in demand for both beets and workers in the 1930s, the return of good weather and increasing consumer/military demand left growers again pining for outside workers. Before 1943, beet growers tried to recruit every available local hand to help in the fields.

In fact, Republican governor John Vivian believed that Arkansas valley residents could help in that regard. Vivian challenged federal policy by freeing eligible farm boys from the draft; he even went so far as to ask for military uniforms that farmers could wear while in the fields so that no one could construe their work as somehow unpatriotic or less vital to the war effort.[86] He wrote Secretary of War Henry Stimson several times to request that the army "grant general furloughs to soldiers to permit them to work on farms" because "our farmers are unceremoniously and, in many instances, shortsightedly taken in the draft." He implored Stimson to release farmers to their homes or "there is likely to be a scarcity of food in Colorado probably never before paralleled."[87] Fortunately, the Extension Service stepped in when it did (less than two weeks after Vivian wrote a final letter to Stimson), as most private companies and even corporations lacked the means to recruit enough workers to satisfy farmers, and the War Department had no intention of releasing all farmers.

The Extension Service approached the labor problem as it had nearly every other issue it faced by sending agents into the field to work with farmers to figure out a solution. Claude Gausman, an agronomist trained at Colorado A&M in Fort Collins, had built connections with the Extension Service and its employees during his time at school, and he entered the field prepared to cooperate with farmers. Once Gausman arrived in Prowers in 1943—he replaced his friend and former college classmate Jack French—he focused on the labor problem. Gausman had been the agent in Baca County since 1939, so he was already familiar with the region and had made connections in the two counties through his work in Springfield.[88] He spoke with 370 farmers about their labor needs once Extension took over the program in May 1943, a number higher than even peak wartime demand in Baca. Unfortunately, Gausman's initial recruits fared poorly; most of the workers did not do much well, and the rate of attrition was high as most were either let go or they left of their own accord.[89] The group included a few American Indians from New Mexico as well as African Americans from Oklahoma in late 1942, but the local newspaper reported several problems with the newly arrived workers, including cases of drunk and disorderly behavior as well as an attempted rape of a local woman. Even though farmers were "critically short" of labor, they looked for alternatives to this round of workers because of such legal issues.[90] Fortunately for farmers, a crop of newly arrived Japanese internees, American Indians, Mexican Nationals, and conscientious objectors combined to meet demand for the harvest that fall, as nearly 3,000 workers took to the fields to bring in the year's beets.[91]

Indeed, by that fall with the camp's construction concluded, nearly 1,000 internees worked the fields to help boost local harvesting efforts. Local beet farmers claimed that the labor dearth would cost them roughly 60,000 tons of beets during the 1942 harvest, and they looked to the camp for workers. The camp's general assembly voted unanimously to aid them and sent 141 volunteers into local fields to help bring in the beets.[92] Local businessmen were quick to thank the internees for saving their harvest. Representatives from the Great Western Sugar Company printed words of thanks in the camp newspaper, attributing the good harvest to internee labor in their time of need. The Great Western labor commissioner claimed that camp workers had a 90 percent efficiency rating in the fields and that the workers deserved the company's genuine thanks for working so hard and so well.[93] In addition, individual farmers placed other notes in the camp newspaper to demonstrate their gratitude, including beet farmers from Holly who appreciated internees' efforts during the 1942 harvest.[94] Internee labor continued to be an important factor in agricultural production during 1943 and early 1944. The camp newspaper claimed that 1,428 Amacheans took seasonal leave in 1943 and celebrated their contribution "towards the 'Food for Victory' campaign during the year in harvesting [the] nation's variety of perishable food products."[95] Another 945 workers left the camp during the harvest of 1944.[96]

These numbers suggest that the camp residents offered significant assistance to Colorado farmers. Yet the slow decline implies that using camp workers became less common after the first year and almost nonexistent by 1945. The steady move to relocate internees to other parts of the country or admit them to colleges or sign them up for military service reduced the number of workers. Other developments compounded that problem, according to Joseph McClelland. A graduate from the University of Missouri where he majored in journalism, McClelland earned a job with the WRA and moved to Denver in July 1942. Within weeks the WRA reassigned him to take a position at Amache, where he served as reports officer and photographer, responsible for distributing information inside Amache as well as to the public as he documented daily activities. McClelland claimed that once area farmers began to appreciate Japanese farmers for their skills and technique ("even though they were in Colorado"), many relocated families entered partnerships with agriculturalists. McClelland remembered several camp families who attained passes and then lived with a local family: "Usually it wasn't a hard labor standpoint, it was a leasing standpoint where they shared the profits" akin to sharecropping. He argued that internees' knowledge and diligence made such arrangements possible and constituted an opportunity that no other group of wartime workers enjoyed.[97]

Growing dislike for the work and for their status on local farms also contributed to declining numbers of internees willing to work for locals. Laboring in beet fields proved especially arduous; even if the worker took pains to protect himself or herself by practicing proper technique, stoop labor took a toll on the worker's body. In addition, the facilities where internees stayed while on leave were often subpar, despite the agreement that farmers signed to offer decent accommodations for workers. Farmers provided "a variety of make-shift shelters—anything from wooden frame houses, to old train coaches, barns, and tents." A survey of Amacheans showed that 83 percent of those on leave had no bathing facilities and 31 percent had no toilet facilities. The same survey suggested that workers had often been confused about the payment system, not only in terms of how much they were due to receive but about the schedule. These conditions frustrated internees and certainly caused some of them to hesitate when considering work outside the camp.[98]

The work requirements on the camp's own acreage constituted another reason for the decline in available labor by 1945. The WRA had hoped that each camp could eventually become mostly self-sufficient by having internees harvest fruit and vegetables as well as work with livestock and swine. The Amache project farm employed more evacuees than any other industry at the camp, and the workers succeeded in using the camp land to meet the evacuees' dietary needs. The variety of fruit and vegetables, coupled with the 1,000 head of hogs, 800 cattle, and 16,000 chickens, offered more than enough for the camp, so the WRA shipped the surplus to other camps. In fact, sixteen railroad cars full of vegetables left the camp for other relocation centers in 1943, when the camp produced a crop worth nearly $190,000. The evacuees produced common commodities like corn, alfalfa, and wheat, in addition to more original and unusual crop varieties like Napa cabbage and daikon, which were new to the region. The camp's success illustrated the internees' agricultural ability, the leadership employed by farm superintendent Ernest Tigges, and the fact that the camp enjoyed junior irrigation rights.[99]

Even the camp faced worker shortages, however, and camp officials became increasingly frustrated that the farm did not reach its full potential. The labor issue represented the culmination of several factors, including the frequency of seasonal leave, segregation of loyal from disloyal evacuees, internal strife, low pay for the work, and cultural beliefs. In addition, the number of internees who relocated to other parts of the country increased after 1943; the total camp population declined, and the number of workers available for any employment dropped as well.[100] The labor scarcity caused such problems, in fact, that the camp sold off portions of its farm acreage to residents when it became clear that not enough evacuees would contribute to production. The manpower issue had become so acute by 1944 that the camp director issued a directive "that required anyone applying for seasonal work to first work two weeks in the center."[101] The farm's output and its ability to sustain the camp population while helping feed evacuees in other camps were remarkable given the labor constraints. By 1943 and 1944, then,

FIG. 22. "Part of the Irish potatoes being grown on the Center Farm," June 4, 1943. Photo by Joe McClelland. War Relocation Authority, Denver Public Library, Western History Collection, x-6582.

the potential camp labor pool slowly evaporated until the camp closed in January 1946, which forced area farmers to find alternative labor sources.

The Extension Service had total control of the labor program by the end of 1943, and it ably supplied farmers and growers with the workers they needed. Farmers also organized into labor organizations designed to funnel workers into the region and then disperse them as necessary on members' farms. For example, Prowers agent Gausman noted that 1943 included negotiations with the Holly Beet Growers Association and the Lamar Beet Growers Association. The newly established East Prowers County Farm Labor Association similarly reached out to Extension in 1944.

The county agent encouraged farmers to utilize the new associations to concentrate their labor demands, thereby making it easier to assess how many farmers needed how many workers and for what work. Yet such associations were often rife with problems, especially in their early stages, because once the associations took over worker management, they had trouble arranging work orders with members, they found it difficult to keep workers on for the whole season, and the process took time and energy since association members were new to such responsibilities.[102] The associations found it easier to reach out to migrants from neighboring states because they could recruit such workers without much help from Extension. But that often proved complicated as well. Per the *Lamar Daily News*, Colorado farmers expected nearly 7,000 migrants to work in the state in 1944. Many of them had previously made connections in Prowers County, so farmers assumed that the migrants would simply follow the same path again from Texas and Oklahoma into Colorado. Unfortunately, problems arose. Heavy rains in the planting season of both 1943 and 1944 meant that farmers had no clear idea of how much they could produce and consequently no clear sense of the contracts they should orchestrate with domestic migrants. As a result, migrants proved reticent to commit and often picked up and left Prowers farmers in search of alternatives in neighboring counties and even other states. The bottom line was that each group posed a series of challenges and promised many benefits, and farmers often found it best to defer to the Extension Service to coordinate the labor regime.[103]

The pattern of combining migrant and guest worker labor continued in 1945, 1946, and 1947, until the final call for wartime production and the dissipation of Extension's labor efforts. Supplemental labor came from German prisoners of war in 1945, when Extension helped repair and replace several buildings and camps that temporarily housed outside labor. Extension funded housing for POWs in two camps owned by sugar companies; the Holly Sugar Corporation owned one facility and American Crystal Sugar Company the other, and combined the two housed 339 Germans in

1945. Extension also contributed to repair migrant camps, again remodeling or repairing hotels, warehouses, and colonies owned by prominent local sugar companies.[104] Problems tended to sprout up when the various groups mingled in the fields, so Extension employees made a concerted effort to separate the disparate workers; in 1945 that meant Mexican nationals focused on east side farms and German prisoners toiled in the western part of the county. Jamaican workers also played a role, albeit a comparatively small one.[105]

This mix of workers toiled in Prowers through 1946 and most of 1947, although 1945 constituted the peak year for utilizing outside labor. The East Prowers County Farm Labor Association operated from 1945 to 1947 and brought in evacuees, Mexican nationals, and prisoners; the West Prowers County Farm Labor Association ran for the same period and utilized Mexican nationals, Jamaicans, and prisoners. These groups looked to Extension to provide labor, but as had been normal in Prowers County, private enterprise took much responsibility for attracting labor for themselves, especially in terms of sugar processors and their contracted growers. They dealt with inclement weather and faced competition from surrounding states. The pool slowly shrank with the quick return of prisoners and the relocation of internees in 1944 and after, and the influx of discharged service people coupled with the migrant and immigrant workers only made up so much slack. In both 1946 and 1947, the state planned with Kansas, Kentucky, and Texas to funnel migrant workers to Colorado to supplement the group of Mexican nationals.[106] In effect, Prowers farmers dealt with their labor problems by capitalizing on federal intervention as well as by organizing into associations devoted to procuring labor as needed by county farmers. The combination worked well enough to meet demand, keep production humming, and create a financial boon to area agriculturalists.

The Extension Service became a jack-of-all-trades during the war by assessing local labor needs, identifying the groups most likely to work well given the circumstances, and procuring the workers to satisfy demand. It also spearheaded two other tasks in securing wartime labor by settling on a pay scale and helping train workers unfamiliar with the crops and conditions found in southeastern Colorado. The county agent then found himself in familiar territory, as he had been performing similar duties of watching the federal purse strings and educating his constituents since the 1930s. This theme held true with wartime responsibilities, including the need to promote conservation techniques and facilitate the genesis and maturation of soil conservation districts in both counties. These various responsibilities combined when the agent cooperated with farmers to pay and educate outside labor. The task tested the core of Extension's raison d'être by requiring patience, an ability to work with locals, and knowledge about effective land use techniques, in addition to demanding familiarity with government protocol and prerogatives. It challenged agents by forcing them to juggle their constituents, the laborers, and the need for conservation while also attempting to meet production goals necessary for the war drive.

Educating workers became a concern once the government recruited and placed foreign workers and migrants in Colorado fields. In their review of the Emergency Farm Labor Program, Hamman and Anderson noted the importance of state-to-state relations and America's ability to negotiate with foreign governments to facilitate the imported labor programs. They rightly emphasized the "concessions" necessary to attract workers, including government payment for transportation and the guarantee that imported workers would receive the same wages that local workers garnered for the same jobs. Additionally, the War Department insisted that prisoners of war also received prevailing wages for similar work, and the WRA set stipulations for internees as well. The payment system relied on county agents to coordinate country wage boards that then deliberated once they heard testimony from local farmers, workers, and anyone else with pertinent information. That county wage board then submitted its findings to the state director of Extension, who turned it over to the federal Farm Labor Office. The board also helped establish adjustment committees that settled disputes between growers and workers about working conditions. The arbitration committees included a representative of the Extension Service, a member of the sponsoring association, and a "disinterested party," in addition to a member of the local Labor Branch to represent the workers, and finally the interested farmer. Again, the county agent helped oversee this and facilitate the necessary intervention by serving as a bridge between federal and local interests.[107]

In the end, prices varied according to the work and the time spent doing the work, such that weeding fetched different rates than planting, and sugar beet work meant different money than a day spent harvesting broomcorn. Additionally, the Extension Service set out a general table for employers and workers as a guideline for hourly work; payment by tonnage also existed, though most of that concerned prisoners.[108] The system seemed generally fair; wages reflected the going rate for agricultural labor, and the workers had several safeguards in place to ensure that they were not being taken advantage of by their employers. Theoretically, all workers had such protection, beginning with the federal government's assurance that their wages and working/living conditions would be acceptable. This parity stirred some unrest among domestic workers, who rarely had any such protection from exploitation, had no contractual agreement for housing, and had no other assurance that they would be retained through the season. Consequently, competition between local workers and contract labor started brewing soon after the guest workers and prisoners arrived in late 1942 and early 1943. Discontented locals conspired against employers, fomented dissent among the workers, and tried to subvert the influx of outside laborers that they believed sabotaged their chance of taking advantage of the wartime boom. Many farmers who declined to tap imported labor because of these problems decided to work themselves and their families harder and longer to negate the need for potentially problematic outside workers. This largely offset the demand for year-round workers but did not assuage the necessity of utilizing seasonal labor, which brought farmers back to the Extension Service looking for assistance in supplying workers.[109]

Agents faced additional problems training the workers that they had placed on area farms. The differences in language and culture between themselves and most of their trainees caused an obvious rift. The work schedule also caused issues; the three-month separation between planting and harvesting seasons meant that workers either had to find employment in the interim, move to another locale, or head back home. Therefore, there was little consistency within the work crews from season to season and even more fluctuation from one year to the next. In response, Extension employees and county agents reached out to workers to promote efficient and safe practices, in hopes that such education could bring new workers up to speed and maintain a steady rate of production despite their lack of experience. The push for efficient production reflected the fact that Extension understood the Emergency Farm Labor Program's fragility. In other words, farmers had access to labor, but there was no guarantee that the workers would be effective or that many might simply stop working without a moment's notice. The number of workers also posed a problem because many farmers had to make do with a minimal labor force. If they had been accustomed to using ten workers, but only five became available during the war, then the five workers had to do their job efficiently and safely. Any injury or lax effort meant a dip in production.

The Extension Service attempted to train both farmers and workers to ensure that the entire system ran as smoothly and effectively as possible. For example, agents held training programs for foreign workers and offered additional training for a crew leader of POW workers so that all parties had some organization and the workers had at least some knowledge about local conditions and practices. The Service put together bulletins and pamphlets, in addition to using motion pictures, as Hamman believed that certain groups learned better through different mediums. Extension even offered informational pamphlets in the workers' home languages to further the learning process. Such methods often proved successful; Hamman claimed that the those who spoke Spanish usually took the pamphlets home with them to their barracks or camp and read the literature in their off time.

The bulletins and leaflets often included basic instructions on labor-saving techniques specific to each subset of hand labor jobs in the region, such as picking string beans, potatoes, or peaches and even topping sugar beets. Extension designed the pamphlets to show growers and workers proper form and to implore that they use the most efficient methods available. Extension also emphasized that farmers could expedite this process by holding meetings to demonstrate how some jobs could be done with fewer workers than normal or how farmers could best pool trucks and machinery with neighbors to save time and man days of work. It was part of a larger scheme to cut down on waste and maximize the resources at hand; in that way educating farmers appeared in tune with repairing vehicles, picking the right crops and seeds, and meeting quotas. Extension held meetings and seminars for farmers to prepare them to supervise and instruct their workers in a productive fashion. Agents also held extra sessions for farmers and companies that presumed to use either prisoners or Mexican nationals so that they could be schooled on the differences in culture, language, and expectations and, in the case with prisoners, how to maintain safe, secure relations.[110] That Extension went to such lengths is an indication of the premium it placed on efficient production that met the federal government's goals.

How each group responded to Extension's efforts, even whether they proved amenable to such instruction, differed from group to group. On the one hand, Hamman claimed that Jamaican workers had no taste for anything approaching agricultural education and bristled at the thought of spending time reading or watching the Extension's films. Hamman noted their apathy with distaste but never ventured a guess as to why they resisted instruction.[111] On the other hand, some workers took advantage of the opportunity to learn new types of farming or techniques; some even used their prior knowledge to their advantage when working in the fields. One example is Rüdiger Freiherr von Wechmar, later German ambassador to the United Nations and chair of the United Nations Security Council, who worked in Kansas and Colorado while stationed at Camp Trinidad.

He remembered his time in the fields fondly: "I learned how to harvest corn, *besenkraut* (broomcorn) and potatoes, and I applied my knowledge for the beet harvest of my farm labor days in Spandau. Between the man-high corn stalks and the *besenkraut* bushes, we learned how to take care of rattlesnakes, which sometimes crossed our paths. The locals explained how to deal with snakes: grab their raised tails and then whirl them like a whip in the air. Thus the bowels of the animals are pushed to their head, they bite and kill themselves with their own venom."[112] Certainly, most workers likely sat somewhere between those who Hamman described as disinterested and the enthusiastic von Wechmar. But Extension provided training to the entire laboring population and then left it to the farmers and workers to put that training to use.

The agent's job continued once the workers arrived, and they started their seasons by conducting "continuous follow-up" with farmers to assess their experience. The Colorado State University Department of Economics and the Experiment Station on campus in Fort Collins regularly contributed to the instruction by doing field research and investigating ways to save money. The Extension Service also resolved to spread new ideas and methods as quickly as they had been discovered and practiced, ensuring that they disseminated the most modern agricultural methods to all farmers. It especially promoted adopting technological advances as another way to maximize production while saving labor. Experiment Station workers modified a power manure loader to load bundle grain and put tools on beet cultivators for cross blocking sugar beets—two methods to cut down on arduous labor. Extension also reached out to private industry and individual inventors in the hope that their efforts might reduce labor and maintain production. It persuaded state sugar processors to organize a foundation to work on new beet implements and join forces with the USDA to experiment with technology to cut labor costs.[113]

A broader public push to use technological innovation to cut into time and labor costs also emerged. Frequent notices of inventions and implementations filled southeastern Colorado newspapers, evidence that folks from across the region experimented with—and then claimed to own—the "next best thing." In some ways, this turn to science and innovation had a lengthy tradition. In one case, C. T. Peacock developed the technology to build a damming chisel and furrow seeder to maximize soil moisture and improve conservation without hampering production—an instrument that Peacock hoped could help farmers maximize profit even when "cooperating with nature."[114] Other instances of this push for innovation occurred during the 1930s and 1940s. Much of this invention dealt with sugar beets, most likely because the beet industry proved both labor-intensive and highly profitable. A new type of beet harvester developed in Wyoming promised to "revolutionize the entire beet growing industry" as it did "everything to a sugar beet but refine it and put it in your coffee."[115] Another tool, dubbed "Schnabel's Machine," could auto-contour one's field, helping conserve soil and save manpower while doing it.[116] The premium on lowering costs and maintaining production continued during the war, although the cause had shifted. The weather and soil exhaustion compelled farmers to think about efficiency during the 1930s, but the dearth in labor prompted a similar focus during the war. Many farmers hoped that technological innovation could solve both problems, but nothing helped as much as their county agents.

Two aspects of the Extension Service's management of the wartime labor program deserve special attention. First, it produced remarkable results and clarified the role that local and state officials played in placing labor successfully. The county agent set meetings and deliberations with local interested parties and relayed his findings to the state committee, and the state representative determined the need as well as the conditions and payment for the various positions. Obviously, the federal government negotiated with other countries to import labor and made the decision to both confine potential subversives and allow prisoners of war to work in the U.S. Extension. This took the consequences of those policies—the able-bodied workers—and translated the motley array into a largely viable workforce that met farmers' demands and helped win the war. Consider the numbers: From 1943 to 1947 Extension averaged 59,210 workers recruited, transported, or placed, and 15,455 farmers served statewide. Furthermore, they did this at a relatively inexpensive rate, averaging $153,878 a year cost to the Service (and therefore taken from the federal allotment for the program).[117] Extension also noted a dramatic improvement in statewide numbers; broomcorn moved from 66,000 acres harvested in 1942 to 104,000 by 1945 while sugar beets jumped from 132,000 in 1941 to 152,000 in 1945. In this case, the total number of acres harvested is set to equate to overall production, and the numbers given by Extension are used to validate its effectiveness. This impact and the contribution to the "Food for Victory" program are impressive.

The labor program reflected a sense of urgency that only national emergencies can elicit. Between 1943 and 1947, the Colorado Extension Service responded to the crisis as it had to the Depression and drought: it relied on county agents to decide what farmers needed, when they needed it, and how best to satisfy their demands. In the process, A. J. Hamman and F. A. Anderson coordinated a dramatic population shift as the Emergency Farm Labor Program recruited workers from four countries and throughout the Great Plains and American West. Indeed, the wartime amalgamation of different races, guest workers and domestics, internees and P O W s, resembled a mixing of peoples that was new to the American home front. The program was largely successful, helping sugar beet, broomcorn, and wheat farmers especially, as it sent needed labor into the fields where the workers helped fuel the war effort. It is fair to contend that this augmented production proved critical to America's ability to fight a two-front war over the course of four years. Problems persisted, however. Racism, bureaucracy, and inefficiency all played a role in souring some workers. Most farmers appreciated workers' efforts, even if some remained reticent to embrace workers or treat them as equal participants in mobilization. It is somewhat surprising that the process went as smoothly as it did, considering the number of moving parts that had to work together to make the system function. When all was said and done, and the Allies had won the war, the Emergency Farm Labor Program had coordinated the placement of more than 250,000 workers on Colorado farms.

Second, we should also view the Extension Service's efforts as a logical derivation of the federal government's recent intervention in rural America. The labor crisis mandated immediate action, and the state's response was made easier because of the strong federal presence already at work. County agents had worked directly with rural Americans during the 1930s and into the 1940s. Their roles certainly changed over that time, but once agents arrived in Prowers and Baca Counties and established themselves they were there to stay.[118]

Indeed, while much of the postwar world certainly looked different than it had in 1929, or even 1939, such variations represented the culmination of New Deal *and* World War II influence on area residents. For example, farmers came out of the war in a tight relationship with Washington DC. Yet the government's role as labor broker proved temporary, and farmers never again had the luxury of having the state funnel workers to them. The internees left, the POWs returned home, and the migrant labor stream slowed after the war. Moreover, the folks who had migrated to cities or to other parts of the country to find work in wartime industries rarely returned to the area. Consequently, farmers had to find new ways to produce, especially in terms of planting and harvesting labor-intensive crops like broomcorn and sugar beets. This trend of outmigration and farmers trying to farm more efficiently started during the Great Depression.

Certainly, farmers looked to technological innovation as a chance to save on labor during the war, but such attention was not new. The wartime boom meant that they had more money to spend on such technologies and mechanical developments. This then contributed to an explosion in production and a new type of farming centered more on highly capitalized farms and the use of science, although the impetus to substitute technology for human labor and animal power had influenced farmers for more than a century. Such improvements cut into the need for labor and made farmers more efficient in preparing the soil, planting, and even harvesting.

Moreover, scientists started to experiment with plant varieties and fertilizers designed to control pests, insects, and weeds; the new combination "had the effect not only of decreasing labor requirements but also of increasing yields per acre."[119] Inexpensive labor was still available to a limited extent, but agricultural production slowly became more manageable by a smaller number of people.

The changing number of farmers represented another trend that accelerated with the war but had been present during the 1930s. Farmers who capitalized on wartime production had money to spend to expand their holdings; in many cases these farmers lived in areas where out-migration led to abandonment or landowners' desire to sell their property at minimal cost. Because of these new opportunities, the average farm's size increased dramatically over the 1940s, even as the number of farmers declined. In other words, the farmers who were left on their land after the war now had more capital, more money to spend on mechanization and expansion, and less competition. Moreover, these large farms tended to have less diversification than their smaller counterparts, as owners quickly identified the most marketable commodities and grew cash crops that could take advantage of growing consumer markets.[120]

The government also continued supporting American farmers by maintaining a safety net to ensure that there was no attrition in postwar America as had happened during the 1930s. The New Deal legacy is evident in this emphasis on providing financial assistance to farmers. Indeed, "since the passage of the first Agricultural Adjustment Act (AAA) in 1933, farm price and income support programs have been the core of agricultural policy in the United States. Federal policy has shifted to some extent since the 1980s when it started to focus more on persuading farmers to produce marketable commodities and more akin to direct government payments instead of subsidies designed to control supply and demand.[121] As it had during the Dust Bowl, Great Depression, and World War II, then, the government continues to prioritize keeping farmers farming even if it means a significant national investment in their well-being.

In addition, the government continued to fund farmers who practiced conservation on their property, another example of New Deal policies remaining an important part of government-farmer relations. The war did very little to upset the maturing conservation state. For example, the Baca County soil conservation districts continued to function, offering members machinery and technological assistance, as well as physical help, to furrow or till the contour, plant shelterbelts, terrace their fields, or use any other method that they employed before the war. Indeed, support for local districts grew on the eve of the war as the Two Buttes district emerged in 1941 to join the western and southeastern districts already in place in Baca.[122] Similar enthusiasm in Prowers led to farmers authorizing the Arkansas Valley Soil Conservation District in 1941.[123] County residents added the Prowers County Soil Conservation District in 1943.[124] Certainly, federal subsidies and assistance in conserving soil and water helped persuade farmers to join, but the district's creation evidenced sustained support for the cause. The federal government's presence in rural America shifted during the war, and part of that happened in response to the weather as the sense of immediacy that accompanied the Dust Bowl evaporated when the rain returned. Yet state and county officials remained on the scene during the war and continually reminded farmers of the need to protect their resources. The county agents maintained their involvement in conservation activities and consistently promoted the need to think about the postwar world. Indeed, through dust, rain, and war, county agents continued to act according to their constituents' needs.

7

An Unquenchable Thirst

In January 1955 Otto C. Lubbers, a farmer on the Fort Lyon irrigation ditch, spent five dollars to purchase a frying pan. On the surface, Lubbers's decision to procure a pan, even if he did pay a considerable amount for it, should draw little interest from historians or, really, anyone not likely to eat from it. This was no ordinary frying pan, however, and Lubbers never intended to use it anywhere near his stove. Lubbers became the first individual in southeastern Colorado to contribute to the "financial campaign for support of the Fryingpan-Arkansas trans-mountain diversion project" then being discussed in Congress by paying his hard-earned money for a tin frying pan that had been gilded. James C. Moore, chairman of the funding drive, sent five hundred pans to Lamar, where several businesses, including the *Lamar Daily News*, offered them for sale to willing contributors. The newspaper celebrated residents who purchased several pans in support of the project and listed the local sites where others could buy them.[1]

The pans became a physical symbol for the project that irrigators hoped would address the water shortage that afflicted most wet farmers in the Arkansas valley, especially during dry years like those that plagued the region during the "Filthy Fifties." In a more abstract way, these cooking utensils also seemed to remind supporters that the project would bolster the agricultural economy and thereby afford farmers more security in their ability to feed their families during the tough years of the 1950s.

Advocates for the diversion of water from the West Slope to the Front Range and the construction of a series of reservoirs designed to capture and hold that water certainly believed that they needed to do whatever possible to ensure their continued (and, ideally, consistently increasing) access to water to protect their interests.

This push for water among irrigators did not reflect a shift in thinking during the 1950s as much as it did an indication of farmers' realization that the southeastern Colorado landscape would never give them all that they desired. That lesson represented an important legacy of the Dust Bowl and became more important as the drought of the 1950s sank in on the western Great Plains. This drought failed to grab national headlines like its predecessor, and comparatively little attention has been paid by historians, although the dry years from 1950 to 1958 proved much drier and hotter than had the span from 1930 to 1938 in parts of the region. Several factors explain the difference. First, the 1950s drought did not cover the same geographical expanse as its predecessor, hitting sections of the southern and western Great Plains instead of impacting the entire region. Second, the Dust Bowl coincided with the worst economic depression in American history, making the plight of plains farmers a fitting symbol for the nation's significant challenges. Conversely, despite a brief postwar recession, most Americans enjoyed financial stability at worst and significant economic growth at best during the 1950s, and there was no sense of despair. Third, while the 1950s drought produced sensational dust storms, nothing like Black Sunday ever occurred, meaning that the type of storm that carried dirt from Colorado and dumped it on politicians in Washington and residents of New York City or Chicago never materialized. No author offered a corollary to John Steinbeck's *The Grapes of Wrath* to dramatize the plight of farmers in the 1950s. The "Filthy Fifties" seemed more regional than national and therefore was not as dramatic as the "Dirty Thirties."

The fourth and most important difference between the two environmental crises, however, is that most of the farmers who faced the 1950s drought did so having lived through the Dust Bowl years better prepared and more confident in their ability to handle sustained dry years. The Dust Bowl forced farmers to adapt in ways that no other impetus could, as an employee of the Extension Service noted in 1938: "Drouth has been the education factor during the past 10 years. There is no question but what nature's method of education has been far more effective than all of the federal agencies and others combined."[2] Fred Betz Sr., editor of the *Lamar Daily News*, considered the drought's lessons in a similar way, suggesting that, while the Dust Bowl was more severe than the 1950s drought, solutions could be learned by looking to the past. "It seems to us," Betz opined in an editorial, "that it would be well for those in charge and federal agencies to re-read what took place [during the Dust Bowl] and take advantage of the information regarding proper land use gained at that time."[3] In celebrating the painstaking gains made during the 1930s, Betz advised that the farmers and federal bureaucrats should cooperate in the face of the new crisis by bringing to bear the experiences each group accrued when facing the Dust Bowl.

Indeed, while the 1930s drought certainly forced farmers to seriously reconsider their practices, it is also true that the federal government and state government changed their roles in helping farmers deal with such crises. It is also the case that no single universal lesson emerged from the Dust Bowl experience. The divergence in how individuals and groups remembered their Dust Bowl years produced an interesting array of approaches to dealing with the "Filthy Fifties." Whether farmers wanted to admit it or even recognized it at the time, however, the Dust Bowl had dramatically affected how they approached their lands and thought about production. Indeed, some researchers have suggested that the 1930s created a sort of memory context that served as a foundation to which all subsequent droughts were compared. In that respect, farmers, residents, and the general public have approached later droughts by explicitly connecting them to the 1930s experience, which has meant that responses to those challenges have never been as urgent but have been more pragmatic.[4]

The indefatigable search for more water, something evident throughout the region during the 1930s, represented the most obvious way agriculturalists reacted to the later drought. They hoped that they might combine efforts with the federal government to expand their access to water. Eager to build on earlier successes in garnering federal financial support and expertise for constructing the John Martin Dam and Reservoir, farmers throughout the Arkansas valley searched for additional projects that might irrigate their farms. Starting in the early 1950s, the Fryingpan-Arkansas Project (Fry-Ark) became the focal point for those in Prowers and throughout the Arkansas valley who firmly believed that federal and local cooperation in erecting the John Martin Dam could be replicated. That they believed it should be repeated exposed the fundamental problem with the dam; John Martin could not guarantee a steady water flow for irrigators. Even though it helped to regulate the flow of the river and several times controlled a rush of water that would have otherwise led to expansive flooding downstream from it, farmers led the charge to increase and make more secure their access to stable irrigation with another project. This motivation produced a rather complex proposal to establish a trans-mountain diversion project that could appropriate water from the West Slope and direct it to the agricultural lands of the valley as well as the fast-growing urban areas of Pueblo and Colorado Springs. The Big Thompson Project, which started at the same time as the John Martin venture, offered a blueprint for the Fryingpan-Arkansas proposal in that it represented a joint federal-state venture to divert water from the West Slope to the Front Range in northern Colorado. It took decades to complete, and construction costs far outstripped estimates, but when southeastern Coloradans started the push for the Fryingpan-Arkansas Project, they did so knowing full well that an activist federal government might be willing to support a similar endeavor. Finally in 1962 John F. Kennedy authorized the bill to fund the project before coming to Pueblo to celebrate with the project's local supporters.[5]

While this legislative victory marked the most sensational example of how farmers applied the lessons they learned to the 1950s drought, it was by no means the only one. Nor was it the sole demonstration of how agriculturalists came to appreciate their continued reliance on water. Indeed, access to water became a primary concern for folks who had previously been unable to reach any, and a dramatic increase in the number of farmers digging wells with hopes of reaching the newly tapped Ogallala Aquifer produced a fundamental shift in the agricultural economy. Farmers who had only known dryland conditions now found themselves able to grow onions, irrigated alfalfa, sugar beets, and other cash crops, which afforded them a new perspective on farming in southeastern Colorado. Access to groundwater gave them a sense of stability that they had not known as dryland farmers, and that security is what they sought after living through the Dust Bowl years.

The pursuit of water certainly influenced land use in southeastern Colorado, but one of the most significant ways that agriculturalists in the 1950s differed from those in the 1930s was that they had already adopted, practiced, and fully supported soil and water conservation programs. The novelty of soil conservation districts no longer existed when drought hit in 1950, but the applied side of community-based policing of land use practices and the widespread support for conservation subsidies meant that the clear majority of farmers in the region faced the onset of the drought having practiced conservation on their lands and stayed steadfast in the continued defense against soil erosion. The Colorado Extension Service, the Soil Conservation Service, the Production Management Administration, and local soil conservation districts provided the kind of resources and infrastructure necessary to promote conservation among farmers and gave them the sort of assistance they needed to protect their lands. Unlike the 1930s, the farmers who faced the onset of drought in the 1950s did so armed with knowledge, techniques, and support to fight erosion. This widespread practice of soil and water conservation largely explains why the "Filthy Fifties" never reached the severity of the "Dirty Thirties" in southeastern Colorado. Farmers were better prepared.

By capitalizing on federal funding for water projects, finding water in new places, and practicing widespread soil and water conservation, farmers fared remarkably well during the 1950s considering they confronted an extensive and sometimes devastating drought. Even then, though, they certainly appreciated the ways that inclement weather affected them and how their inability to control the weather remained at the center of their existence. The 1950s drought broke in 1958, but in 1960 the rains stopped again, leaving farmers with yet another reminder of the climate's inconsistency. Fortunately, by the time Kennedy authorized the Fryingpan-Arkansas Project, they had become adept at using government assistance to their advantage and countering the effects of sustained drought. The lessons they learned from the 1950s reinforced what they took away from the 1930s and prepared them to face future droughts, a prospect that seemed inevitable to folks who had lived through decades of dry weather and hot winds.

The "Filthy Fifties" hit Great Plains farmers hard and marked an incredible reversal of fortune. While it proved more devastating in other plains locales, it nonetheless struck southeastern Colorado and reminded many people of the 1930s. The *Springfield Plainsman Herald* reported several highway accidents caused by blinding dust storms that left visibility near zero. One of the more serious cases involved a six-car collision when high winds and dust left motorists unable to navigate the roadway.[6] It also recounted the health problems of men who suffered through "severe" attacks of pneumonia that doctors believed had been caused by the dust.[7] The newspaper shared stories of a destructive storm, "the worst dust storm since the depression days of the 1930s," that combined dust and snow to cause millions of dollars in property damage in Baca County.[8] Mrs. M. H. Young cited several examples of comparisons between the two periods, including the need to call prayer meetings for folks to request heavenly intervention into the weather as well as the presence of dust mounds that covered fence rows as in the 1930s.[9] The 1950s drought also hampered production, leaving the state's crop production in 1954 as its lowest since 1935.[10]

Some residents refused to accept a growing narrative that gained traction nationally that emphasized the need for farmers and ranchers to reconsider their chosen livelihoods in an area that succumbed to such devastation.

In the *Lamar Daily News*, Fred Betz openly chastised the author of a *Life* magazine article that drew connections between the Dust Bowl and the 1950s drought. Betz argued that the author dramatized the situation by only describing the dust storm that hit early one morning and not admitting that the rest of the day was "one of those salubrious January days" that locals enjoyed. Betz continued that editors of the *Denver Post* and the *Pueblo Chieftain* deserved condemnation for providing negative coverage and exaggerating.[11] A self-proclaimed "Old Timer" presented his case more bluntly, arguing that Prowers County remained the best place where he could live despite the weather. He admitted, "No, we don't like the dust," but "Does it faze us? No! Great Scott, we are no sissies." He claimed that a little dust would not be fatal and suggested that residents would get through these years just fine despite the challenges.[12] Baca residents announced their faith at the 1954 state fair as well, producing a booth with the title "We'll Beat the Drouth and Dust" and assuring Coloradans that the weather proved no match for their resolve.[13]

Regardless of this bravado, however, the 1950s drought represented a significant climatic event that impacted the nation in a profound way, leading to dust storms throughout much of the Southern Great Plains as well as high temperatures and water shortages. It marked "one of the more severe [droughts] on record, and it affected more than half the agricultural land west of the Mississippi River." It also warranted significant government involvement through relief, production reduction, and conservation subsidy programs. Such policies resulted in federal expenditures of nearly $400 million in direct aid and $250 million in loans to peoples living in drought areas throughout the Plains and Southwest.[14] The scale and scope of federal assistance during this period did not compare to what it had done in the 1930s, but significant government involvement in the agricultural economy began quickly after it became clear that the drought would stick and that people needed help.

Some of this attention came to Colorado, as the state and federal governments helped in the form of conservation subsidies, low-interest loans, and programs designed to better support farmers in the drought areas. Most of the focus centered on the southeastern corner as Baca and Prowers became two of the harder-hit counties in the state. Governor Dan Thornton approached the drought with a sense of immediacy and activism, working with state legislators and representatives from federal agencies to design programs to assure and bolster farmers. On a tour through the counties in June 1953, Thornton stressed the need for Colorado to extend drought aid and claimed that "this is no funeral" for farmers on the Colorado plains.[15] The state drought disaster committee also met in Lamar in 1953, showing their support for the area and focusing their attention firmly on the parts most impacted. The Colorado delegation left that meeting with a sense of urgency to help residents and pressured Governor Thornton to step up the state's efforts.[16] Baca quickly earned recognition for its troubles, becoming one of nine counties included as "disaster areas" in need of concerted state assistance.[17] In a demonstration of federal responsiveness to eastern Colorado, Secretary of Agriculture Ezra Benson toured the area in 1955 and emphasized the government's efforts to keep farmers afloat during lean years. Benson utilized a speech in Lamar to ensure his listeners that his department would not facilitate widespread land retirement or the removal of farmers from the area, and he would instead work with residents to stabilize the agricultural economy and keep farmers farming.[18]

While residents may have enjoyed these words of encouragement, and while the dramatic increase in federal and state spending that accompanied the rhetoric certainly helped, they knew that the only surefire way to get out of the drought was to have it rain. The major newspapers publicized weather forecasts and updates on their front pages, demonstrating both the conviction that every storm signaled the end of the dry years and the prominent role that weather had in residents' lives. The *Springfield Plainsman Herald* noted the connection between residents' emotional well-being and the weather, offering a short update entitled "Rain Brings Hope to Dust Bowl Counties, Southeast Colo." and suggesting that the rain produced "a new look in Baca county today."[19] The *Lamar Daily*

News celebrated a storm that fell on the "western tri-state area," calling it a "million dollar rain."[20] The Springfield paper likewise touted a "million dollar rain" that made "everybody happy," suggesting that "the drought is definitely broken, now the citizenry of the several states can go back to work with new hopes for the future, as the moisture came as a blessing on people that were oppressed by the thought of losing their farms and businesses."[21]

The optimistic prognostications about the end of drought showed both the residents' belief that the drought could break that easily as well as their understanding that rain would stabilize their financial outlook and their mental health. But they connote a sense of helplessness that never held sway among locals. Rather, inhabitants attempted to take matters into their own hands and did so in a variety of ways, some of which seemed unconventional. One of the more remarkable trends to emerge during the 1950s drought combined the long-held conviction that water could solve their problem with a more recent appreciation for technocratic solutions. Residents began to search for technological options for increasing rainfall, relying on the expansion of technology that accompanied American involvement in World War II and continued into the postwar period. This adoption of technological means to address agricultural problems became much more common after the war, as more companies and government agencies devoted resources to the research and development of new seeds, fertilizers, machinery, and other elements of "modern" farming.[22] In that way, the push by residents of the Great Plains to pursue technological solutions to aridity—in addition to engaging these other methods—reflects the zeitgeist of the era.

In addition, the rise of such precipitation-increasing theories, including the popular "rain follows the plow" and adoption of "pluviculture" more broadly, predated the 1950s drought, but the key difference between these earlier periods and the Filthy Fifties is the manipulation of new technology to modify the weather.[23] This difference emerged as early as the late spring of 1950 and at a point when few believed that drought would last at all, let alone for nearly eight years, when residents looked for ways to increase rainfall by utilizing chemical compounds and airplanes. A group of farmers convened in April and combined their money to employ Reed Crandle of Holly to fly into the clouds to drop "dry ice and other chemicals" to create rain. Proponents believed that clouds "laden with moisture" too often move east beyond Colorado before they condense and deposit rain on the arid plains, leaving Kansans to benefit and Coloradans to suffer. As the *Springfield Plainsman Herald* noted that the investment might prove worthwhile since "If such a thing could be worked out, and if it could do just partially what is expected, the benefits would far off-set the time and expenditure involved in the experiment."[24]

This initial trial afforded some traction for rainmakers and their services, as many observers came to believe that such weather modification might bear fruit. The pursuit gained a sense of credibility such that a variety of organizations and associations proved willing to test the theory by financing rainmakers' endeavors. The Lamar Chamber of Commerce agreed to meet with rainmakers in the fall of 1950 and listen to their pitches, eventually expressing some doubt about their impact but willing to finance a project to increase rain.[25] A private association of farmers cooperated to execute a fund-raising campaign to raise the $45,000 it needed to hire rainmakers. The group looked to contract a rainmaker for the calendar year so that they could capitalize on cloud formations during all seasons, thereby increasing both rain and snow output.[26] Another group of at least 100 people, primarily farmers from Baca and Prowers, organized the Southeast Colorado Weather Research Association to raise additional money and employ rainmakers through the winter. They aimed to raise $35,000 to hire the Weather Research Association of Pasadena, California, to devise a good plan for augmenting precipitation and then to facilitate it.[27] The shift from proponents employing a local pilot to a California firm suggests that residents felt increasingly confident in the prospect of modification. It also points to the development of the industry, since companies like the Pasadena-based one made themselves available to plains farmers and had the capacity to fulfill a contract.

Unfortunately for both the farmers and the rainmakers, this initial enthusiasm for the prospect of weather modification became increasingly hard to sustain when the rainmakers proved unable to consistently produce results. For farmers, the return on investment for rainmaking efforts sometimes justified the cost—or at least they believed that it remained worthwhile—especially during the early years of the drought. When rain fell in September 1950, the Pasadena-based firm claimed responsibility and suggested that their efforts saved 45,000 acres from further devastation. Locals promptly agreed to raise money for his services throughout that fall and into the winter.[28] Gerald E. Brown penned an article for the *Springfield Plainsman Herald* that celebrated the rainmaker, arguing that "there is little doubt that precipitation can and will be increased anywhere it is desired." Brown contended that this would change the area from "a land of 'almost rain'" to a land of rain.[29] Brown's support might have resulted from his position as resident rainmaker, however, since he purchased a used rainmaking generator that season and concocted his own mixture of chemicals to make rain. His efforts were in vain, but his support for the rainmaking prospects continued.[30]

Resident support for rainmaking fluctuated according to the perceived results and the depth of the drought. On the one hand, the science of weather modification gained more credibility with each passing year. In 1954 the federal government funded a Weather Modification Committee that it tasked with reviewing the science, evaluating the impact of various techniques, and researching solutions to problems associated with uncontrollable weather. Fred Betz believed that this was a necessary step to manage the shift in climatic conditions that resulted from a general warming of the Earth; this surprising interpretation of global warming demonstrated the perception that science could provide a solution to such problems.[31] On the other hand, though, residents less often invoked the cause as much as the result of having little rainfall, and they pushed for rainmaking when the drought seemed to reach its nadir. Proponents celebrated the increased funding that came with federal recognition of the problem and support for additional research, but the newspapers only really included information on rainmaking in brief bursts of attention and generally when they believed drought had just begun or worsened considerably. For example, no mention of rainmaking in either the *Lamar Daily News* or the *Springfield Plainsman Herald* emerged between 1954 and 1964, even though the 1950s drought lasted through 1958 and another drought started in 1960. Headlines from 1964 again referred to weather modification as a sort of saving grace, promoting the decision to hire a Denver-based "cloud master" to make rain that resulted from a more general, renewed interest in stabilizing precipitation levels through scientific intervention. No mention of such options emerged during the first three years of drought.[32]

This rather inconsistent support for rainmaking shows a level of skepticism among residents, but it might have also been due to the high cost of such experimental efforts to modify weather. Residents became much less ambiguous about the programs when the federal government offered to fund research and development, but when they had to pay for the service, they showed a reticence to commit to the practice. This lack of commitment likely reflected the fact that the agricultural economy declined during the 1950s. Farmers in Baca County, for example, tried to raise money to fund projects in cloud seeding but failed to draw the necessary $8,000 to start it. In a rather gentle admission that the economy had declined, the *Springfield Plainsman Herald* noted that "money seemed very scarce" to explain the decision not to start the project early in 1952.[33]

In a more telling example of both their ambiguity and their financial challenges, farmers contracted with the Weather Research Association for their services, but then failed to pay them on time or in full. The *Springfield Plainsman Herald* completely supported the rainmaking service, suggesting farmers owed rainmakers for the "fairy tale" turnaround in their weather fortunes.[34] The editor expressed his gratitude after a healthy 6–8 inch snowfall in January 1951: "All of this sounds like 'hooey' to some people of Baca county, but if it had not been for a large number of desperate farmers who were willing to try something in order to get moisture, and spend a large sum of money in getting the rainmakers to seed the clouds, Baca county would have had a very, very dry season last year."[35] And yet it took months to raise the money to satisfy the contract, and there is no further mention in this newspaper of this company or its services, or any other company like it, for the duration of the 1950s drought. The sense of urgency emerged during tough years but then dissipated when it became clear that private funding alone had to procure the weather modification services. It is also the case that even when they find their access to water more stable and the amount of water they can access increasing, farmers never seemed wholly secure with the water levels as they stood. The effort to combat aridity by manipulating the weather—quite literally by looking to the skies to solve their problems—represents an important continuity in how farmers approached their environments.

As some farmers looked for answers up in the clouds, others looked down under the soil, finding a potential solution to their problems in the form of the Ogallala Aquifer, an incredibly large and seemingly inexhaustible underground reservoir that stretched across much of the Great Plains. This reliance on groundwater is certainly "an enduring legacy of the Dust Bowl," though efforts to try and tap into the aquifer began in the 1930s.[36]

The drive to develop groundwater sources became especially pronounced in Baca County, in large part because access to the aquifer constituted the first time that farmers had an opportunity to address aridity somehow other than waiting for the rain to come. The momentum to search for and exploit groundwater gained steam in 1953 and 1954, with the *Plainsman Herald* first noting that "Irrigation Wells Become Numerous in Baca County" on October 22, 1953. Another article in the same paper indicated that a huge windfall for hopeful irrigators might be on the horizon when it relayed the story of a man outside of nearby Las Animas. The Bent County man had been randomly digging on his property when he tapped into a "'tremendous' underground river of water" that he believed ran under the whole county and much of the region.[37] A further indication of growing support for well drilling and irrigation from the aquifer comes in the form of advertising, specifically the ads for Silas E. Camp that ran regularly in the paper during the mid-1950s. The ad promoted Camp for his abilities in drilling test holes to assess aquifer depth and the likelihood of tapping into it on one's property. Camp celebrated his nine years of experience, his overall expertise, and his recent investment in new equipment, all of which made him the best option in the area for farmers looking to capitalize on the underground source.[38] Another advertisement, one for Dreiling Drilling that appeared in the *Holly Chieftain*, offered well-drilling expertise and reminded the reader that "Water Is Great Stuff and We Can Get It for You."[39]

Baca residents witnessed the slow but gradual shift in support among agencies, both federal and state, that started to take notice of these irrigation possibilities as well. These entities, most of which had already turned some attention to the region because of the continued drought, began to share farmers' confidence in pump irrigation to help the region become more drought-resistant and productive. Governor Dan Thornton returned to the region in the spring with the intention of looking for ways in which farmers might capitalize on underground sources. Thornton promised to help farmers by providing government assistance to research potential sources and do the necessary exploration to help farmers tap into them.[40] Not coincidentally, Thornton had expressed serious concerns over the quickly deteriorating situation in the region because of the drought and increasing severity of accompanying dust storms, so his call for irrigation in the county reflected one element of his broader agenda to address the crisis.[41] Whether he compelled federal action or it developed of its own accord is difficult to know for sure, but shortly after his visit to the region, the federal government published its intentions of at least considering how to utilize loans to farmers who had an interest in and the opportunity to develop pump irrigation. The *Springfield Plainsman Herald* printed an article about the potential for a Soil Conservation Service or even a Federal Housing Administration loan to farmers who might need assistance in attaining the capital to invest in irrigation projects on their property.

It is not surprising that this combination of public and government emphasis on pump irrigation was most profound in areas like Baca. In some ways, though, it is a shock to note that the county agents took a bit longer to come around to the idea. Certainly, farmers and county agents alike came to understand that the aquifer was a game-changer; it could certainly help buttress farmers against the impact of cyclical droughts, but it also marked a chance for them to evaluate how and what they grew for the market. Many farmers adopted the view that the potential for tapping and utilizing the water represented the best way to combat the problems of the Filthy Fifties, but the first explicit mention of county agent support for farmers considering pump irrigation only comes in 1956. By then they seemed completely on board, however, as agents undertook extensive studies of what water Baca farmers might use, how much they would need under most circumstances, and how the aquifer might address their demand. Arwin Bolin, agent in Baca throughout much of 1956, organized a subcommittee of county farmers that became responsible for compiling information and needs and access that he then used as part of a paper he wrote, "Water Resources in Baca County."[42] By the next year, Bolin had become more active in trying to assist farmers looking to irrigate and sponsored a "pump irrigation clinic" for interested residents.[43]

His efforts to implement sound irrigation practices came when the public embrace of pump irrigation and the widespread use of the systems to extract aquifer water began to spread significantly. In a general way, the use of groundwater by Coloradans came under scrutiny in 1957 with the passage of the Colorado Ground Water Law that required a permit from the state engineer to drill new wells and that mandated well owners register existing wells.[44] The passage of that law suggests a statewide shift in attention toward groundwater that accompanied increased use, a shift that some farmers embraced but others tried to limit. Groups of farmers had been fighting for a law like this one that might control the development of groundwater irrigation in hopes of protecting the water table. Some restriction advocates noted that the drop in water might have been a consequence of weather—that they offered this acknowledgment seems unnecessary given the fact that drought had just ravaged most of the state for the preceding several years—but they also blamed farmers who had abused the system. The advocates most likely never mentioned limiting their own use, as there is no sense of their role in reducing the table in this story.[45] The criticism certainly indicates a level of self-interest and the advocates' desire to protect their own access to water, but the fact that legislation and the broader push to control use emerged during the dry 1950s and when drilling became much more widely pursued further reinforces the notion that there can never be enough water for everyone who wants it and who demands access.

Most farmers in an arid region would relish the chance to access supplemental water, and for those who had never been able to tap into that resource, the prospect of drilling proved incredibly enticing. This seems particularly true in southeastern Colorado, where farmers demonstrated a remarkable level of enthusiasm behind exploring the aquifer that might suggest few places matched their level of optimism about these new possibilities. To be sure, Prowers farmers took every opportunity to use groundwater. The number of pumps used for irrigation tripled between 1947 and 1953, and the average depth of those wells came in at 35 feet, which meant that folks could reach water cheaply and quickly.[46] Similar growth happened in Baca, though it started a bit later. Every annual report done by Baca agents during the period from 1957 to 1963 noted the expansion of irrigation, often citing the total number of wells in use and the acreage covered under pump irrigation. This expansion occurred at a time when the population dropped considerably and the average size of farms increased dramatically, a trend that suggests the possible correlation between the growth of individual acreages and the ability to tap into water sources.[47] By 1965 Baca farmers were operating 425 wells, meaning a high proportion of farms had some sort of well given that census data shows that only 773 farms existed in the county in 1959.

The continued digging of wells only slowed in 1965, according to agent Thomas Doherty, because a new round of legislation had been passed in the state that limited the expansion of groundwater irrigation and did so after farmers had faced another round of drought in the early 1960s. The 1965 Ground Water Management Act stiffened the regulations in place to control the construction of new wells. It also authorized the creation of "local groundwater management districts" designed to monitor and restrict the digging of wells in areas that already faced shortages. Perhaps most important, it imposed the doctrine of prior appropriation as the basis for "economic development through the maintenance of reasonable pumping levels."[48] In other words, the state realized that the potential for over appropriation of groundwater existed and became increasingly likely given the explosion in the drilling of wells, so it utilized its authority to control groundwater like it had flowing sources. The timing also corresponded to the weather, though, as the dryness of 1960–63 influenced the number of people who wanted to drill more and more deeply. For example, as of 1963, the Baca agent noted, "There was considerable well drilling activity in the spring and early summer due to the droughty conditions."[49] The benefits of looking to pump irrigation to counter the aridity became clear in Prowers during these dry years, too, as agent Fred Fitzsimmons noted that when stream flow from the Arkansas proved limited, the supplemental pump irrigation helped those who had access to the river.[50]

Pump irrigation allowed farmers the chance to experiment with new cash crops. Farmers in Baca had an opportunity to dramatically expand their crop base because of irrigation. As Arwin Bolin noted in 1957, Baca was making "somewhat of a come-back" due to improved weather but also new crops—sugar beets, onions, watermelons, wheat, alfalfa, and potatoes—that became available to farmers once they had water.[51] Area farmers seemed to understand these new opportunities and showed immediate interest in producing these crops. Onion farming started in earnest in 1956 with the harvest of 30 acres, due to the help provided by irrigation water. More important, the turn to sugar beets occurred quickly. Farmers met in December 1956 to discuss with beet company representatives the prospect for contractual beet growing.[52] By the next January, 75 farmers attended a meeting to sign up for 1954 acres of beet production for 1957, and the *Springfield Plainsman Herald* noted the connection between irrigation and the new crop by suggesting that "the beet raising possibility has created an interest in additional irrigation wells, and well drillers are busy in the Walsh area at this time."[53] The Holly Sugar Company and American Crystal encouraged Baca farmers to enroll and also hoped to attract "experienced beet farmers" who might rent land around Walsh to help production.[54] The companies' attention paid off that fall, when Baca farmers harvested 2,100 acres of beets for Holly Sugar; by that point the infrastructure had developed to deal with the harvest as the Santa Fe Railroad provided receiving stations in the county to pick up beets and deliver them to Holly.[55]

The development of a beet industry allowed Baca farmers to experience a style of cash crop farming that had become familiar to Prowers farmers, and the formerly dryland agriculturalists seemed to enjoy it. Prowers farmers had always been aware of how water buttressed their crops against aridity and how much the irrigated crops brought in for their county. Even in 1958, generations after beet farming had ascended in the county, the *Lamar Daily News* noted that beet production, including the workers, contract farmers, and the companies, had an undeniably positive impact on the region. The financial benefits alone fueled a variety of regional developments and made Lamar a desirable place to live. When the paper spent all of 1960 celebrating the fact that Lamar had been voted an "All American City," it often referenced how sugar helped make that possible.[56] With the development of irrigated crops in Baca, these same kinds of sentiments emerged that celebrated the benefits that irrigation provided. A story in 1962 claimed that irrigated wheat would contribute $2 million to the local economy, money that was especially important as the dryland areas struggled with drought.[57]

The new irrigation systems affected farmers' psyches as well, giving cause for celebration, pride, and a sense of independence. Baca celebrated its first "sugar per acre" contest in 1961, a competition that drew about 10 percent of the sugar beet growers and named Tony Ochoa as the Baca champion grower for the year. This effort paled in comparison with some other festivals, including the broomcorn celebration in Baca and melon and beet festivals in towns along the Arkansas River. But it proved that farmers had started a new period of production.[58] That new period promised other changes, too, including the start of a new relation between local farmers and the federal government. While some farmers embraced federal intervention, guidance, and assistance—and most did so when it meant significant financial investment in local farming—others chafed under the renewed federal management of agriculture that had begun during the 1930s, sagged during the war years, and returned in the 1950s. This level of involvement produced a level of dependency that left some lamenting their loss of freedom to choose what to grow, when, and where. The prospect of irrigation symbolized a shift in that trend, though, as the "installation of wells" marked a movement "close to the top of the list of progressive movements" that assured a breaking of the dependence on "Uncle Sam" that most dryland farmers had come to expect. Water helped diversify the economy, developed local and county tax bases, and benefited all residents, not just the irrigators, which helped everyone become more independent from government control.[59]

This rather awkward dance between independence from and reliance on the federal government had become part of agricultural life in the arid West, as most irrigators developed some relationship with the state in the process of attaining water rights in the first place. That relationship becomes even trickier during times of crisis and especially when farmers come to believe that there is not enough water to go around. The thing about water for farmers in arid regions, of course, is that demand on the resource will never balance with supply, and the system in the West has largely accepted the fact that over appropriation of rights is a common occurrence when no one can reasonably predict water levels from year to year.[60] If everyone drawing from a supply understands that the demand will not be met, then each stakeholder will try to either protect his/her claim or find more water. This is demonstrated by both the continued search for additional access, the willingness of some irrigating Coloradans to cut off the supplies to others, and the push by Colorado farmers to ensure that the water from their rivers stayed in the state instead of ending up in the fields farmed by Kansans.

Despite postwar efforts to provide more water through deep wells or by encouraging precipitations, however, many farmers realized that a "natural deficiency in precipitation" would continue to challenge them. That sense of vulnerability did not emerge only in the 1950s, of course, but severe drought tends to expose those sorts of challenges in ways that wet years tend to conceal. Drought also motivates people in unusual ways, adding a sense of desperation during periods already rife with uncertainty. The 1950s drought years forced this kind of distress in many farmers and convinced irrigators to more fully appreciate their dependency on the Arkansas River for their well-being. That combination of vulnerability and reliance produced enough momentum to push the Fryingpan-Arkansas Project through Congress and gain federal support for the distribution system that would bolster their access to water. It was "imperative," promoters argued, that the project address the "critical shortage of water . . . that menaces the economic stability" of the entire valley because the project represented "the only hope for new water."[61]

While project promoters later contended that valley residents had "dreamed of a major transmountain diversion of water from the Western Slope to augment existing supplies" since the 1910s and 1920s, the initial grassroots efforts to garner federal support for such a project began in earnest during the 1930s.[62] That such a push started during the 1930s drought and under the New Deal administration should not be a surprise, of course, but it is somewhat curious to note that the first emphasis on this project coincided with the struggle to obtain federal support for the John Martin Dam. It seems likely that advocates of the two projects believed that neither would address the urgent need by itself; it might have also been the case that proponents realized the unprecedented opportunity to garner federal funding that existed under FDR and sought to strike while the iron was hot. The drought, dust storms, and overwhelming sense of crisis certainly added further conviction that it was the right time to push for support.[63]

In his explanation of why the Fry-Ark project marked a necessary step in the nation's development of its resources, James Ogilvie focused on the crisis of the 1930s. Ogilvie, who had a long career in the fields of irrigation and water management, worked with the Bureau of Reclamation on several projects before becoming the project manager for Fry-Ark once it gained congressional approval. He had done a lion's share of promotional work in getting support for the project. In his speech to advertise the project's benefits, Ogilvie demonstrated how important the lessons of the 1930s were in helping people within and outside of Colorado understand the challenges of living and farming on the Colorado plains during periods of drought. He argued, "Everyone will recall the shock that came to the people of America in 1934. In that year, a vast transcontinental dust laden windstorm, darkening the sun, broadcast the fact that large, once fertile portions of five western states, Kansas, Texas, Oklahoma, Colorado, and New Mexico, had become a desolate dustbowl. The great majority of us have forgotten that catastrophe, but tens of thousands of people have not." The major takeaway is the need to adapt humans' approach to the earth, as "man can no longer plunder the earth, but must work with nature, rather than in conflict. The turning point of recovery and reclamation has been reached in the Arkansas River valley." Ogilvie also connected the 1960s movement to transfer water to the 1930s by reminding his listeners that the Bureau of Reclamation had already proven up to task of moving water through the mountains. The successful Colorado–Big Thompson Project indicated that the project could work and provided a model for supporters to reference when they advocated for a similar effort with the

Arkansas. Supporters of the Arkansas valley plan seemed to understand that the federal government would likely only support one such project at a time, however, so when Congress announced its support for the Big Thompson, it took some of the momentum out of the push for an Arkansas River project.[64]

At the same time, however, Ogilvie and other promoters attempted to sell the project based on a reading of present problems and eventual challenges to residents throughout the Arkansas Valley. They did so in part by casting this as a multiuse project that would benefit all inhabitants, especially in terms of promising an economic impact that would help everyone. Ogilvie emphasized the role that agriculture could play in providing economic prosperity to all Coloradans. The Southeastern Colorado Water Conservancy District published a pamphlet in support of the project that celebrated ways the project would improve the local economy. By providing water it would stabilize farms, which, in turn, would strengthen the agricultural economy. The farmers would then enjoy more success, allowing them to buy from local retailers because of their increased purchasing power. That spending would invite new industries to relocate into the region, which would provide additional jobs and wages for current workers. These changes would then produce an increase in land values. It would also provide many benefits for the state and the nation. Higher property taxes and income taxes would contribute to higher state and federal revenue, which would allow those governments to function more efficiently. The project would also provide for national defense through its capture of ample water power and ability to provide laborers willing to work in defense industries that might capitalize on hydroelectricity.[65]

The proposal also identified other benefits that might assuage concerns that only really developed in the postwar period. While the history of the project suggests that it began in earnest during the 1930s and 1940s, the emphasis within the project for the ways that the project would benefit an increasingly urban and industrializing population indicates a change in focus. In selling it as a multiuse project, the proposal highlighted aspects that reflect the long-held goals the Bureau of Reclamation had for these kinds of diversion and irrigation projects, including irrigation, flood control, and even power. Yet the inclusion of municipal water, new recreational opportunities, and the preservation of fish and wildlife represent a shift away from purely Progressive Era concerns.

This longer list shows that promoters had developed an understanding of the political landscape by the 1950s that perhaps they had lacked in the 1930s and 1940s. Indeed, the Fry-Ark proposal that politicians debated in Washington ruffled the feathers of many folks in Colorado and throughout the nation who believed that the project would do more harm than good. For example, the notion that the project might have unanticipated consequences of stream flows, wildlife, fish, and even the quality of outdoor recreation both in the mountains at the tip of the project as well as along the Front Range compelled many organizations to weigh in. While some groups like the Sierra Club refused to get involved, others jumped in when they believed the project represented just another reclamation project designed to commodify western resources. The Wilderness Society recruited the Sierra Club to oppose the project, and the Isaak Walton League of America similarly voiced its concern over the need to reconsider the project as early as 1953. The environmental movement was rather immature and without the groundswell of support it would eventually enjoy in the later 1950s and 1960s. So this marked an important interest group in questioning the project but one that lacked much credibility or power to fundamentally challenge it. The Sierra Club had been leading the charge against the Echo Park Dam and decided not to split its attention, while the Isaak Walton League eventually felt comfortable that the project did not pose an immediate threat to recreational opportunities and dropped its opposition.[66] The ability for project promoters to adapt and in doing so to meet the demands and answer the questions posed by stakeholders helped it eventually win support in many circles.

That adaptation is evident in multiple shifts that advocates made in the lead-up to the 1961 congressional review that eventually led to its passage.

Starting in 1953, the bill earned review in 1954, 1955, 1956, 1958, and again in 1960 only to gain support by the Senate and refusal by the House. This demonstrated that the proponents needed to address the conflicts engendered by the bill not only in the state but across the country. As they developed this more comprehensive understanding of the postwar political landscape, they adopted a more complicated approach to earning support. For example, organizers started a program for people who lived in the state but had business interests in other states to encourage them to write their representatives in that state in search of support. They also identified more influential people to send to Washington as lobbyists to further their cause and establish a broader constituency. Finally, they tallied the votes levied in previous reviews of the bill so that they could track its supporters and detractors more clearly and work with those who might be persuaded.[67]

The promoters also found themselves acting in a more conciliatory manner toward opponents to ensure their eventual support for the bill. For instance, in the face of criticism over how the structures might interrupt fish and wildlife, proponents addressed requests by members of the Department of Fish and Wildlife to protect them. Frank Milenski, who had been attached to the project in some form since the first Gunnison Arkansas Project meeting organized at the Elks Club in Rocky Ford during 1951, recalled that a variety of issues arose that troubled Fish and Wildlife, but that in every case the proponents of Fry-Ark assuaged their concerns. Milenski, who attended the congressional hearings in Washington, lamented that "you have to pay a price to a lot of interests" to get things passed. To compromise with Fish and Wildlife demands, Milenski indicated that the project would provide elevated crossings over power canals in order to protect big game crossing the water, it would consider funding a fish hatchery on the eastern slope, and it would help promote lake fishing instead of stream fishing in order to meet demands across the region.[68] The combination of rhetoric and concessions apparently did enough to convince Secretary of the Interior Stewart Udall that the project would protect these resources. Udall glowingly suggested that "fish and wildlife resources will be conserved and developed, and opportunities for outdoor recreation will be created and stimulated." The conscious use of such action verbs indicates his belief that the project would do more than merely protect access and would instead allow outdoor recreationists to enjoy new opportunities to engage the natural world.[69]

An exchange between Colorado representative Wayne Aspinall and Secretary of Agriculture Orville Freeman demonstrates another concession made by proponents, namely, the decision to allocate additional water to farmers with established water rights and not to promote the expansion of new farms in the region. Supporters of John Martin had employed the same rhetoric to assuage concerns in the 1930s. Referencing an address by President Kennedy, Secretary Freeman suggested that the USDA had been instructed to use tremendous discretion in authorizing reclamation projects. Freeman cited the president's concern as evidence of a broader sense of urgency to try and limit the expansion of agriculture and instead encourage a turn toward the use of fewer lands for production. Aspinall deflected those concerns by assuring Freeman that the Fry-Ark would serve as a "rescue operation for the existing irrigation lands." He supported the project for attempting to make available a water supply that would support the private agriculture then being undertaken in the valley and, in the process, make the distribution of water more efficient. Aspinall also sold it as a way to give farmers more flexibility in crop options, which would allow them to "move away from those crops which have contributed to the surplus problem."[70]

Secretary Freeman responded by throwing the weight of the USDA behind the project, giving it another prominent ally in government. Freeman introduced his letter to Aspinall by emphasizing that he understood how the Fry-Ark proposal represented a concerted move toward efficiency, which was a necessary step for farmers who had to develop a more effective strategy to deal with years of "short water supply." The USDA hoped "to provide appropriate and needed services for improvement of the family farm pattern and of farm and rural living." While Freeman worried about the potential imbalance of supply and demand that might result from increasing production, he committed his department to the idea "that reclamation and irrigation have a highly necessary role to play in the wise present and future use of national land and water resources for the economic growth of the Nation."[71]

It is not at all surprising that a project like this one invited criticism given its scope, cost, and the belief that many detractors held that this represented just another way to steal water from folks who had a right to use it. Indeed, even when the project eventually won congressional approval it did so in large part because proponents agreed to a voice vote rather than a roll call vote. A roll call vote would have required each member to go on record in a show of support for or in opposition to the legislation, and according to Representative J. Edgar Chenoweth, whom some considered the "Father of the Fryingpan-Arkansas Project," many congressmen friendly to the project were nonetheless unwilling to publicly support it in fear of retaliation from their constituents back home. Somewhat expectedly, members of Congress from other states voiced their concerns that depleting the water from the Colorado River basin would adversely affect the total output of that river, which would then impact users in their home states. The staunchest objections came from Californians, who believed that folks in Colorado were effectively abusing their rights to Colorado River water to the detriment of California's irrigators and cities that relied on the waters. Chenoweth recalled Congressman Craig Hosmer, Republican from California, levying attacks against the project that included a variety of "ridiculous things" that had nothing to do with the actual project. For example, Chenoweth remembered when Hosmer commented on the floor of the House that "Colorado is here, they want to take water so they can raise bananas on the top of Pikes Peak."[72] In other cases, like that of Representative John P. Saylor of Pennsylvania, some politicians simply did not like the project and took personal interest in fighting it. Chenoweth, who considered Saylor "a very fine gentleman and a very good friend," failed to explain why Saylor "took great delight in needling the witnesses who would come back on the Fryingpan Project."

What is surprising, however, is that one of the most vocal champions of the project who played a role in getting it adopted, Wayne Aspinall, had once been a prominent detractor. Aspinall, a member of the House from Colorado's Fourth District that sat primarily on the West Slope, served as chairman of the House of Representatives Subcommittee on Irrigation and the Reclamation Committee on Interior and Insular Affairs. Frank Milenski remembered clearly the scene in the House when Aspinall and the committee reviewed the proposal in 1961. Milenski noted that Aspinall sat tall in his chair facing a "fine oil painting" of himself that hung on the far wall. He conjectured that his constituents had paid for the portrait, which Milenski took as a show of Aspinall's power and position. Aspinall, Milenski recalled, "didn't lack for starch."[73] Harold Christy, longtime proponent of the project and supporter of making more water available to the Arkansas valley, served as a witness in front of Aspinall's committee in 1953. He did not enjoy himself, recalling that "it wasn't a pleasant occasion" in large part because Aspinall peppered him with questions about the project. Christy stated that Aspinall "was not too friendly to the project" and spent so much time trying to get Christy to expose the push to exploit the Colorado for the benefit of the eastern slope and detriment of the western slope.[74]

Aspinall's principal reason for objecting to the project shows his attention to his constituents, many of whom felt that federal attention to the Fry-Ark would negate federal support for other projects in Colorado. Aspinall worried most about the proposal for a Colorado River Storage Project, which would provide five reclamation projects and three hydroelectric dams for the western slope. He seemed to think that, somewhat akin to how supporters of the Fry-Ark became concerned that federal support for the John Martin project would negate the chance they had to gain federal support for two projects, federal promotion of the Fry-Ark would supersede federal attention to his project. That might explain in part why he remained very sensitive to the Fry-Ark project's cost and often criticized the plan for abuse of funding.[75] Two developments seem to have changed his view, however, and left him not only supporting the plan but currying favor among his peers to lend their support as well. First, Congress passed legislation to initiate the Colorado River Storage Project in 1956, giving Aspinall a significant victory for his constituents. Second, interests on the eastern and western slopes agreed to establish Ruedi Reservoir and use it instead of the smaller Aspen Reservoir to hold water from the Colorado. The larger Ruedi could hold more water, which would better defend western slope residents' water rights and keep more water in Colorado rather than letting it work its way out of the state.[76] With his concerns addressed, Aspinall became a vocal supporter of the project and steered it through his committee and through Congress.

With the entire Colorado contingent of congressmen on board, vocal enthusiasm from the governor, and an incredibly long and diverse list of organizations lending support to the project, Congress finally agreed to it and put into motion. Frank Milenski recalled that the decision led to much excitement among both the politicians and supporters in the Arkansas valley, especially those in the Southeastern Colorado Water Conservation District (SCWCD). He remembered that Aspinall asked the District to throw a cocktail party to celebrate and then suggested that it pay for the whole endeavor. Milenski noted that it "wasn't too much," though he seemed upset that the SCWCD had to fund a victory party that "was more to help the Congressmen's egos" than anything else.[77]

The SCWCD also paid the tab when President John Kennedy inaugurated the Fry-Ark project upon his visit to Pueblo in August 1962. Milenski identified the District's responsibility for the visit to amount to $832.92, but he certainly understood that the president's symbolic visit to the region, his vocal support for the project, and the kind of credibility that support afforded the project warranted the expense. In his speech, President Kennedy celebrated the dam for the way that it represented the potential greatness that a postwar America still sought to achieve. Kennedy remarked that anyone who flew over some of the "bleakest" lands in the United States only to find themselves alongside a flowing river like the Arkansas could fully appreciate the value of the dam. He hoped that his contemporaries might have the forethought to develop programs like this and emphasized how its most strident proponents had been pushing to gain support for more than thirty years. He suggested that the dam represented a foresight that would become more necessary with increasing population and more demands on the nation's natural resources that he believed would come by the turn of the century. This project, he thought, also represented a "rising tide" that would benefit all Americans; the program required a local, state, and national effort that marked an "investment in the growth of the West" and showed the way for other states and regions to develop resources. Even as he intoned the need to develop the nation's resources by harkening back to the conservationist mantra of protecting resources to ensure future generations' prosperity, however, Kennedy capitalized on the occasion to pressure Congress to pass stringent clean air and clean water laws, expand existing and create new national parks, and finance extensive conservation efforts to plan for future protection. At the close of Kennedy's speech, a representative for the grateful citizens of Colorado presented the president with a frying pan to symbolize their appreciation and to demonstrate his support for the project.[78]

It is highly doubtful, of course, that the president ever realized that he had something in common with Otto Lubbers, but Kennedy's visit to the region marked the culmination of a long push by locals to bring more water to the valley, to which Lubbers had contributed. The president's comments aligned well with how many supporters had viewed the project: it was a legitimate use of federal assistance to promote a conservation agenda designed to benefit Colorado, the West, and the nation. It supported urbanization along the Front Range, outdoor recreation in the mountains, agriculture in the Valley, and the entire state's economy. Few Coloradans voiced much concern over the transmountain endeavor once Aspinall came on board with the proposal.

That the project came to fruition when it did is significant for several reasons. It represented one effort in a much larger struggle to secure water for agricultural production throughout the Arkansas valley. The

FIG. 23. President John F. Kennedy at the dedication of the Fryingpan-Arkansas Project, Colorado, August 17, 1962. National Archives and Record Administration, Rocky Mountain Region, Denver CO, Record Group 115, Bureau of Reclamation.

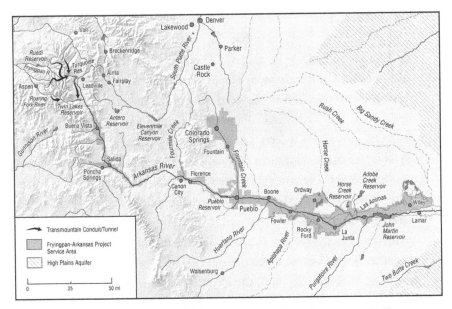

FIG. 24. The Fryingpan-Arkansas Project represents the culmination of efforts by both urban and rural residents to draw water from the West Slope to meet increasing demand in southeastern Colorado. Map created by Erin Greb.

impulse to draw water from wells, to capitalize on federal largesse, and to improve water conservation techniques to maximize returns all exemplify the continued search for more water. In addition, the project marked the developing ideas of postwar urbanization along the Front Range, including industrialization in those communities, and outdoor tourism in the Colorado Rockies.[79] These stakeholders came at the notion of a transmountain diversion project from various perspectives but each found something worthwhile in the project, allowing farmers to get what they wanted in the end.

At the same time, however, agriculturalists wanted the same thing in the 1960s that they had wanted in the 1930s, which was a safeguard against drought and dust storms. That sense of urgency only became more pronounced after the Filthy Fifties settled in, leaving farmers feeling once again exposed and apprehensive about their future production. That the 1950s drought had less of an impact than the 1930s version is partly a testament to the legacies of the Dust Bowl in that fewer people faced the 1950s drought and those who did were more prepared to call upon the techniques and strategies that they had employed during the earlier crisis. They certainly understood that water would not make them immune to drought, but their willingness to rely on federal help, to search for water in unusual places, and to lobby for the development of water resources helped them in the 1950s as it had before. The water they receive will never be enough, of course, in part because the water is so heavily allocated that holders of water rights rarely see their full shares. Beyond that, though, farmers need to realize that water has never been the kind of cure-all that they hoped it would become. Beginning with area boosters in the 1880s and continuing into the present, folks hold on to the idea that water represented a foolproof insurance policy against the ravages of drought. And while it has certainly helped ease the burden of folks who have access, the ideal security for which many have spent their lives searching has never fully materialized.

8

Back to Work

A story in the *Springfield Plainsman Herald* from November 1945 stands apart from most newspaper coverage of the end of World War II. While the paper celebrated the conclusion of the war, it also included a blurb that presented the view held by some Baca farmers who actually wanted the war to "continue"—not out of a malicious sadism but because they knew that the war had been incredibly good for business and they hoped for continued prosperity.[1] These farmers were not alone in identifying the connection between mobilization and prosperity. The front page of the *Lamar Daily News* celebrated Japan's surrender by printing "VICTORY," demonstrating the kind of patriotism and pride that seemed ubiquitous in the nation at the conclusion of World War II.[2] A day later, the paper's editor offered a sobering reminder that the nation now faced the challenge of reverting to a peacetime economy and that such a transition demanded residents' full attention. He claimed that farmers must have confidence in the country's leaders to ensure continued, balanced prosperity to thus avoid "The Problems of Peace" that had plagued the nation coming out of World War I.[3] Instead of rushing headlong into the postwar economy, farmers should be smart and disciplined about their choices, which was a sentiment common to many in the region.[4]

Despite these calls for tempered excitement, however, the newspapers exhibited no memory of the hard times of the 1930s and no real concern for falling into old habits, demonstrating a kind of selective amnesia that had come to typify how residents often thought about the past. As they had before, residents seemed to correlate weather and economics, suggesting that "Peace and a General Rain" promised prosperity going into the postwar period. Indeed, headlines in 1945 and 1946 demonstrated a sort of blind faith in future gains that resembled efforts by nineteenth-century boosters to encourage migration to the region. Just as they extolled the beauty of the land and the productivity of its soil, the 1940s version looked to "sell Baca county to the outside world" through an attempt to "show its excellent advantages by giving figures and facts, and to prove their contention backs them up with pictures."[5] A few weeks later, a similarly enthusiastic article claimed that 1945 had marked the most prosperous year in the county's history and that they hoped 1946 could be better still.[6]

Yet, farmers approached these new opportunities through a lens colored by their experience in the 1930s, whether that meant searching for water or producing in a more sustainable way. They leaned on the history when thinking about the present. By the middle of 1945, for example, farmers started to lobby for continued federal support in providing the two things that reflected both the good years and the bad: soil conservation assistance to help correct destructive practices and migrant laborers who could help them reap the benefits of the wet years and high demand. Ostensibly, these two issues seem unrelated, and perhaps antithetical, in that many observers believed that the use of temporary workers exacerbated problems of soil erosion. They suggest two important legacies engendered by the challenges of the Dust Bowl years which both derived from the fact that drought so greatly affected the agricultural economy that old practices no longer worked and old workers were no longer available.

On the one hand, the continued emphasis on conservation is not surprising because the government provided farmers with everything they needed when they made the soil conservation districts available. And when the going got tough in the 1950s, they returned to the federal government. While wartime demand meant an easing of some of the conservation restrictions on production, the end of the war brought a renewed focus on soil and water. That attention to conservation became even more pronounced when the drought of the 1950s set in, leaving farmers recalling their efforts to combat the earlier drought to prepare themselves for the new challenge. On the other hand, several shifts in the postwar agricultural economy required farmers to adapt to new circumstances. For example, the increased availability of water allowed them to grow crops that they had previously been unable to sustain. Furthermore, the demographic collapse of rural communities across the plains during the 1930s and continuing through the war left many farmers facing an extensive labor shortage. The government stepped in to meet wartime demand, but farmers had no real sense of what they might do to meet their labor needs once prisoners, internees, and other government-provided workers became unavailable. Thus, regardless of what they hoped to grow for market, they needed to address the immediate priority of finding help to bring in their harvests, which was in and of itself a tricky proposition.

The federal government and state government, primarily through the Extension Service, helped sustain this new agricultural economy, and as farmers maintained their conservation regimes and solidified their access to workers, they came out of the 1950s drought in much better shape than they had in the 1930s. For most farmers, however, survival was not good enough even in the face of such significant challenges. They regularly criticized the government for not doing exactly what they had requested. They also established a system of labor that often left workers underpaid, overworked, and critical of their employers for failing to meet the agreements they had signed to come work in Colorado. The government responded in both respects by trying to assuage them and eventually capitulating to much of what the farmers demanded and by trying to soften the harshest elements of the labor regime by supplementing what farmers provided workers with additional support services. The result was an agricultural economy that by the early 1960s little resembled that of 1929 and instead manifested as the culmination of three decades of federal and farmer cooperation to keep farmers working on the Colorado plains. That system remains largely in place now and represents one of the more pronounced legacies of the Dust Bowl.

The first years after the war found southeastern Coloradans feeling at best ambiguous and at worst almost schizophrenic about the necessity of implementing soil and water conservation techniques during times of good precipitation. Debates about the value of conservation and the place of the Extension Service raged on the front pages of local newspapers. A chorus of vocal resistance to state intervention emerged over the returning emphasis on soil conservation in the face of forecast dry weather and the increased possibility for dust storms. Evidence from newspapers and the declining influence levied by the Extension Service show that the late 1940s became another period, much like the mid-1920s, when residents chose sides and deliberated over the viability of continuing support for the state's conservation regime.

Articles in the *Denver Post* and *Life* magazine that prognosticated another Dust Bowl inspired a local response that promoted residents' knowledge and chastised national coverage of the weather as missing the mark on conditions and dramatizing the situation. The front page of the *Springfield Plainsman Herald* in September 1946 showed this resistance clearly when the editor refuted the *Denver Post*'s article on the coming drought. The editor charged the *Post* with acting as the vehicle for soil conservationists who only wanted to ensure themselves of having a "soft" job that came at the expense of taxpayers. He blamed them for digging "trenches that will take years for nature to fill back up again, destroying nature [*sic*] grass in the process, count[ing] blades of grass and import[ing] them back to Washington" for further study.[7]

The implication that outsiders had little knowledge of current conditions and that the federal government caused more harm than good also surfaced in a note by John Neal, president of the Western Baca County Soil Erosion District, that the *Springfield Plainsman Herald* published in August 1946.

Neal suggested that members of the Soil Conservation Service (scs) had actively been spreading rumors about continuing problems in the region to justify its expansive budget. He argued that the scs has done little to help farmers and in fact had done too much to restrict them from doing what they did best: grow crops. "If scs had been in existence 60 years ago," he argued, "no sod land would have been plowed west of the Missouri line, leaving west Nebraska, Kansas, Northern Oklahoma, North New Mexico, and east Colorado out of production." Fortunately, the farmers made their own decisions about expanding into those regions, and by their "initiative" the region turned into the "greatest wheat belt in the U.S." He continued a month later, again invoking the farmer and his "initiative" by saying that without it the United States would "not have been in a position to help win World War II" and that because "the farmers did not wait for some Government scs technician to come out and tell him how to get the job done," the farmers could meet the demands brought on by mobilization. The emphasis on local rather than federal knowledge became a common refrain in these criticisms of the scs. The editor of the *Plainsman Herald* echoed Neal's view, suggesting that the soil conservationists deserved no credit for building "this great country" and that the farmers should receive the praise. Neal claimed that the scs "never did stop any wind eroded farm lands" and that "control of blowing land has always reverted back to the men of the soil." Neal also suggested that the federal government had overpaid for scs expertise, and the government should have given that money directly to residents to establish a reserve from which they could draw to combat problems.[8]

The ironic part of this is that Neal fully embraced the role of the local districts in mitigating the potential for soil conservation, and in doing so he demonstrated the fact that most of this vitriol proved more rhetorical than actual. Neal's diatribe against the scs closed with his vocal support for his local district, which, with him as president, continued the conservation agenda they had established a decade earlier. In his view, the local districts were prepared to handle any "ordinary wind hazard," and the state law from 1937 had put them in a perfect position to defend against another Dust Bowl.[9] In another article about the potential for severe dust storms, the *Springfield Plainsman Herald* printed the views of John Bird and others from various conservation agencies to demonstrate that these experts refuted the belief that another Dust Bowl was imminent. In Bird's view, the "big operators are belligerently positive about their ability to take care of their lands" because "there is a much better brand of farming on the Plains today. Many farmers have put unstable acres back into grass; summer fallowing has a widespread following; contour cultivation is practiced, and strip cropping is used."[10] Even as the editor of the *Lamar Daily News* lambasted the *Denver Post* for predicting dust storms, the paper included front-page coverage of Extension workers and head soil conservation- ist encouraging farmers to conserve their soil. The article recommended farmers remain proactive in their efforts to ensure continued production and compliance with state and federal conservation mandates to earn subsidies for such activity.[11]

This frustration with the intimation that Coloradans faced another Dust Bowl also translated into how residents proffered rather mixed reviews of federal intervention during the early 1950s. The Office of Price Administration (opa) took the brunt of it from those who wrote into local newspapers to express their concerns. Once the war ended, most argued, the federal government had no reason to dictate production or to determine which farmers should abide by federal policy. Much of this condemnation represented a pro-business, anti–New Deal viewpoint that many conservatives promoted. An advertisement for the Jett Hardware and Lumber Company criticized the opa for anti-American policies like "importing binder twine from Mexico" instead of buying from American producers. It also suggested that federal regulation of the farm implement business, which was itself a product of too much intervention in the agri- cultural economy that was no longer necessary after the war, meant that their business suffered and that local farmers could not purchase what they needed to prosper.[12]

The *Springfield Plainsman Herald* also disparaged the OPA as a corruption of state power. The paper printed several articles on the abuses of the agency and argued that it was an extension of New Deal policies that had long gone after "business people" who deserved better treatment. New Deal policies had taken money away from the "ones who have and give[n] to the ones who have not," even though the have-nots are "the ones who get practically all of the soft jobs." Many of those workers have made themselves a career by moving within and between federal agencies, and their advancement has resulted in a great "power over the citizenry at much larger salary."[13] The paper entered the political arena with its criticisms when it offered its full-throated support for J. Edgar Chenoweth in his fight against the OPA and its intervention in the agricultural economy. Again, the editor launched an attack against these "Communistic-Radical" policies that marked the continuing influence of New Deal policies. Asking the readers if they had "had enough," he advised them to use their influence to "Vote 'Em Out" and replace liberal Democrats with conservative Republicans who would tear down the walls of regulation that threatened postwar prosperity.[14]

Just as in the 1930s when drought hit and erased virtually all resistance to Extension programs that existed in the 1920s, so too the 1950s drought purged opposition to the Service that had surfaced in the 1940s. Somewhat surprisingly, Guy Robbins, agent for Baca County in 1946, noted that the Extension program had "dwindled to a point where both the county commissioners and the state office questioned the value of continuing an extension program in the county."[15] The numbers for Prowers County also showed a drop-off, with agents in 1946 and 1947 noting a significant decline in the number of farms changing practices, first-time users of Extension assistance, and total numbers of farms influenced over that period. At the same time, though, Prowers agent Stanley Boyes referenced a meeting with Prowers resident Calvin Stalford, "the state soil conservation champion" and a reporter from the *Denver Post*. Prowers agents also celebrated a "soil conservation week" and showed conservation films as part of a broader public relations plan to encourage county support.[16] This suggests more ambiguity than animosity, but all of that disappeared as drought returned. Indeed, the 1950s witnessed among farmers the kind of enthusiasm for Extension Service programs that had become commonplace in the late 1930s, and the positive attention to conservation represents how farmers returned to the tried and true techniques used in the Dust Bowl to respond to problems in the 1950s. After a temporary drop in participation, residents again attended Extension demonstrations, toured the demonstration farms, established and expanded the Extension-assisted soil conservation districts, and prioritized soil and water conservation by cooperating with state and federal programs. Baca County agent Chester Fithian's experience is a good example of how residents and agents grew closer as the drought became more severe. He had no illusions about the fact that people were much more likely to work with Extension because of it. As he stated in 1952 when he described the turnout at the regular meeting of the Western Baca soil conservation district, while 108 people attended, and that was a solid number, "if it had rained two inches before the meeting there wouldn't have been twenty-five people here."[17]

Crucially, however, farmers now had such organization and associations in place to utilize during drought, and they started well ahead of their predecessors as a result. This meant not only a reliance on and expectation of federal intervention as farmers saw fit; it also led to a more positive relationship between agents and farmers. As Fithian stated in early 1953, "To the writer's knowledge, there is no opposition to existing programs, nor ill-feeling towards the personnel, service, and programs, either active or in formulation, which gives the County Extension Service full rein towards unlimited progress in many areas of activity." While agents were never immune to optimistic statements about future growth, the Narrative Summary from that year contained several examples of Fithian's faith in community-based activism in response to the drought. He expressed confidence that farmers, the Extension Service, and county committees could cooperate in trying to deal with erosion and other challenges, and he explained in detail how they went at the problems.[18] Indeed, everyone seemed to contribute. Baca farmers established their County Emergency Drouth Committee to organize conservation programs in the county well before the state or federal programs began, indicating both local initiative and the greater sense of urgency among residents than those in Washington. Fithian roundly applauded local leaders for their leadership and urgency in coordinating these efforts during the early years of the crisis. He implied that residents had taken advantage of state help but that they felt responsible for ensuring that these policies helped those who needed it by taking command of their enforcement.[19]

We can also see this in how farmers greeted new federal programs like the Production and Marketing Administration (PMA), designed in part as an agency responsible for balancing supply of and demand for agricultural products, with a lukewarm response. They grew concerned about the consequent levels of bureaucracy and the sheer number of programs at work in the region, so they celebrated local leaders for tackling conservation problems. Fithian again emphasized local effort by announcing the organization of a Baca County committee within the PMA apparatus that could then meet regularly and work together to control wind erosion. A corresponding group emerged in Prowers County at the same time, in part due to the same problems of deciding how to implement federal policy. Prowers agents Vernon Carter and Bruce Young noted that farmers had trouble understanding the policy and how it affected them, leaving them confused and frustrated. As they had in the 1930s, however, agents like Carter and Young took on the challenge of having to translate federal policy and encourage participation on the local level. In Prowers, this meant that Carter and Young met weekly with the county and U.S. Department of Agriculture Drought Committee to decipher policy changes and to determine eligibility for the programs and subsidies offered by the federal government.[20]

Farmers relied on the Extension Service in other ways throughout this period as well, particularly as agents took on community responsibilities on boards and committees designed to combat soil erosion and land degradation. This existed prominently in agent cooperation with the SCS, as agents in both Prowers and Baca worked closely with the SCS to encourage farmers to continue and even increase their use of conservation considering the dust storms. Agents also recalled working with the USDA to implement feed programs; Chester Fithian served as secretary of the county committee created to offer emergency provisions to ranchers. Baca agents facilitated subcommittees on water needs and conservation to ensure that users employed the best methods to save water they drew from their new wells. Similarly, Prowers agents orchestrated a Water Users Association to work with both rural and urban interests and devise a viable strategy to meet the needs of both populations. Agents maintained relations with other groups as well, principally communities of farmers that organized within distinct associations like the Farm Bureau. They also formed connections with some newly created associations like the Holly Veterans Agricultural Group and the Lamar Veterans Agricultural Group to help veterans get established in the region upon their return to the home front.[21]

The agents' most important effort in trying to encourage and sustain conservation practices in the area came in the form of their leadership in numerous soil conservation districts in the two counties. The three major districts in Baca, all formed during the first five years after the Colorado law to allow for such districts passed in 1937, remained prominent features in the fight against erosion until they eventually merged in 1982.[22] Prowers County districts experienced more fluctuation during the period, in part because they had not embraced the district model as strongly during the 1930s and in part because they joined the fight against erosion with gusto in the 1950s. They even considered establishing a new district to correspond to the Fry-Ark project area and ensure that the new water was being put to the best use.[23]

The Prowers districts had very much come into their own by the end of the 1950s, and the county won the state award for "best effort in Soil Conservation" for 1961, an award that largely represented work done by the Northeast Prowers Soil Conservation District.[24] While only formed in 1947, and therefore a much younger district than its counterparts in the region, the district took the lead in working with the community to build a strong conservation agenda, and their efforts paid off. Indeed, the district won a contest that the radio station KLZ and the *Denver Post* organized to recognize the most active and influential district in the state. The contest rewarded them for their outreach programs, which the district excelled in. For example, the group organized a "District Conservation Context" that tested the conservation work of boys (most of them presumably associated with the 4-H or the Future Farmers of America program at Holly High School) and awarded prizes. Perhaps more important, the district took to the pages of the *Holly Chieftain* to publicize its efforts and advertise its conservation message to the broader readership. In so doing, the district invoked three central themes to articulate the value in protecting natural resources. First, it explained why a district formed, how it operated, and what it intended to accomplish for farmers and rural residents. This effort to explain its role in the community helped build a bridge with readers in working to gain their support.[25] Second, it celebrated a "Conservation Pledge," suggesting that members "pledge as an American to save and faithfully to defend from waste the natural resources of my country— its soil and minerals, its forests, waters, and wildlife." By connecting the natural to the national, the district continued the tried and true method of utilizing a call for patriotism to justify conservation.[26]

The third element of the district's public relations campaign came in the form of promoting economic growth by linking business and conservation together, which showed how the district worked with others in the community to achieve its goals. For example, one sequence of headlines promoted conservation as "good business" and emphasized that farming with responsible conservation methods led to better production rates and that a stable farm economy meant economic prosperity for everyone. One advertisement suggested that conservation "pays big dividends, not only now but in the future . . . in greater production, better crops, [and] more soil inches." An adjacent picture caption included words like "rich," "richest," "prosperity," and "saving" to describe the need to conserve soil, which they reminded readers marked a "never-ending battle that has just begun." Notably, these advertisements had been sponsored by local businesses that included their names and phone numbers at the bottom of the page. This kind of marketing that associated their business's future with the future of responsible farming in the area showed the connection that many in the business community made between their livelihood and more sustainable farming. It also represented the marriage between those businesses and the district, since each advertisement also included information about the next district meeting time and location to remind readers of upcoming events to promote participation.[27]

This focus on the districts as the main conduit for the conservation message is one legacy of the Dust Bowl because districts throughout the state became central to the government's fight against soil erosion in the 1950s just as they had in the late 1930s. Indeed, district members participated on local committees and boards, organized support for government directives, worked closely with the Extension Service, and kept a watchful eye on farmers to ensure they practiced sound conservation. In many respects, the response to the 1950s drought by the government, the Extension Service, and the districts proved remarkably akin to how they combatted drought in the 1930s. This makes sense, of course, for the many experts and citizens who remembered how the cooperation between these groups developed in the face of crisis and recalled how vital that teamwork was to dealing with those issues. Yet policymakers and residents also pushed new programs that might better protect against future droughts and marked a more reasonable approach to long-term planning than anything previous. Even these programs capitalized on districts and agents, and those two entities remained irrevocably tied to the conservation program.

Several 1950s policies reflected the kind of thinking that led to reforms passed in the 1930s, and most of them required agency cooperation as well as unity between federal, state, and local branches of those agencies. For the most part, policies supported the notion that farmers should be paid to conserve their resources, an idea that matured during the 1930s with the development of conservation subsidy programs. Much of this money came from state and federal emergency programs established to offset farmers' costs to combat erosion. Undoubtedly, these programs succeeded in encouraging farmers to comply with conservation programs and practice techniques on their own land to protect their resources. Millions of dollars in subsidies certainly can motivate even the most skeptical conservationists. Once farmers in the two counties became eligible for emergency drought aid, they embraced the opportunity to capitalize on funding. For example, nine hundred Baca farmers enrolled in the wind erosion component of the Agricultural Stabilization and Conservation program in 1954 and received thirty cents an acre for their efforts. The government paid out $240,000 for the 800,000 acres prepared by participating farmers. Furthermore, the programs designated certain payouts for specific activities, just as they had during the Dust Bowl, to inspire farmers to adopt a variety of approaches to conservation. Thus farmers could employ terracing, plant cover crops, farm on the contour, or retire segments of their land to satisfy the requirements and tap into funding streams.[28] The government also utilized tax refunds and offered tax breaks for farmers who could prove that they had practiced these techniques before the money became available. This form of retroactive support certainly reinforced the notion that conservation during the good years could still help farmers' bottom lines.[29]

The 1950s drought did engender a few new adaptations from policy-makers and farmers who seemed to finally grasp the cyclical nature of drought in the region, which allowed for a more aggressive push toward conservation that incorporated earlier policies but with additional enthusiasm. For example, the Soil District Conservation Law from 1937 created districts and allowed them some level of control over neighbors' decisions by giving them legal means to compel other farmers to keep their land from blowing. By 1952 the Prowers County agent had worked with the county commissioners, the county attorney, and others to devise a plan to police residents' practices that resembled the earlier effort. This plan took on the added significance of incorporating several branches of county government to compel user compliance. The group outlined its authority, justified its action based in the Colorado House Bill from 1951, codified a scale for financial penalty, and informed the guilty party of a timeline for completing necessary erosion control. It proved effective and seemed to entail less controversy than similar attempts during the 1930s. In one case, C. D. Buckner approached the county commissioners and filed a complaint against his neighbor Henry Deihl over Deihl's apparent failure to control soil blowing on his ranch. County agent Vernon Carter assessed Deihl's farm, found it lacking erosion controls, and informed Deihl that he had twenty-four hours to address the problem or the county would work it and charge him for the labor and supplies. Deihl agreed to that timeline and put into practice an erosion program on his land.[30] In addition to this local mechanism to control blowing, the state got more involved as well when Governor Dan Thornton authorized the State Soil Conservation Board to manage soil erosion issues, particularly as they related to watershed protection but also generally to help coordinate state efforts.[31]

The Extension Service also adopted a more focused approach to dealing with dryland farming problems by installing a research station in Baca. The research station, dubbed the Southeastern Colorado Branch Experiment Station, included an extensive and diversified farming operation devoted to primarily dryland farming techniques and conservation practices to protect against soil erosion. Initially on an expanse of 3,840 acres, the experiment station conducted research projects on issues like moisture control, crop rotation schedules, variety tests with wheat and other products, and reseeding programs designed to determine how best to recover land lost to erosion. It represented the third station run by the Extension Service on the eastern plains and combined with the irrigated farming station in Rocky Ford to provide two outposts for research and development in southeastern Colorado. Somewhat ironically, the dryland station gained funding and support precisely when more farmers in Baca had access to water in deep wells. Nonetheless, this is another example of the already considerable attention afforded to plains farming by Extension that became more intense during periods of crisis.[32]

Two other changes came to fruition because farmers and policymakers capitalized on the 1950s drought to strengthen or broaden approaches that began in the 1930s. First, the federal and state governments teamed up to create a program that rewarded long-range planning by cooperating with farmers and ranchers who would agree to a ten-year conservation program designed to protect against erosion and overproduction. Second, the land retirement programs that started during the 1930s enjoyed a rebirth during the 1950s and culminated in the formation of national grasslands that functioned under the protection of federal agencies. These two differences showcase how conservation policy matured in the face of another round of destructive erosion. These two adaptations also demonstrate how the legacy of the Dust Bowl caused farmers and policymakers to refocus their efforts on instituting more sustainable practices in the 1950s considering their collective memory of what had worked earlier.

The long-range conservation planning policy took time to build momentum, but it eventually gained enough political support by 1956 to earn passage as Public Law 84-1021, otherwise known as the Great Plains Conservation Program. The new law amended the Soil Conservation and Domestic Allotment Act of 1936 to expand that program in hopes of securing more sustainable agriculture. The secretary of agriculture effectively entered a contract with producers in the Great Plains. Those contracts, "not to exceed ten years," meant to "assist farm and ranch operators to make, in orderly progression over a period of years, changes in their cropping systems and land uses that are needed to conserve the soil and water resources of their farms and ranches and to install the soil and water conservation measures needed under such changed systems and uses." Farmers cooperated with experts to create a plan "to be practicable for maximum mitigation of climatic hazards" and that could thus defend against further drought.[33] The program generated a series of regulations designed to capitalize on cost-sharing between the individual and the USDA. Once the individual agreed to effectively reserve some of his/her land, then the government contributed to the costs to execute the long-term plan that the farmer devised. Such measures included the use of contour strip cropping, terracing, grassing waterways, leveling land, and planting windbreaks to encourage conservation, as well as providing irrigation for pasture and forage and fencing to facilitate the transition from farming to ranching.[34] Indeed, the program allowed for a liberal interpretation of how farmers might better take care of their lands. Roy Hanes, who signed the first contract with southeastern Colorado farmers in August 1958, put nearly 10,000 acres under protection as part of a plan that included changing fertilizers, studying irrigation trends and implementing new practices, and even utilizing wildlife conservation strategies to protect animals on his farm.[35]

This approach marked a new strategy, as the program did not allow for spontaneous changes or quick withdrawals by farmers wanting to plow up land that they had set aside. It also enjoyed considerable support from Coloradans as well as residents in other Plains states, who signed up by the thousands to participate. By late 1959, less than two years after the bill's passage, more than 3,000 contracts covering 8,597,385 acres had already been signed across the region. By the early 1960s the program had become so popular that most of the states' financial obligations were filled quickly, leaving many interested farmers without the option to participate. Congress responded by extending the contract length and allocating additional money to support more contracts.[36] This trend suggests an overarching support for conservation that existed among farmers even if they decided against, or did not earn, the ten-year contract. The *Springfield Plainsman Herald* noted that 1961 numbers showed 170 plans covering roughly 500,000 acres in the Arkansas valley, plus an additional 625,000 acres of soil survey work accomplished and 262,000 acres under very specific "complete conservation jobs."[37] More specifically, the paper found thirty-nine

Baca farmers signed up for various conservation programs, with another ten opting for the ten-year plan.[38]

Similar support emerged when Congress resuscitated land retirement programs designed to remove land from production, either for a period of years or permanently. The Soil Bank program, a product of the Agricultural Act of 1956, introduced government policy to reduce crops, conserve soil, and maintain farm income through two services. First, the Acreage Reserve Program encouraged farmers to take highly productive land out of production to control the supply of six basic crops, including wheat. More substantially, the Conservation Reserve Program (CRP) sought "an enduring change from cropland to pasture, range, forest, and wildlife uses." It allowed participants to sign up for 3-,5-, or 10-year contracts, and during that period the farmer and the government entered a cost-sharing program that paid farmers to remove lands from production or to redirect them away from agriculture.[39]

The CRP proved especially popular among farmers in southeastern Colorado, principally because it gave them a buffer against drought and dust storms by limiting production and facilitating government-funded conservation. As Chairman Branom of the County Agricultural Stabilization and Conservation Committee explained, farmers who participated in the program cut costs dramatically by easing demands on labor, equipment, seed, fertilizer, and other necessities, and they improved demand by limiting supply. Branom encouraged farmers to put as much acreage aside as they could by noting that the payments increase if farmers put all "eligible" land in the program.[40] By that point, in late fall 1958, farmers in the region had already jumped on board. Indeed, by June 1957 Baca farmers had already set aside 167,578 acres, earning $14 or $15 per acre for a total of $2,371,925 in federal funding—quite the financial windfall.[41] Nearly 200 farmers asked to be involved in the Soil Bank program in 1958, a number that amounted to nearly 25 percent of the county's farmers and that showed how much support these programs garnered even as the 1950s drought had subsided.[42] Furthermore, 193 farmers in Prowers applied for the Soil

Bank starting in 1959.[43] Part of the Soil Bank's appeal may have been that farmers could determine how long they could reserve their acreage and could then plan their next several years of production accordingly. It acted as a compromise between advocates for continued maximal production and those more concerned with erosion, overproduction, and the need for drastic changes to land use.

The other land retirement program that came to fruition during the late 1950s and early 1960s represented less of a compromise and more of a permanent response to the Dust Bowl that culminated in the creation of Comanche National Grassland. Starting with the land utilization programs of the New Deal years, the federal government made a concerted effort to buy up submarginal land across the plains to remove it from production. Indeed, the National Grassland serves "as a landmark to a great experiment in state planning and soil conservation during a time when the grass was not always green nor the sky always blue."[44] Policymakers experimented with that program to determine whether taking the most vulnerable lands out of production and bolstering the local economy by paying farmers for that acreage could alleviate erosion problems. To some extent, their experiment worked, but it happened on a rather limited scale and never enjoyed enough funding to buy up all the submarginal land in the region. Moreover, once rain returned and the war brought demand, most of the momentum behind buying up additional acreage abated, and the program lost support.

Even after the war, however, federal experts failed to fully buy into the program, and most of the concern associated with these submarginal lands (then called Title III lands) revolved around who should manage the acreage. By the early postwar period, the Soil Conservation Service managed most of these lands and used them as demonstration plots to show new conservation practices to local audiences. It also executed a broader plan to restore those landscapes to protect against further erosion. Though successful in its stewardship, the SCS eventually gave way to the U.S. Forest Service, which took control over these lands in 1953 to ensure that they were used per the "multiple use" philosophy of the administration.[45] Considerable debate also existed about whether to open up this land to the public for resale or to continue federal ownership, and this debate came to a head in the late 1950s when many farmers believed that they could utilize this acreage to their financial advantage. For many observers, though, private ownership would only ensure a repeat of the misuse of the landscape that would engender further erosion and crop failure. For example, John F. Douglas, forest supervisor of the San Isabel National Forest and the Title III land in southeastern Colorado, spoke with local leaders in Baca about the potential return of federal land to private farmers. According to Douglas, Baca leaders, including the county assessor, editor of the *Springfield Plainsman Herald*, and president of the Springfield Bank, noted that, without exception, federal ownership had become necessary to protect denuded lands. They felt confident that most of the county would similarly criticize the disposal policy of returning land to sale and informed Douglas that he and the Forest Service should maintain control.[46] The election of John F. Kennedy granted them their wish to keep the Title III lands under federal control as the president pushed the Multiple Use Act in 1960 that designated all such lands as national grasslands. This included a stretch of roughly 419,000 acres in Baca, Las Animas, and Otero Counties that garnered the name Comanche National Grassland with its formal recognition in 1960.[47]

Undoubtedly, the classification of these lands as national grasslands irrevocably changed how farmers and residents used it as well as how the landscape itself became divided between public and private land. In total, about 669,000 acres within Baca County are enclosed within the Forest Service's administrative district, which means that private owners often use public lands, and the protected areas are open for visitation. Most use of these acres was not by private individuals, however, as the population within the Comanche boundary dropped considerably over time, from about 1,300 in 1930 before it became established to about 300 in 1980 when the policy had been in place for twenty years. The Forest Service has continued its mantra of multiple use and offers grazing contracts for local ranchers. Most use is based in outdoor recreation, with the formation of picnic sites, hiking trails, hunting, and the establishment of tours for visitors to enjoy the wonders of Picketwire Canyon and Picture Canyon located in southwestern Baca County. The Forest Service currently offers "dinosaur digs" in Picketwire Canyon because the canyon became a hotbed for dinosaur tracks, bones, archaeological resources and artifacts, and other paleontological findings. The Service also runs guided tours through Picture Canyon, aptly named because of the ancient hieroglyphics that have brought a measure of fame.[48]

The formation of the Comanche National Grassland indicates very clearly that the Dust Bowl's legacy affected federal policy and local practices. Certainly, the checkerboard style of the Comanche acreage has allowed for inconsistent land use, meaning that some private owners will continue to engender erosion even when their plots stand adjacent to or across the street from grassland that has remained out of production for more than fifty years. Yet the establishment of federal stewardship has alleviated many of the stresses that led to the dust storms that plagued the 1930s and 1950s. Rather than threatening fines against neighbors who refused to control erosion on their lands, for example, owners could be confident that federal managers would conserve water, revegetate problem areas, and nourish wildlife. The recreation options also increased tourism to the area and gave residents more opportunities to enjoy the open spaces.

In some ways, then, the 1960 policy marked a significant turning point in farmers and policymakers turning the page on the kinds of activities that had helped make possible the dust storms of earlier decades. Land use surely looked different in the 1960s than it had even in the early postwar period, largely because of the continued migration of people out of the region and the increased use of federal conservation policies among farmers. The shift away from exploitive tendencies and overproduction has never been total, and there are certainly still indications that farmers continue to tempt fate with soil erosion, but clearly the Dust Bowl stimulated a response that culminated in the 1950s as a more conservation-minded and sustainable agricultural program that better reflected the kind of climatic challenges that farmers on the High Plains face.

While the postwar period engendered a shift in *how* people worked the land, it also produced an important change in *who* worked the land, as the move away from tenancy and wartime regimes led farmers to search for new ways to address their labor needs. These adaptations became even more important as the farmers shifted their growing patterns and their products as some transitioned to irrigation and as the farms them-selves grew in acreage. The demographic shift that began during the 1930s continued into the 1950s and beyond, leaving the mostly rural counties on the Colorado plains with declining populations generally and fewer farmers more specifically. As the average size of operations increased, the demand for labor did as well in some respects, but the pool of available workers that the Extension Service recruited and the federal government provided during the war steadily dried up with the war's end. To make matters worse, most farms no longer employed family workers to meet demand, which left them scrambling a bit to find available labor. Farmers turned to two major sources of labor after the war, both of which proved somewhat familiar especially to the old-timers who had been employing people since the 1920s. First, the bracero program marked a continuation of the importation program that began in earnest during the war and continued until 1964. The program enjoyed some success and experienced some controversy; it never fully satisfied employers' demands and often left both the worker and the employer unhappy with their relationship. Second, farmers rekindled their old practice of recruiting migrant workers from neighboring states, particularly Texas, as they had done before the start of the Great Depression after which migrant worker streams shrank. This combination met employers' demands but required a whole new level of government participation, as farmers looked to the state to provide imported labor, monitor migrant workers, and supplement private efforts to address migrants' needs once they arrived in Colorado. The resulting system, largely built on the backs of Latino workers, represented another important way that the agricultural economy in southeastern Colorado changed after the war.

The motivation to solve the labor problem through mechanization remained an important component of postwar farming in sugar beets and broomcorn. As they had before the war, farmers and inventors conspired to find alternatives to hiring out by constructing cutters, bailers, and harvesters that might alleviate the need for workers and by designing better tools that would make the field workers that farmers employed more efficient. A quick snapshot of developments in broomcorn is illustrative. Ralph R. Woods of Springfield applied to patent his broomcorn cutter in 1950; his drawings depict an automobile with a cutter resembling a windmill built off the front bumper and reaching several feet into the air to cut the tallest stalks. Harve Adamson of Holly and Roy Haney of Walsh (the same Roy Haney who was the first in the region to cooperate with the ten-year Great Plains Conservation Program) designed another harvesting machine and sought a patent for it in 1955. Work in plant genetics also aimed to make the harvesting process easier by limiting the number of seeds, widening the plant but shortening the stalks, and producing plant hybrids that might grow faster and be more durable.[49] These attempts reflect the desire not only to grow more but to allow for more efficient harvesting and therefore a smaller labor cost. Overall, though, none of these adaptations and inventions accomplished much before the 1970s, leaving farmers still largely reliant on humans willing to do the kind of stoop labor required to harvest beets and the physical work needed to produce a broomcorn crop.

The bracero program represented one of the two main sources of workers, although the program often proved problematic for both workers and employers. Wartime agreements between Mexico and the United States to provide labor and to ensure that the workers enjoyed necessities succeeded in helping farmers supply the war effort. While some considered the program a temporary fix, the two countries continued the importation program until 1964 and orchestrated the movement of nearly 4.5 million Mexican workers into the United States to address domestic labor shortages. Though designed with the intention of protecting both braceros and American workers from exploitation, the agreements often included loopholes that allowed for a variety of abuses on the part of farmers who sought to limit their labor costs by not providing adequate housing or food, insurance, transportation, or decent wages. It may have been the case that Mexico lost its negotiating power over time and could no longer protect its workers as it had during the war. It may also be that the U.S. agribusiness lobby became so powerful that it started to dictate terms for the agreement, thereby allowing farmers more opportunities to exploit those loopholes and lower costs.[50]

While the reasons for its mixed record certainly matter, the consequences of the program's inability to provide a consistent labor stream for Colorado farmers demonstrate how the postwar period produced a different work relationship between employer and worker. For example, unlike the relative harmony that the two sides enjoyed during the war, problems arose almost immediately after the war ended. The Mexican government refused to send braceros in February 1946 because it found "discriminatory treatment and unsatisfactory earnings" on Colorado farms, leading it to boycott the state as it had Texas on other occasions. Despite Governor John Vivian's best efforts, Mexico withheld its workers for part of that season.[51] Camp conditions became an issue again in 1953 when Mexico stopped the program because braceros had subpar accommodations in the state.[52] Indeed, the availability of workers—both in terms of whether they might come and how many might travel—seemed to vary by the year.

In 1958 the Mexican consul for the region, Hector Jara, assured Coloradans that Mexico would participate in the program and that farmers could take some comfort in knowing that workers would be ready. Jara noted that domestic workers would have the first opportunity to fill need but that the program had become more stable, conditions had improved, and the workers would fulfill their contracts. As Jara explained, "The old problem of Mexicans coming into this area without proper restrictions and working for a dollar a day and so on was solved a long time ago." He continued that "Colorado has had only 15 Mexican deportees in the last 10 years," promoting the idea of consistency and selling the viability of the program while also assuaging concerns about the quality and character of the workers.[53]

By the late 1950s, however, even Jara knew that the program might not last much longer because other workers had become more appealing than braceros. "Spanish American migratory workers," as he called them, could come from bordering states and work for next-to-nothing, giving farmers a cheaper option than braceros that also allowed them to avoid the kinds of rules and regulations in place to protect Mexican workers.[54] This chance to improve the bottom line and still have some state support for bringing in workers persuaded many farmers in the area to shift from bracero to domestic migrant labor, a transition that reinvigorated some of the labor streams that had existed before the Great Depression. Migrants became the backbone of the agricultural economy in the region, far surpassing the influence that local workers had or even what braceros could contribute.

The numbers of migrant workers and their importance to seasonal agricultural production led Colorado to establish several distinct agencies and programs to house, educate, and tend to migrants and their families. Whereas private companies, and especially sugar firms, had done this in the 1920s, the state adopted these responsibilities to a much larger degree after the war. In that way, the state again played an important role in providing workers for farmers who needed the help and became a mix of its hyperactive role in the war and its relatively hands-off approach before it. Once the state started to intervene in migrant labor issues, it adopted an approach that tended to emphasize its role in addressing the needs of the migrant community and move away from functioning as a labor broker. One of the best demonstrations of this increasing state support was the growth in state-funded attention to education. The Colorado Department of Education took charge in terms of both providing educational opportunities for migrant children and taking on a sociological approach to determine the social and cultural aspects of migrant children's lives in their home states as well as in Colorado. Thus their work with the migrant community is a window into how that community developed in the 1950s and how the state responded to its presence. A good starting point to identifying characteristics of that community is a Migrant Education Research Project study from 1959 that compiled demographic information from migrant parents and their children who were enrolled in migrant schools throughout the state.

The study, headed by the director of the Migrant Education Research Project, Alfred M. Potts II, represented a "compilation of the responses from a questionnaire that was completed by teachers for the families of Colorado's four 1958 pilot schools for migratory children."[55] In total, 86 percent of the 393 students, who came from 160 families, responded to the questionnaire. The pilot school program started in 1955 with instruction in Wiggins and eventually expanded to seven summer schools around the state, with a school in Rocky Ford serving children in southeastern Colorado. The Colorado Department of Education funded the program in various ways. For example, it orchestrated a five-week workshop at Adams State College in 1957 that taught teachers how to better deal with migrant children. It educated instructors in how to have a "more complete understanding" of the children by appreciating their experience and transition into the state and having patience with potential problems associated with the "attitude and value formation in the migrant child." The state also shared costs to run the schools, pay the teachers, and coordinate with migrant families to address children's needs by sending experts into their communities. The federal Office of Education also contributed to the program and worked with the state Department of Education to expand the summer school program and construct a year-round migrant program that might better respond to the needs of children.[56]

To be clear, the report demonstrates the assimilative impulse at work among education reformers and others aligned with the program. Potts was at times heavy-handed in his assessment of the program's goals, so that, while it is certainly the case that he and others in the program sought to help these students in what they believed to be the best ways, he also proclaimed his goal of finding a way to teach the children "the ways of our society" and to "best prepare these children to live in our society." He continued: "These children are ours. They are of our society. If we teach them wisely, both the children and society stand to gain from the ringing of school bells that call these boys and girls, too, into classrooms." By laying claim to "our society," Potts effectively renounced the migrants' inclusion in it, which is troubling when we consider that most of these students were born in the United States. The report and Potts's reading of it also contain racialized language and depictions of the children, both of which reflected biases held by many white Americans about Latinos during that period. Yet he forwarded an agenda that eventually helped migrant families deal with life in Colorado.[57]

He was also not alone in voicing this tension between disparaging the migrant community and trying to help them. Educators knew why the "nomads" had come to Colorado and, as the principal at the Rocky Ford summer school, Lynn Gulmon, noted, "These people are indispensable in their work" and "The crops raised in the Arkansas valley demand stoop labor, and Anglos do not do this type of work." Gulmon echoed Potts's perspective, suggesting that it "may be true" that migrants "are careless with property and do not wish to be more clean and sanitary" and that "we" (presumably the white American teachers) "are the ones who should teach parental responsibility and self-discipline to the children" in addition to the basis curriculum. Gulmon strove to teach students how to comply with the rules and regulations of the school to prepare them to abide by laws demanded of American citizens. In a way, then, he implied that their status as citizens afforded them a chance to better fit in the United States as adults. Yet he also claimed to have seen the living and working conditions that most migrants experienced and claimed that "some of it is bad beyond description," obviously not befitting citizens who toiled on Colorado farms. He believed that employers were getting away with substandard pay and housing when dealing with the migrants ("citizen labor") but would have to meet a higher standard if they wanted to use Mexican workers. For that reason, Gulmon claimed that "we do contribute to their condition," and he felt a compulsion to help them get through school on their way to full citizenship.[58]

One of the more interesting elements of the studies and their deeper examination of migrants' lives is that the migrant community largely reflected broader social and cultural changes to be found throughout the country, making them appear more like stable populations than many might have guessed. In some respects, of course, any similarities would be immediately dismissed because of the common stereotypes laid on migrants as well as the impression held by many that they were more different than alike. As Gulmon's explanation makes clear, even those working to help migrants often laid bare their assumptions and injected racialized language in describing their situation. But the review done by Potts and his associates in 1959 betrays many of those assumptions and demonstrates stability among the migrant population.

The report began by noting that most respondents had come from Texas and had "Spanish American" surnames but that not all migrants came from this same group. According to their number, about 76 percent of the families came from this Spanish American classification, so they were "people of mixed bloods, whose primary or secondary cultural background was essentially the Southwestern Spanish-influenced culture. Spanish or Spanish dialects were often a primary or secondary language, but this was not a necessary requisite. The names and physical appearance may have been the primary means of identifying those who were classified in this group." The report counted 21 percent "Anglo," meaning "the white stock of mixed racial background other than Spanish mixtures." While the authors offered no explanation for it, they noted that location of the school led to remarkable diversity in the numbers. For example, 97 percent of the population at Fort Lupton was Spanish American while 54 percent of those at Palisade were "Anglo." Regardless of where they ended up, many had started in a similar place. The report concluded that 150 of the 339 children had a "home base" in Texas, with Colorado coming in second with 90 children. Somewhat surprisingly, 34 students gave no answer to the question. The report also suggested that many of those from other states, which included Utah, South Dakota, and Arizona, remarked that they had an "intention to settle" in Colorado. The authors tied this to a broader "northern movement" of workers who had come to the state in hopes of finding steady work and making a home for themselves here.[59]

One important difference between migrants and Mexican Nationals brought in through the bracero program is that most migrants to Colorado arrived with their families. Per the Potts study, most of the children lived with both parents (89 percent) and another group lived with a parent and stepparent (4 percent), which led the authors to discount the idea that the family unit remained somehow unstable. Most of the children's parents were American citizens, with more than 42 percent born in Texas. The study cited 15 percent of respondents having parents from Mexico, giving them a chance to earn citizenship as first-generation Americans. The households sometimes consisted of extended family members, and most families had several children, making it hard to generalize about family dynamics among those who participated. In most cases, however, only the father worked, primarily in agricultural labor, while the mother stayed home as a "housewife." The authors contended that one path for migrants hoping to stay in Colorado was to earn a job in a semiskilled trade and to demonstrate their ability to do it well. Most families, however, seemed content to have the father work to provide for the family, and that path seemed to be the surest to both maximize their earnings and potentially cultivate a path to stay put.[60]

The data regarding migrants' finances paint a complicated picture because they show elements of middle-class consumerism as well as problems faced by those in poverty. The authors noted that about 43 percent of the families owned their homes while 55.7 percent rented their dwellings, below the national average of 60.4 percent home ownership and the regional average of 58.3 percent ownership in areas "outside standard metropolitan areas." Not surprisingly, the home ownership numbers varied by location, such that only 20 percent of workers in Fort Lupton owned a home compared with 48 percent of migrants in Rocky Ford. More specific numbers about the home suggest that migrants remained well below what might be considered average living standards. For example, a full 15 percent of permanent homes consisted of adobe. Most of the homes had more than three rooms, but these buildings housed an average of 8.13 persons, with an average of almost 6 children in each home. The authors noted that "climate allows a great deal of out-of-doors living" and that "overcrowding in an adobe house in South Texas hardly creates the same problems as overcrowding in a New York or Chicago tenement district." The lack of modern conveniences may have exacerbated some of those issues, especially considering that only 40 percent of the homes had running water, 16 percent a telephone, 62 percent refrigeration, and 33 percent a bath. A full 90 percent of the homes had electricity and 78 percent contained a radio with another 50 percent have a television. These numbers, considered "by modern standards . . . very low," demonstrate that many of the migrants were not enjoying the postwar wave of prosperity that is so often emphasized when talking about the 1950s and 1960s.[61]

This relative poverty factored in with other elements of migrant children's lives to challenge teachers. The study emphasized a few aspects of family life that likely made instruction more difficult, but nothing seemed to pose an obstacle bigger than the language gap. Potts and his team also pointed to low usage rates of newspapers and to the few families that owned magazines or books to suggest that literacy among the adult population also represented a challenge to get students excited about school. Students failed to get involved in many social organizations like the Boy Scouts, and few parents joined parent-teacher associations. While the clear majority of students claimed that they intended to finish school, the authors believed that many would fail, due to their transience more than anything else.[62]

Because of that central issue, programs designed to meet the needs of migrant workers and families never successfully combatted most of the problems that resulted from unstable living situations. Indeed, the education program expanded somewhat dramatically over the 1960s. Not only did the program employ new techniques like vocational training to reach more students but they also utilized migrant education mobile units to better meet children's physical and medical needs by moving into camps instead of only conducting exams at schools. Moreover, the number and geographic reach of the schools in the system also grew. The program started to run year-round in the late 1960s, with twenty-two programs existing in the fall and spring of 1968 and twenty-four during the summer. The schools also emerged in new places, including one facility in Baca County and two in Prowers.[63] This growth suggests that considerably more students needed help from the state and local districts, which resulted in higher costs, more people working in the programs, and greater attention to the problem of educating migrant children.

The education program eventually became only one element of a broader approach that the state used to meet migrant workers' needs, which by the late 1970s represented a considerable piece of the state's budget. States across the nation as well as the federal government started to become more sensitive to the plight of workers in response to the *Harvest of Shame* documentary that Edward R. Murrow and Frew W. Friendly put together to address the ongoing problems of poverty and deprivation. In Colorado this meant the genesis of the Colorado Migrant Council (CMC), a group organized to build on the work done by Potts and others to better understand the problems that continued to face farmworkers. Rather than dealing solely in education, however, the CMC devised a program designed to work on behalf of the neediest workers as their advocates and by helping with medical service, education, family planning, and direct welfare like food stamps.

The CMC study, based on findings in 1975, shows that many of the tendencies that emerged in the 1950s continued in the decades to come.

If anything, those trends became more pronounced. For instance, 75 percent of the migrants living in the Arkansas valley claimed to hail from Texas, and 96 percent of the population considered themselves "Chicano," numbers that show an increase in the percentage of migrants claiming both identities. The study also found that the community was largely impoverished, with little education, low home ownership numbers, and fewer skills that would allow them to climb the agricultural ladder or take on more responsibility in the fields. Most of the workers came from families where at least the father also worked in agriculture, meaning that many had been accustomed to this transitory life. Yet there was only so much that one could truly become used to given the uncertainty of that experience. Even as of 1975, 23 percent of workers in the Arkansas Valley had no idea when they might return home and 11 percent had no idea where they would go.[64]

Despite these numbers, however, Robert McCracken, member of the CMC and author of this study, reminded his readers that things are not always as they seem. "There is a tendency among well-intentioned liberal thinkers who have not looked at farmworkers' lives in depth, or have not understood their lives, to characterize the farmworker's lifestyle as a sort of living hell," he lamented. "Farmwork is a way of life; it is a culture with beliefs, values, and prescriptions for living" and "no attempt should be made to understand or deal with them in terms of a generalized morphic mass." He cited the 83 percent of respondents from the Arkansas valley, even those needy enough to qualify for CMC assistance, who reported being "satisfied" with working as farm laborers. Unlike many reformers who argued that the entire system of farmwork should be eradicated, McCracken believed that the CMC needed to contribute to the "design of programs that provide farmworkers with options in their lives, giving those who want to get out of farmwork the opportunity to do so and at the same time giving those who want to stay in farmwork the means of improving their standard of living." Even though 61 percent claimed that they would welcome participating in the U.S. Department of Labor's policy to retrain and resettle farmworkers, McCracken contended that many within that population would continue their work if the system better attended to their needs.[65]

The local side of this broader debate about the viability of supporting migrant workers and calculating basic demographic data to identify trends and problems shows a much more immediate and, in some ways intimate, relationship between residents and migrants. The two major newspapers in Baca and Prowers did not fully reflect the findings in the state-sponsored studies, and the exceptions illuminate another side of the complex system. To be sure, the mention of workers most often included some reference to their migrant status. It is also the case that several articles covered the issue of wages and conditions for workers, since that economic relationship lay at the heart of the system. This usually meant that there had been some disagreement about wages throughout the county or that the conditions provided on major operations like the Neal Ranch were not ideal.[66] In some cases, the newspaper lamented the plight of the workers. One article specifically referenced the 90 percent of migrant workers who had arrived late for the harvest and had to "mooch" food or money to buy food. While "orderly," they had to be under "police control" while they stayed in the county waiting for work.[67] Many articles made mention of the violent relations between migrants or other issues related to lawlessness, most often drunkenness, which painted a picture of depravity and immoral behavior.[68] Yet the newspapers also pointed out that workers should work with state and federal programs and agencies to ensure that they found placement and could defend their best interests.[69]

This often meant an allusion to their home base or less frequently an identification of where they might have come from specifically. While some of these references mentioned Texas, as one would guess according to the statewide studies, several noted other home states. Unfortunately, when newspapers mentioned migrant workers by name, it was often because of tragic consequences. For example, the *Springfield Plainsman Herald* noted that John Thomas Keene of Prairie Grove, Arkansas, had passed away in the county.[70] Similarly, the paper explained injuries to Roy Lee Adams and Everett Marcomy, two middle-aged men from Oklahoma.[71] Russell Hummingbird, from near Tulsa, Oklahoma, had been working on Curtis Porter's farm for two years before he died in a traffic accident in 1958.[72] The Baca papers also uniformly addressed workers employed in the broomcorn industry and made no real mention of those in other fields even though the statewide studies suggested that sugar beets were almost exclusively the draw for workers to southeastern Colorado. These discrepancies, though minor, show that the Colorado studies failed to capture the whole picture.

One of the more interesting elements of the relationship between locals and migrants was the increasing attention that the *Lamar Daily News* paid to Latino history and culture. Beginning in 1962, the paper ran regular sections entitled "Lightest Mexico" that featured stories about Mexico to bring awareness of and connection to that country. Indeed, the paper expressed interest in the Vaqueros fiesta to be held in town to celebrate the tradition of Mexican vaqueros and how that history related to the founding of the county and the region more generally.[73] The newspaper also shared the story of a group of new citizens who threw a party to thank their patrons. The paper celebrated a small group of folks who worked with the local couple to study English and American history to pass the citizenship test, and the paper's emphasis on their success shows an appreciation for the challenges that many migrants faced and an approval of their decision to become Americans.

That the local population worked with and even eventually embraced agricultural workers should not be terribly surprising given the region's history. Between 1900 and 1970, locals who needed workers found them, employed them to work their lands, and tried to find a way to stabilize the workforce to benefit their bottom line. As they had since the 1930s, farmers worked with the federal and state governments to find sufficient labor. During the postwar period, this meant primarily individual and company-led recruitment methods designed to bring migrant workers from nearby states into Colorado during the peak seasons. By then they had become accustomed to using outside labor of all kinds, so reuniting with recruitment boards or reaching out to migrants to entice them to come into the state did not represent a serious challenge. While the newspapers often mentioned shortages, most years the migrants satisfied employers' demands and did the kind of physical stoop labor that they required. The two communities never fully gelled, with members of residential communities sometimes failing to hide their belief that the workers were more "outsiders" than fellow Americans. Mistreatment of workers sometimes grew from this same prejudice. Overall, though, both worker and employer entered the labor contract with the intention of planting, pruning, or harvesting goods for sale. The tens of thousands of workers who stayed in the area at least temporarily made the development of postwar agricultural economy possible. They never enjoyed the level of prosperity that many of their employers did, but they added a vital piece to the regional economy that allowed it to post some of its finest years on record. Migrants provided the kind of stability within the labor market that farmers had long sought, and they continue to toil in southeastern Colorado fields following the rhythms of the seasons.

It might seem difficult to connect the Dust Bowl of the 1930s to a doctor in a state-sponsored mobile unit driving around Prowers County in 1958 meeting with and examining migrant workers. But the Dust Bowl set in motion a series of changes that irrevocably changed how and where farmers farmed. The decision to conserve resources or to hire migrant laborers reflected the importance farmers placed on steadying the bottom line and protecting their economic well-being. That emphasis on stabilizing the agricultural economy became more intense during the Dust Bowl years, but it has been on the minds of farmers throughout the region.

The most important change caused by the Dust Bowl in affecting how dryland farmers worked was the shift in the number and size of farms. That is really the starting point for everything else regarding conservation or labor. As the number of farms declined and their size increased, farmers found themselves looking more at options to move away from the kind of farming that had led them into the mess of the post–World War I era and the Great Plow Up. At that point, though, the federal and state governments combined with the Extension Service to work with agriculturalists to design a system that would better ensure sustainable production and avoid the sort of speculation that had contributed to the economic downfall of the 1920s and the Dust Bowl. The increased government involvement, while it produced some resistance among farmers, nonetheless made this possible. The dramatic increase in federal intervention that began during the 1930s and continued during the war laid the groundwork for the kind and scope of state involvement in agriculture after the war.

The use of federal money, the expansion of federal programs, the retirement of submarginal lands, and the focus on conservation to keep farmers productive all reflect the ways that state involvement engendered postwar agriculture. The 1950s drought and dust storms convinced most observers that they needed to do more—from a federal, state, and local level—to forestall the chance that the worst years could return. The onset of a ten-year conservation plan for water and soil conservation showed this shift. It is likely that the federal subsidies swayed farmers to do it, but regardless of their motivation, they regularly let land sit fallow, rotated crops, planted shelterbelts, and contoured their plots. The conservation itself made sense, and with the federal government paying the bill, this was a win-win situation for most farmers in the region. The 1930s were insufficient in fully bringing about that shifting mentality, but the Dust Bowl established a foundation from which those changes could be made. Not many people would have been so eager to embrace federal intervention had they not been through the 1930s, and even though they sometimes resisted that influence, locals capitalized on federal largesse and Extension assistance to survive the 1950s drought relatively unscathed. Indeed, the Filthy Fifties seemed to do just enough to finally convince farmers, policymakers, and observers that they needed to strive for a more sustainable and stable agricultural system.

The ready and willing pool of migrant laborers that allowed farmers to continue farming certainly contributed to this growing sense of security. Despite some issues related to workers, the shift from wartime, government-supplied workers to a more consistent and readily available stream of workers from neighboring states provided farmers with a potentially larger and more skilled set of workers. Farmers also benefited from having the state, and especially the Colorado Department of Education, establish an extensive infrastructure capable of reaching members of the migrant community and meeting their needs. In many cases, while some migrants lived on large-scale farming operations while in the state and others lived in places like the retired Civilian Conservation Corps camp, individuals and families also found themselves trying to navigate living in a new area, finding new jobs, and figuring out ways to support themselves. By stepping in to help, the state removed much of the potential burden from the shoulders of employers and instead inserted itself into a prominent role helping migrants get established for the season. While problems existed, the migrant labor stream provided farmers with cheap workers willing to do the very physical and demanding work necessary to produce crops like sugar beets and broomcorn. That farmers had an opportunity to rely on them year in and year out instead of worrying about the number of internees or prisoners of war that might become available made the system work much more smoothly than it had since at least the 1920s.

The combination of a more conservation-focused approach and the availability of a sustainable labor force gave most farmers the chance to capitalize on postwar markets and sustain themselves during lean years. Their larger farms and heavier investments in their productivity gave them greater resiliency in the face of drought and dust, and the government stepped in when crisis hit to help keep them afloat. Even that potent combination could not forestall the kind of drain from the countryside that Extension agent James Read had envisioned. Read, who served in Prowers County during the late 1950s, lamented the potential loss of community in the county and in other rural areas in his narrative summary from 1958.

He understood that the growth of urban areas, the creation of new jobs and opportunities in those cities, and the steady decline of population in rural areas would leave residents at a crossroads. He worried that local programs, schools, and churches would be "gradually disintegrating" and leave residents with only a facsimile of the vibrant and close-knit communities they once enjoyed.[74] The number of farms, acreage in farming, and relative importance of agriculture to the state's economy all slowly dropped after the 1950s, leaving fewer farmers to produce the goods that had been such an important part of Colorado farming after the war. The slow erosion of the broomcorn market and steady drop in sugar beet production in the area further compounded these issues in southeastern Colorado by the turn of the twenty-first century. While some observers certainly saw that trend developing, many farmers and residents believed that they would be ready for anything after they survived the Filthy Fifties.

CONCLUSION

There and Back Again?

The *Denver Post* published an article lamenting the long hot summer and invasion by unwanted pests: "Colorado farmers already plagued by a debilitating drought are now fighting the arrival of crop-eating insects who like the hot, dry weather that has settled over the state and elsewhere. 'It's to the point where we just feel beat up,' said Harry Strohauer, who has already let 500 acres of corn on his 3,500-acre Weld County spread die to conserve water." The article also quoted Rick Davis's opinion on conditions. Davis, a farmer on 600 irrigated acres near Julesburg, felt that "no matter what man does, Mother Nature has a way of getting around it."[1] The *Post* printed the article in July 2012.

The cycle of drought and economic instability that has plagued the Great Plains over the last century and a half continues today. For all intents and purposes, southeastern Colorado remains inhospitable country. Settlers quickly understood the challenges that the semiarid environment posed; current residents have long appreciated those challenges and face them daily. Many who still live in Baca or Prowers County were born in the area; some inherited farms or moved onto farms near their parents and other family members. As a longtime Prowers resident noted, most folks stayed because they had ties to the area or had no viable alternatives. Census information for Prowers shows a freeze on population since 1950, and Baca has lost almost 33 percent of its level since that time. The main streets in both counties resemble what they may have looked like during the 1930s, when drought and depression combined to wreak havoc with agricultural production and area residents' lives. Things are not likely to change dramatically soon, as the Great Recession has certainly taken a toll, as has the most recent drought. The familiar cycle of boom and bust that has seemed to define the area and has often fluctuated according to weather is not budging too far from "bust." While most residents refuse to say it aloud, a down economy, lack of water, and hot temperatures are scary omens, reminiscent of the 1930s.[2]

That is a bit simplistic, as much has changed since the New Deal years. Consider the landscape. The Comanche National Grassland comprises almost 444,000 acres in the region, and more than half of those acres sit within Baca County. The land represents a legacy of the New Deal effort to buy up submarginal land and retire it from production, an effort that gained federal support within the Resettlement Administration and the Soil Conservation Service. It reflects the New Deal emphasis on stabilizing the agricultural economy by helping farmers choose acreage more selectively rather than farming wherever, whatever, and whenever. With the help of county agents, the federal government identified which denuded plots warranted retirement and offered the owner a passable rate for the acreage. In many cases, farmers sold land that could have passed as arable but chose instead to cut losses and escape crushing debt. In others, however, the land proved so problematic, so devastated, that the federal government helped protect the farmer by buying him out as well as the farmer's neighbors to cut down on erosion. The grassland, formally created in 1960, is now managed by the U.S. Forest Service and offers several campgrounds, hiking trails, and fishing spots. It embodies the shift in land use policy from the production-first mentality to the embrace of recreation as a viable alternative. It also represents the success that New Deal purchasing programs had in removing acres from agriculture and rectifying the mistakes that unmitigated production had on the soil.[3]

Another characteristic of the area that suggests change is the number of windmills that dot the landscape. Some of the windmills, particularly in the east end of Baca and the southeast end of Prowers, are devoted to pulling water up from the Ogallala Aquifer. As William Ashworth writes, "Groundwater is the glass slipper that has transformed this Cinderella landscape [the high Great Plains] into a princess. Under the sand hills, under the shortgrass prairie, under the rich harvest of corn and wheat and cotton, lurks an ocean: the Ogallala Aquifer."[4] While scientists and hydrologists knew of the aquifer and the first attempts to capitalize on the water for irrigation date back to 1911, use expanded dramatically after World War II. At that point, the federal government widely subsidized or directly funded drilling projects designed to tap the aquifer for agriculture. Additionally, irrigation technology improved and made it more likely that farmers could institute personal pumps at minimal charge and with relatively small investment. The inexpensive machinery allowed farmers in southern Prowers and throughout much of Baca to access water for the first time and opened the possibility for more stable and secure production. It seemed an answer to their prayers, but, alas, even an ocean has a finite amount of water, and farmers are draining the aquifer quickly. By 1950 the aquifer irrigated 4.5 million acres; as of 2000 that number sat at 16.5 million acres. Irrigators are draining the source at more than ten times the natural rate of recharge, meaning that the aquifer cannot keep up with demand and will be sapped in the near future. Moreover, farmers' devotion to fertilizer is jeopardizing the aquifer water by introducing several harmful chemicals into the groundwater reserves and polluting the reservoir.[5] Therefore, while farmers who did not have access to irrigation now do, the aquifer seems less like a viable long-range alternative and more like a temporary boon. David Danbom notes that nearly half of the aquifer is now gone after only forty-five years of sustained use, "a natural resource millions of years in the making was in danger of being frittered away in less than two generations."[6]

Irrigated farmers in Prowers County also face distinct supply issues. The Arkansas River and its tributaries reflect the drought that flared up during the mid-2000s. In 2006 the reservoir hit its lowest level in thirty years.[7] In fact, as recently as 2009 the Colorado Division of Wildlife and Colorado State Parks jointly purchased an additional 3,000 acre-feet of water from the city of Colorado Springs to ensure that the John Martin Reservoir had a permanent pool worthy of sustaining fish and wildlife. The drought and users' demands had so depleted the reservoir that officials worried about the fishery and the animal population that relied on the reservoir to survive.[8] This was not the first time that area residents worried about the impact that irrigation and drought had on dwindling water levels. Indeed, a group of conservationists and recreationists gathered several times during the 1950s when thousands of fish died simply because the reservoir held no water. These enthusiasts successfully lobbied the Army Corps of Engineers to maintain a permanent pool to preserve the fish population and invite wildlife to the water's edge. Even as locals viewed the water in new ways, as a place for outdoor enthusiasts as well as a container for irrigation water, they quibbled over how to utilize the water and found that they never had quite enough to keep everyone happy.[9]

Other windmills have also sprouted up across the region, though they are designed to amass power instead of water. The Colorado Public Utilities Commission decided in 2001 that a wind-powered electric plant should be added to Xcel Energy's efforts to meet growing demand and looked to Lamar as prime real estate for a new power plant. Residents, especially the Emick family who owned large swaths of land on the eastern and western sides of Colorado Highway 50, then leased acreage to the company for the erection of wind turbines to fuel the plant. Such green energy alternatives have gained in popularity during the last decade, and the steady winds of the Colorado High Plains made Prowers an ideal locale for the extensive wind farm, later named Colorado Green. The turbines illustrate how Baca residents have taken to changing their resource use. Baca resident Joe Rosen-grants, for example, decided to leave the farming to his sons and opened

Sand Arroyo Energy, a company devoted to marketing wind turbines and solar panels. Rosengrants told a reporter from the *Pueblo Chieftain* that he made the move once he realized that "he couldn't beat the sun and wind, so [he] joined them."[10]

Perhaps the most telling characteristic of life in Baca and Prowers Counties is the sustained federal presence, which explains how people continue to farm and live in the region. The names of the programs and the agencies involved have changed since the Dust Bowl days, but the federal government's intervention in the rural economy during the 1930s set the stage for much of today's agriculture. Subsidy programs remain crucial to maintaining the local agricultural economy. Per the Environmental Working Group (EWG) database, the combination of wheat, corn, and sorghum subsidies effectively have kept farmers afloat since the war. The EWG tallied the number of recipients of various subsidies and the money spent on the subsidies by calculating totals provided by the USDA (see tables 1 and 2). The subsidy programs reward growers of certain commodities that the USDA decides are most valuable to the market. The number of farmers who primarily or even only grow those crops suggests a level of soft coercion from the federal government. Yet the farmers continue to farm in large part because of such federal involvement.

The other Dust Bowl legacy of note is the Conservation Reserve Program, which the USDA instituted in 1985. Per the USDA, the CRP "is a voluntary program available to agricultural producers to help them use environmentally sensitive land for conservation benefits. Producers enrolled in CRP plant long-term, resource-conserving covers to improve the quality of water, control soil erosion, and develop wildlife habitat. In return, FSA [Farm Service Agency] provides participants with rental payments and cost-share assistance." The principal agency in charge of the program, the Natural Resources Conservation Service (NRCS), is the present-day incarnation of the Soil Conservation Service. Employees of the NRCS join with members of local soil and water conservation districts to promote enrollment. Again, we see the federal and local entities working together

Table 1. Federal subsidies, Prowers County, 1995–2011

PROGRAM	NUMBER OF RECIPIENTS	TOTAL SUBSIDY
Conservation Reserve Program	867	$87,795,183
Wheat subsidies	1,336	$67,214,577
Disaster payments	1,007	$30,619,881
Corn subsidies	659	$25,188,223
Sorghum subsidies	1,156	$16,730,080

Source: The Environmental Working Group offers a database for such federal payments, searchable by state, county, and even congressional district. See "Prowers County, Colorado Farm Subsidy Information," accessed August 25, 2017, http://farm.ewg.org/region .php?fips=08099.

Table 2. Federal subsidies, Baca County, 1995–2011

PROGRAM	NUMBER OF RECIPIENTS	TOTAL SUBSIDY
Conservation Reserve Program	1,583	$122,426,727
Wheat subsidies	1,916	$100,260,721
Corn subsidies	521	$52,278,320
Sorghum subsidies	1,721	$46,390,610
Disaster payments	1,420	$44,230,787

Source: "Baca County, Colorado Farm Subsidy Information," accessed August 25, 2017, http://farm.ewg.org/region.php?fips=08009.

to protect natural resources by enabling farmers to retire parts of their acreage by compensating them for their decision.[11]

The figures suggest that the CRP made quite a difference in both land use regimes and farmers' willingness to remove acreage from production. Moreover, even the temporary retirement of such lands and the replanting of native grasses helps keep another Dust Bowl from recurring. Baca farmer Rosalie Bitner agrees, suggesting that "CRP has most definitely been instrumental in controlling erosion, preserving wildlife, and generating a healthier economy for Baca County's agricultural community." The USDA also argues that the rehabilitation and retirement program helped maintain the topsoil during the first few years of the current drought cycle. When "Baca County faced one of the worst drought periods on record" from 2001 through 2004, "the landscape was blown and dry, but no soil drifted into the neighboring fence rows and no blowing sand darkened the skies." "The land was prepared to weather the harsh conditions" because of the agricultural conservation efforts pushed by the USDA [12]. The CRP is also a perfect representation of how changes enacted during the 1930s continue to influence land use on the Colorado plains. Even then, however, not everyone is willing to participate in such programs, so their acreage will always be susceptible to blowing and drifting (see fig. 25). Neither local nor federal proponents can force people to opt into the program, just as farmers had to choose to participate in the New Deal programs that aimed to accomplish the same goals. There is no doubt, however, that the return to native grasses that has resulted from the CRP marks an important adaptation toward soil conservation that was not present in the 1920s (see fig. 26).

It is unclear whether the region will ever truly reach a point of stability. Drought and economic downturns continue to wreak havoc on locals. Population decline is evident, especially in downtown Springfield, where many shops have closed, leaving the county seat looking relatively desolate and depressed. Lamar seems much more bustling, and with good reason, as its population is more than double that of Baca County. The town has a much more diversified economic base, is home to Lamar Community College, and has a downtown full of restaurants, bars, and local shops. Judging by such superficial qualities, one might assume that Prowers residents are much more secure than their neighbors to the south. Yet there is no escaping the fact that they are still subjected to the fluctuations of a

volatile agricultural market and an unpredictable climate. Prowers residents are quick to point out that times are tough all over, that irrigation does not matter as much during a drought, and that the federal government has more to do to keep farmers solvent.

Perhaps more than ever, residents of the Colorado plains rely on federal intervention. That relationship and the amount of money that the federal government spends every year to keep farmers on the land in southeastern Colorado are difficult to characterize. In some ways, agricultural production is the backbone of the nation's economy, providing commodities for domestic and international consumption. Yet the level of federal involvement in that production, whether through price stabilization or resource conservation subsidies, calls into question the value of farming in such an arid environment. Perhaps early explorers like John Wesley Powell and Zebulon Pike had a point about the viability of agricultural settlement on the High Plains. The environment may not be suited to intensive cultivation.

To their credit, and although it has taken a long time to move in that direction, residents have moved away from the type of production frenzy that contributed to the Dust Bowl. Conservation districts play important roles in both counties by helping members make land use decisions based on sustainable production. The NRCS maintains a presence, and the county agents continue to patrol area farms, lead community meetings, and work to stabilize the population and the economy. Farmers and the federal

FIG. 25. "The difference between CRP (*top*) and cropland in Baca County is visible in the effects of wind soil erosion. In these fields, fine, sandy soil erodes off the cropland and accumulates in the center foreground. Vegetative cover on the CRP acreage catches the blowing soil as well as holds in place the topsoil under the cover." "Dust Settles on Baca County Thanks to CRP," Farm Security Agency, U.S. Department of Agriculture.

FIG. 26. This picture celebrates the impact that returning acreage to native grass has had on the landscape. "Dust Settles on Baca County Thanks to CRP," Farm Security Agency, U.S. Department of Agriculture.

government have worked to retire hundreds of thousands of acres and are taking better care of their property than perhaps ever before. Growers still face challenges. Water sources are finite, and even the flowing Arkansas will never meet users' demands. The wind will always blow and the sun will always beat down on the soil. Many residents seem stuck in a perpetually marginal existence, unable to make a lot of money but secure enough to stay on their land. It is doubtful that the plains will ever reach a point of stasis given the circumstances, so farmers may have to handle the ups and downs if they choose to stay in the region. In that respect, some things never seem to change.

NOTES

1. See Burns, *Dust Bowl*. Timothy Egan's book, *The Worst Hard Time*, increased attention to the region. A smattering of coverage in textbooks has also helped in that regard.
2. On Lange, see Gordon, *Dorothea Lange*.
3. A number of works on the Dust Bowl have informed this study, but they almost universally address only the 1930s. See Worster, *Dust Bowl*; Hurt, *Dust Bowl*; Cronon, "A Place for Stories"; Riney-Kehrberg, *Rooted in Dust*; Fearon, *Kansas in the Great Depression*; Nelson, *Prairie Winnows Out Its Own*; Stock, *Main Street in Crisis*; Cunfer, *On the Great Plains*. Cunfer's work builds on earlier studies of agriculture on the Great Plains such as Webb, *Great Plains*, and Malin, *Grassland of North America*.
4. The interwar years have yet to garner enough attention among environmental historians. See Paul S. Sutter's examination of this dearth in "Terra Incognita." See also Sheflin, "Environment and the New Deal." One of the better recent attempts to rectify this is Phillips, *This Land, This Nation*.
5. The most attention given to labor in the region deals most directly with the war years. See Wei, "Strangest City in Colorado"; Schrager, *Principled Politician*; Takahara, *Off the Fat of the Land*; and Harvey, *Amache*. Prominent works that deal with elements of farm labor during this period primarily deal with Hispanic workers. See, e.g. Markoff, "Sugar Beet Industry"; Reisler, *By the Sweat of Their Brow*; Gamboa, *Mexican Labor and World War II*; and Galarza, *Merchants of Labor*.

6. Sarah Deutsch gives a good explanation of this general shift in the region in *No Separate Refuge*.

7. A few exceptions exist. See, e.g., W. Rasmussen, *Farmers, Cooperatives, and USDA*; W. Rasmussen, *Taking the University*; Scott, *Reluctant Farmer*; Reid, *Reaping a Greater Harvest*; and Firkus, "Agricultural Extension Service and Non-Whites in California."

8. This emphasis on the Extension Service goes against most histories that show FDR at the center and his administration's agencies and policies as integral to the New Deal agricultural policy. See Henderson and Woolner, *FDR and the Environment*; Schuyler, *Dread of Plenty*; Baldwin, *Poverty and Politics*. The exception is Neil Maher's work on the Soil Conservation Service (SCS) and its impact on farmers in Kansas. See Maher, "'Crazy Quilt Farming on Round Land,'" 319–39.

I. EARLY LESSONS

1. J. L. Farrand, "Annual Report, Extension Service, Baca County, April 15, 1929, to November 30, 1929," 5–11, folder 44, box 8, Records of the Colorado Cooperative Extension.

2. Many historians have discussed the transition following World War I toward expansive landholding and mechanization on American farms, which proved especially costly when prices dropped and farmers were stuck with debt through the 1920s. This, by that reasoning, made 1929 almost another year in a line of many instead of the start of the Great Depression. See, e.g., Grant, *Down and Out on the Family Farm*; Hurt, *Dust Bowl*; Nelson, *Prairie Winnows Out Its Own*; and Stock, *Main Street in Crisis*. Conversely, a smaller group of historians has downplayed the decline within the agricultural economy during the 1920s. Most notable is Hamilton, *From New Day to New Deal*.

3. The term "Valley of Content" was first used by irrigation promoters in the late nineteenth century. Sherow, "Discord in the 'Valley of Content.'"

4. Osteen, *A Place Called Baca*, 169.

5. *Lamar Daily News*, October 24, 1929. The *Springfield Democrat Herald*, the largest Baca County paper, made no mention of the Crash and seems much less connected to national and even to some regional news than the *Lamar Daily News*.

6. Dillon, "Stephen Long's Great American Desert."

7. West, *Contested Plains*, 13.

8. Ubbelohde, Benson, and Smith, *A Colorado History*, 56–62, 326.

9. West, *Contested Plains*, 110–13.

10. Berwanger, *Rise of the Centennial State*, 5–8.

11. Stoll, *Larding the Lean Earth*. Much of his second chapter explains the relationship between soil degradation and emigration, but see 143–50.

12. Taylor, "Town Boom in Las Animas and Baca Counties," 120–23.

13. *Thirteenth Census of the United States Taken in the Year 1910: Population*.

14. Gillis, "'We Traded Your High Chair,'" 38–47; Durrell, "Homesteading in Colorado," 93–114.

15. Gutmann, Great Plains Population and Environment Data.

16. Examples of company pamphlets and brochures are available at the Colorado Historical Society. See box Arkansas Valley: [ephemera] 1894–1976, Stephen H. Hart Library and Research Center, Denver. Other examples include *The Arkansas Valley Truth*.

17. Emmons, *Garden in the Grasslands*, 128.

18. Quoted from Emmons, *Garden in the Grasslands*, 159. Original quote from Syndicate Land and Irrigation Company, *Valuable Data*, 42–43. A more recent explanation of this phenomena and its impact on settlement is Wishart, *Last Days of the Rainbelt*.

19. Van Hook, "Settlement and Economic Development," 134–40, 127. Van Hook offers specific settlement numbers for what will become Prowers County, but does not include numbers for Baca; the rough estimate of 2000 is based on his assessment.

20. Gillis, "'We Traded Your High Chair,'" 39, 40, 43.

21. Gillis, "'We Traded Your High Chair,'" 40–46.

22. Durrell, "Homesteading in Colorado," 98–99. Durrell explains that he was party to conversations about soil conservation and fallowing as a way to ensure stable returns, but few farmers cared enough or had the luxury to remove a portion of their acreage from production.

23. Van Hook, "Settlement and Economic Development," 128–30.

24. U.S. Congress, 44th, 2nd sess., "An Act to Provide for the Sale of Desert Lands in Certain States and Territories."

25. U.S. Congress, 57th, 1st sess., "An Act Appropriating the Receipts from the Sale and Disposal of Public Lands in Certain States and Territories to the Construction of Irrigation Works for the Reclamation of Arid Lands." A

number of works deal with the act, most significantly Worster, *Rivers of Empire* and Hays, *Conservation and the Gospel of Efficiency.*

26. *Fifteenth Census of the United States 1930: Agriculture.*

27. "Floyd M. Wilson and the Alfalfa Milling Industry," 100.

28. *Sixteenth Census of the United States, 1940: Agriculture,* 269. This document and all further census citations come from http://www.agcensus.usda.gov/Publications/Historical_Publications/index.php.

29. Markoff, "Beet Sugar Industry, 33, 35–38; Sherow, "Discord in the 'Valley of Content,'" 78–82.

30. Markoff, "Beet Sugar Industry," 30.

31. Markoff, "Beet Sugar Industry," 1–30.

32. Markoff, "Beet Sugar Industry," 1–24.

33. See Arrington, "Science, Government, and Enterprise in Economic Development," 15–16; Markoff, "Sugar Beet Industry in Microcosm," 6–7, 261–62.

34. Markoff, "Sugar Beet Industry in Microcosm," 224–98.

35. Blinn, *Development of the Rockyford Cantaloupe Industry,* Bulletin 108, March 1906, 1–17.

36. Bowman, "History of Bent County," 879–83.

37. Van Hook, "Settlement and Economic Development," 199–243.

38. Gillis, "'We Traded Your High Chair,'" 43.

39. Lockwood, *Locust,* xvii, 67–86.

40. Abbott, Leonard, and Noel, *Colorado,* 167–69; Riney-Kehrberg, *Rooted in Dust,* 2–3; Hurt, *Dust Bowl,* 6–14.

41. Abbott, Leonard, and Noel, *Colorado,* 168–69.

42. Cottrell, *Dry Land Farming,* 4–7, 20–27.

43. Abbott, Leonard, and Noel, *Colorado,* 169.

44. Hewes, "Early Suitcase Farming in the Central Great Plains."

45. *Thirteenth Census of the United States Taken in the Year 1910: Agriculture; Fifteenth Census of the United States 1930: Agriculture.*

46. *Thirteenth Census of the United States Taken in the Year 1910: Agriculture; Fifteenth Census of the United States 1930: Agriculture.*

47. Farrand, "Annual Report, April 15, 1929, to November 30, 1929," 11.

2. THE COUNTY AGENTS

1. Joe T. Lawless, The Unofficial Observer, *Lamar Daily News,* May 8, 1933.

2. *Springfield Democrat Herald*, November 10, 1932.

3. *Congressional Record*, 63rd Cong., 2nd sess., 1932.

4. *Congressional Record*, 1937.

5. *Congressional Record*, 1932.

6. Causley, "Agricultural Experiment Stations."

7. Hays, *Conservation and the Gospel of Efficiency*.

8. *Congressional Record*, 1937.

9. Peters, "'Every Farmer Should Be Awakened,'" 190.

10. Senate, *Report of the Country Life Commission*, 26–28.

11. Senate, *Report of the Country Life Commission*, 24, 9, 17.

12. Senate, *Report of the Country Life Commission*, 60–65.

13. Postel, *Populist Vision*, 282.

14. Postel, *Populist Vision*, 50.

15. Senate, *Report of the Country Life Commission*, 38–39.

16. *Congressional Record*, 63rd Cong., 2nd sess., 1935.

17. Kelsey and Hearne, *Cooperative Extension Work*, 117–18; Sanders, *Cooperative Extension Service*, 417.

18. Osteen, *A Place Called Baca*, 169.

19. See *Lamar Daily News* and *Springfield Democrat Herald* from the middle of October to beginning of November 1929. The only day that the *Daily News* did not include trial coverage as its central story was the day after the Crash. Ava Betz also covers the robbery, murder, and trial in her brief pamphlet on Prowers County history, *Prowers County, Colorado*.

20. Hamilton, *From New Day to New Deal*, 5–7, 27.

21. Shideler, "Herbert Hoover and the Federal Farm Board Project," 711–18.

22. Osteen, *A Place Called Baca*, 170.

23. "Sources of Farm Income, Baca County, Colorado: Comparison of Crop Values for Years 1920, 1925, 1930–1935," folder Baca County, box 3 Colorado, SP10(S) Records Relating to Land Utilization 1936–1939, Records of the Soil Conservation Service (hereafter SCS).

24. R. Johnson, "Part-Time Leader," 539–40.

25. "The Co. Agent Racket," *Springfield Democrat Herald*, March 29, 1929.

26. Marlon D. Lasley, editorial, *Springfield Democrat Herald*, April 19, 1929.

27. Joe T. Lawless, The Unofficial Observer, *Lamar Daily News*, May 8, 1933.

28. Frank R. Lamb, "Annual Report, Extension Service, Prowers County, November 30, 1931, to December 1, 1932," folder 1, box 67. Records of the Colorado Cooperative Extension, Prowers County Annual Reports, 1931–1932.

29. "The County Agent Matter" from The Editor Speaking, *Lamar Daily News*, November 11, 1933.

30. Mrs. C. L. Nickelson, "Letter to the Editor," *Lamar Daily News*, December 8, 1933.

31. George B. Long, "Letter to the Editor," *Lamar Daily News*, December 8, 1933. Long also demonstrates a certain anxiety about federal intervention by claiming that he was in fact unsure whether federal involvement was a good thing for locals or if it sacrificed independence.

32. R. E. Frisbie, "Annual Report, Extension Service, Prowers County, January 15, 1934, to June 25, 1934," folder 5, box 67.

33. J. L. Farrand, "Annual Report, Extension Service, Baca County, April 15, 1929, to November 30, 1929," folder 44, box 8 Also see W. Rasmussen, *Taking the University*, 86–88.

34. Lamb, "Annual Report, November 30, 1931, to December 1, 1932," 88–89.

35. Frank R. Lamb, "Annual Report, Extension Service, Prowers County, December 1, 1929, to November 30, 1930," 1, folder 62, box 66.

36. W. Rasmussen, *Taking the University*, 78–79. Also see Headley, "Soil Conservation and Cooperative Extension," 293.

37. W. Rasmussen, *Taking the University*, 92–94.

38. Lamb, "Annual Report, November 30, 1931, to December 1, 1932," 88–89.

39. Fausold, "President Hoover's Farm Policies," 376–77. Fausold argues that Hoover became too reliant on individual compliance and voluntary action by farmers, thus refusing to offer any enticement to make sure the system remained afloat. It seems that the federal subsidization also helped persuade farmers to abide by the AAA in higher numbers and with less reticence than they had with the Agricultural Marketing Act and consequent Farm Board.

40. Saloutos, "New Deal Agricultural Policy," 396.

41. Rasmussen, *Taking the University*, 97.

42. R. E. Frisbie, "Annual Report, January 15, 1934, to June 25, 1934," 6–7, 19–20, folder 45, box 8.

43. A. J. Hamman, "Annual Report, Extension Service, Prowers County, January 15, 1934, to November 30, 1934," folder 5, box 67.

44. *Lamar Daily News*, October 28, 1933.

45. Hamman, *My Long Journey*, 120, 129.

46. *Springfield Democrat Herald*, June 15, 1933. That date included some of the most glowing support for the CCC, but various editions of that paper throughout the summer of 1933 likewise argued in favor of the program.

47. Colorado Preservation, "Depression and Dust Bowl," http://www.coloradopreservation.org/crsurvey/rural/baca/sites/baca_resources_depression.html (accessed March 10, 2018); Santa Fe Trail Scenic and Historic Byway, www.santafetrailscenicandhistoricbyway.org/ArenaDustBowlTours.

48. Christman, *Legacy of the New Deal on Colorado's Eastern Plains*, 5.

49. For a more detailed discussion of the New Deal's impact on the built environment, see Cutler, *Public Landscape of the New Deal* and J. Smith, *Building New Deal Liberalism*.

50. Colorado Preservation argues that roughly half of Baca residents were on relief in the early 1930s. See Colorado Preservation, "Depression and Dust Bowl."

51. *Lamar Daily News*, January 8, 1934.

52. Van Hook, "Development of Irrigation."

53. Fred Betz, The Editor Speaking, *Lamar Daily News*, July 27, 1933.

54. See statistics from Frank R. Lamb, "Annual Report, Extension Service, Prowers County, December 1, 1929, to November 30, 1930," folder 62, box 66, and "Annual Report, Extension Service, Prowers County, November 30,1931, to December 1, 1932," folder 1, box 67; J. L. Farrand, "Annual Report, Extension Service, Baca County, April 15, 1929, to November 30, 1929," folder 44, box 8; and R. E. Frisbie, "Annual Report, Extension Service, Baca County, January 15, 1934, to June 25, 1934," 6–7, folder 45, box 8.

3. DIRT

1. Worster, *Dust Bowl*, 13–14.

2. T. G. Stewart, "Historical Review of the Soil Conservation Problem in Colorado," in "Annual Report of T. G. Stewart, Extension Soil Conservationist, December 1, 1938, to November 30, 1939," 1, folder Soil Conservation Specialists Report 1939, box 127, folder 7, Records of the Colorado Cooperative Extension.

3. T. G. Stewart, "Annual Report December 1, 1938, to November 30, 1939."

4. Stewart, "Annual Report, December 1, 1938, to November 30, 1939."

5. Stewart, "Permanent Agricultural Program for the Southern Great Plains," 26, folder Soil Conservation Specialist Reports 1937, box 127, Records of the Colorado Cooperative Extension.

6. *Lamar Daily News*, June 16, 1934.

7. *Springfield Democrat Herald*, July 26, 1934.

8. See, e.g., *Springfield Democrat Herald*, May 2, 1935, May 9, 1935, and April 2, 1936.

9. *Lamar Daily News*, June 16, 1934.

10. *Springfield Democrat Herald*, August 30, 1934.

11. *Springfield Democrat Herald*, April 18, 1935, April 25, 1935; Osteen, *A Place Called Baca*, 176; Hurt, *Dust Bowl*, 52–53.

12. Gutman, Great Plains Population and Environment Data.

13. *Springfield Democrat Herald*, October 27, 1938.

14. Robert T. McMillan, "Social Problems of Farm Families in Baca County, Colorado," Report I, Sociology Section, Land Use Planning Division, Resettlement Administration, Region XII, Amarillo TX, 1937, 13, folder Baca County Sociological Study, box 1 General, SP 10 Records Relating to Land Utilization 1936–1939, Records of the SCS.

15. McMillan, "Social Problems of Farm Families," 10–11.

16. Great Plains Drought Area Committee, *Report of the Great Plains Drought Area Committee, August, 1936*. Sarah T. Phillips's examination of the relationship between poor land and poor people, and the New Deal focus on rehabilitating the land to restore economic stability, was an important template for this book. She does an excellent job tying economic recovery to conservation and then illustrating how such thinking developed during the 1930s. See Phillips, *This Land, This Nation*.

17. F. A. Anderson, "The Editor Speaking," *Lamar Daily News*, August 29, 1938.

18. Opie, *Ogallala*, 100.

19. Gray, "Federal Purchase and Administration of Submarginal Land in the Great Plains," 123. See also Gray, "Research Relating to Policies for Submarginal Areas."

20. Gray, "Federal Purchase," 130.

21. Gray, "Federal Purchase," 130–31.

22. For a concise explanation of the economic interpretation of submarginal land, see Johnson and Barlowe, *Land Problems and Policies*.

23. Black, "Notes on 'Poor Land,' and 'Submarginal Land,'" 361.

24. For a brief biography of Wilson, see the historical note included in the introduction to his papers at Montana State University via http://www.lib .montana.edu/collect/spcoll/findaid/2100.html.

25. Wilson, "Problem of Poverty in Agriculture," 10, 13.

26. McMillan, "Social Problems of Farm Families," 38.

27. McMillan, "Social Problems of Farm Families," 9.

28. McMillan, "Social Problems of Farm Families," 63.

29. McMillan, "Social Problems of Farm Families," 26.

30. McMillan, "Social Problems of Farm Families," 43–44; quote from page 43.

31. McMillan, "Social Problems of Farm Families," 38–39.

32. McMillan, "Social Problems of Farm Families," 45.

33. Cotton, "Regulations of Farm Landlord-Tenant Relationships," 520. The issues of tenancy and tenants' treatment of land are more fully covered in chapter 5.

34. Sarah Phillips's work on the "New Conservationists" of the 1920s demonstrates the growing importance of land use planning as a way to reconcile land abuse and the depressed agricultural economy. See Phillips, *This Land, This Nation*, 9–12, 21–45; Phillips, "FDR, Hoover, and the New Rural Conservation."

35. Cannon, *Remaking the Agrarian Dream*, 2–6.

36. Jack N. French, "Annual Report, Extension Service, Prowers County, November 30, 1939, to November 30, 1940," folder 10, box 67; French, "Annual Report, Extension Service, Prowers County, November 30, 1940 to November 30, 1941," folder 11, box 67; *Lamar Daily News*, January 18, 1939.

37. *Lamar Daily News*, October 18, 1935.

38. *Lamar Daily News*, August 18, 1936, and January 21, 1937.

39. Cannon, *Remaking the Agrarian Dream*, 19–22, 59–62. See also J. E. Morrison, "An Evaluation of the Farmers Home Administration," 21, in folder Reports 1935–1950 (2), box 155, Records of the Colorado Cooperative Extension.

40. Division of Project Organization, Bureau of Agricultural Economics, USDA, "Land Acquisition Plan for Southeastern Colorado Land Utilization and Land Conservation Project, Part IV," May 1, 1938, 2–3, box 3 Colorado, folder Land Acquisition Plan—Southeastern Colo., SP 10 Records Relating to Land Utilization 1936–1939, Records of the SCS.

41. Division of Project Organization, "Land Acquisition Plan," 6.

42. Division of Project Organization, "Land Acquisition Plan," 7–8.

43. Raymond H. Skitt, "Annual Report, Extension Service, Baca County, December 1, 1937, to November 30, 1938," 40–42, folder 50, box 8.

44. Fiege, "Weedy West," 24.

45. *Lamar Daily News*, October 26, 1936.

46. Griffin and Stoll, "Evolutionary Processes in Soil Conservation Policy," 30.

47. Helms, *Readings in the History of the Soil Conservation Service*, 17, 21, 136.

48. "A Policy Statement on Relationships between the Soil Conservation Service and the Extension Service," March 29, 1940, box 6, folder Erosion Control Work—Co. Agents Assembly, SP 5 Records of Regional Conservator H. H. Finnell, 1934–1942, Records of the SCS.

49. Underwood, "Physical Land Conditions," 11. Douglas Helms has also written about these surveys, though he argues that they were often subjective. See Helms, "Development of the Land Capability Classification."

50. Underwood, "Physical Land Conditions," 11, 14, 17–18.

51. Underwood, "Physical Land Conditions," 28–29.

52. H. H. Finnell to Dr. Austin L. Patrick, n.d., folder CO-38-22 S E Colorado, Springfield, Colo., box 9 LU Corres.–1941–42—KA-38-21-Elkhart, Morton County, Kans., SP 5 Records of Regional Conservator H. H. Finnell, 1934–1942, Records of the SCS.

53. H. H. Bennett, "Statement by H. H. Bennett," 1, untitled folder, box 1, SP 19 General Correspondence 1935–1936, Records of the SCS.

54. Colorado State College, Extension Service, *Keeping the Farm at Home*, 4.

55. The explanation about FDR's involvement and the push to get local boards involved comes from "75 Years Helping People Help the Land: A Brief History of NRCS," accessed August 25, 2017, http://www.nrcs.usda.gov.

56. Colorado Soil Conservation Act, Soil Erosion Districts, Colorado House Bill 258 (1937), 1169–70.

57. "Memorandum of Procedure for the Organization of Soil Conservation Associations and the Development of a Soil and Moisture Conservation Program in Colorado," n.d., 32, 33, folder Digest of Federal Farm Programs in Colorado, box 155, Records of the Colorado Cooperative Extension. The loose paper is not dated, but its content and location in the box suggest it was written in 1935.

58. Colorado Soil Conservation Act, 1169–70.

59. Colorado Soil Conservation Act, 1173–77, 1189–90.

60. Colorado Soil Conservation Act, 1179–80.

61. Colorado Soil Conservation Act, 1179–80.

62. Colorado Soil Conservation Act, 1186, 1188.

63. "All Eyes on Baca County," *Garden City Daily Telegram*, n.d.

64. "Dust Bowlers Gird for Action," *Pueblo Star Journal*, n.d.

65. "Self-Help in Baca County," *Rocky Mountain News*, n.d.

66. A. J. Hamman, "Soil Conservation Specialists Report, 1938," folder Soil Conservation Specialist Reports 1938, box 127, Records of the Colorado Cooperative Extension.

67. A. J. Hamman, "Soil Conservation Specialists Report, 1941," folder Soil Conservation Specialist Reports 1941, box 127, Records of the Colorado Cooperative Extension.

68. Office of Board of Supervisors, "Western Baca County Soil Erosion District," 57–58.

69. Office of Board of Supervisors, "Western Baca County Soil Erosion District," 65.

70. Stewart, "Annual Report, Extension Soil Conservationist, December 1, 1937 to November 30, 1938," 96, folder 7, box 127, Records of the Colorado Cooperative Extension.

71. Raymond H. Skitt to Roy I. Kimmell, "Proposal Western Baca County Soil Erosion District," March 28, 1938, folder Baca County Soil Conservation District, box 15, November 19, '36–Dalhart, Tex.—Soil Cons. Dists. Colo., SP 18 Records of the Coordinator 1936–1942, Records of the SCS.

72. Skitt to Kimmell, "Proposal Western Baca County Soil Erosion District," 6–7, 8.

73. Leo E. Oyler, "Annual Report, Extension Service, Baca County, December 1, 1935, to November 30, 1936," 1–6, folder 47, box 8.

74. Leo E. Oyler, "Annual Report, Extension Service, Baca County, December 1, 1936, to November 30, 1937," 25–27, folder 49, box 8.

75. Raymond H. Skitt, "Annual Report, Extension Service, Baca County, December 1, 1937, to November 30, 1938," folder 50, box 8.

76. "Wind Erosion Control Program, Colorado Report, November 30, 1936," 5, folder 3, Report Wind Erosion Control Program, box 156, Records of the Colorado Cooperative Extension.

77. A. J. Hamman, "Annual Report, Extension Service, Prowers County, November 1, 1935, to October 31, 1936," folder 7, box 67.

78. Jack N. French, "Annual Report, Extension Service, Prowers County, December 1, 1937, to December 1, 1938," folder 9, box 67.

79. Jas. O. Dougan to Lauriston Walsh, April 12, 1940, folder Baca County Soil Conservation District, box 16 Western Baca Co. Dist. Rep—Water Facilities, Colo., SP 18 Records of the Coordinator, 1936–1942, Records of the SCS.

80. Walsh to Dougan, April 11, 1940, folder Baca County Soil Conservation District; box 16 Western Baca Co. Dist. Rep—Water Facilities, Colo., SP 18 Records of the Coordinator, 1936–1942, Records of the SCS.

81. Walsh to Dougan, April 16, 1940, folder Baca County Soil Conservation District, box 16 Western Baca Co. Dist. Rep—Water Facilities, Colo., SP 18 Records of the Coordinator, 1936–1942, Records of the SCS.

82. Dougan to Walsh, n.d., folder Baca County Soil Conservation District, box 16 Western Baca Co. Dist. Rep—Water Facilities, Colo., SP 18 Records of the Coordinator, 1936–1942, Records of the SCS. Italics in original.

83. See, e.g., Raymond H. Skitt, "Annual Report, Extension Service, Baca County, December 1, 1938, to November 15, 1939," 5–6, 44, folder 50, box 8; *Springfield Democrat Herald*, April 26, 1929, and June 13, 1935.

84. "Public Hearing of Southern Great Plains & the President's Great Plains Drought Committee," Dalhart TX, November 18 and 19, 1936, folder Southern Great Plains Region, box 1 General, SP 10 Records Relating to Land Utilization 1936–1939, Records of the SCS

85. W. K. Chalmers to Roy I. Kimmel, January 6, 1938, FD Baca County SC District: 1937–1939, Colo., box 16 Western Baca Co. Dist. Rep—Water Facilities, Colo., SP 18 Records of the Coordinator, 1936–1942, Records of the SCS.

86. J. H. Neal to H. H. Bennett, July 21, 1938, FD Baca County SC District: 1937–1939, Colo., box 16 Western Baca Co. Dist. Rep—Water Facilities, Colo., SP 18 Records of the Coordinator, 1936–1942, Records of the SCS.

87. W. O. Brown, transcript of telephone conversation with Al Hurt, July 2, 1938, FD Baca County SC District: 1937–1939, Colo., box 16 Western Baca Co. Dist. Rep—Water Facilities, Colo., SP 18 Records of the Coordinator, 1936–1942, Records of the SCS.

88. "Memorandum of Understanding," no author, n.d., folder Baca County SC District: 1937–1939, Colo., box 16 Western Baca Co. Dist. Rep—Water Facilities, Colo., SP 18 Records of the Coordinator, 1936–1942, Records of the SCS.

89. "Wind Erosion Control Program, Colorado Report, November 30, 1936," 5.

90. F. R. Stansbury and Norman Fuller to W. R. Watson, n.d., 2, folder 731-Agronomy, box 1, SP23 (BG) General Records, 1939–1941, Records of the SCS.

91. Two important examples of recent works on this developing thirst for outdoor recreation are Maher, *Nature's New Deal* and Gregg, *Managing the Mountains.*

92. *Springfield Democrat Herald*, May 11, 1939.

93. Cornebise, *CCC Chronicles*, 38.

94. A check of the CCC Collection illustrates these two themes among enrollees. Thanks to Gerald E. Sherard for taking the time to extract such information from CCC enrollment records and to the Colorado State Archives for making his lists available to the public. See http://www.colorado.gov/dpa/doit/archives/ccc/state_ccc.htm for additional information.

95. Jas. O. Dougan to Secretary Henry A. Wallace, August 16, 1940, folder Cooperation with C.C.C.—Colorado, box 3 Colorado, SP 3 Records Concerning Cooperation with Other Agencies 1937–1941, Records of the SCS; Hayden K. Rouse to A. E. McClymonds, August 14, 1936, untitled folder, box 2, SP 19 General Correspondence 1935–1936, Records of the SCS.

96. John J. Underwood, "Physical Land Conditions in the Western and Southeastern Baca County Soil Conservation Districts," 30–31.

97. Office of Board of Supervisors, "Western Baca County Soil Erosion District," 59.

98. Geographical and Geological Survey of the Rocky Mountain Region (U.S.), Powell et al., *Report on the Lands of the Arid Region of the United States*, 3–29.

99. T. G. Stewart, "Permanent Agricultural Program for the Southern Great Plains Region," in "Annual Report, December 1, 1936, to November 30, 1937," folder Soil Conservation Specialist Reports 1937, box 127, Records of the Colorado Cooperative Extension.

100. Underwood, "Physical Land Conditions in the Western and Southeastern Baca County Soil Conservation Districts," 28–29.

101. Great Plains Drought Area Committee, *Report of the Great Plains Drought Area Committee, August, 1936*, 6.

102. Stewart, "Permanent Agricultural Program," 21–23.

103. Claude E. Gausman, "Annual Report, Extension Service, Baca County, November 16, 1939, to November 1, 1940," folder 52, box 8.

104. Hurt, "National Grasslands," 256.

1. *Sixteenth Census of the United States: 1940.*

2. McHendrie, "Early History of Irrigation in Colorado," 14–16. See also Holleran, "Historic Context for Irrigation and Water Supply Ditches and Canals in Colorado."

3. Holleran, "Historic Context," 10–11. Additionally see Sanchez, *Forgotten Cuchareños of the Lower Valley.*

4. Betz, *Prowers County History*, 127–43.

5. Miles, "Salinity in the Arkansas Valley of Colorado," 3.

6. Betz, *Prowers County History*, 127–43; Athearn, *Land of Contrast*, chap. 8.

7. Hall, *History of the State of Colorado*, 282.

8. Sherow, "Utopia, Reality, and Irrigation," 174–75.

9. Sherow, "Utopia, Reality, and Irrigation," 4–7.

10. Sherow, *Watering the Valley*, 31.

11. Sherow, *Watering the Valley*, 30–32; Miles, "Salinity in the Arkansas Valley of Colorado," 12.

12. Van Hook, "Settlement and Economic Development," 285–304.

13. Van Hook, "Settlement and Economic Development," 304.

14. Miles, "Salinity in the Arkansas Valley of Colorado," 6–7

15. Baca County Historical Society, *Baca County*, 14.

16. Code, "Construction of Irrigation Wells," 4.

17. Code, "Construction of Irrigation Wells," 6.

18. Office of the State Engineer, *Fifteenth Biennial Report of the State Engineer to the Governor of Colorado for the Years 1909–10*, 125–26.

19. On the Carey Act, see Bonner, "Elwood Mead"; Gease, "William N. Byers and the Case for Federal Aid to Irrigation in the Arid West."

20. Hill, "A History of Baca County," 119–23, Baca County Historical Society, *Baca County*, 52–53

21. *United States Census of Agriculture 1925; United States Census of Agriculture 1935.*

22. *Fifteenth Census of the United States 1930.* Like the agricultural census data, population census information also comes from websites devoted to housing the documentation. See http://www.census.gov/.

23. John Martin to Lt. Col. E. Reybold, November 25, 1936, folder WPA–Arkansas River Prowers County, Colorado–Horse Creek Dam, box 75 WPA Projects

Arkansas River/Cucharas River/Huerfano River/Purgatoire River, Records of the Office of the Chief of Engineers. See also "Works Progress Administration Project Proposal" in the same folder.

24. Lowitt, *New Deal and the West*, 81.

25. Phillips, *This Land, This Nation*, 80–107.

26. "The Editor Speaking," *Lamar Daily News*, January 8, 1934.

27. "The Editor Speaking," *Lamar Daily News*, June 25, 1934.

28. "The Editor Speaking," *Lamar Daily News*, July 1, 1933.

29. *Lamar Daily News*, July 21, 1933.

30. "Ickes Irrigation Policy Flayed by Rep. Martin," *Lamar Daily News*, October 31, 1933.

31. John A. Martin, telegram to Harold Ickes as quoted in "Ickes Irrigation Policy Flayed by Rep. Martin."

32. "Ickes Irrigation Policy Flayed by Rep. Martin."

33. Harold Ickes, quoted in "Ickes Hails National Planning," *New York Times*, October 14, 1934.

34. Arkansas Basin Committee, "Request for Approval of the Caddoa Dam and Reservoir Project on the Arkansas River in Colorado," 1, FD 800.12 JMD Request for Approval of the Caddoa Dam and Reservoir Project, 15, box 4, Records of the Office of the Chief of Engineers.

35. Arkansas Basin Committee, "Request for Approval of the Caddoa Dam," 1–2.

36. Arkansas Basin Committee, "Request for Approval of the Caddoa Dam," 8–10.

37. Floyd E. Brown, "Annual Report of Floyd E. Brown, Extension Specialist in Irrigation Practice, December 1, 1938, to November 30, 1939," 90, folder Irrigation Practice Specialist Reports 1939, box 121, Records of the Colorado Cooperative Extension.

38. Brown, "Annual Report, 1938," 93.

39. Brown, "Annual Report, 1938," 94.

40. Floyd E. Brown, "Annual Report of Floyd E. Brown, Extension Specialist in Irrigation Practice, December 1, 1939, to November 30, 1940," 39, folder Irrigation Practice Specialist Reports 1940, box 121, Records of the Colorado Cooperative Extension.

41. Brown, "Annual Report, 1939," 7, 14.

42. Brown, "Annual Report, 1939," 14.

43. Brown, "Annual Report, 1939," 4.

44. Sherow, "Discord in the 'Valley of Content,'" 326–29. Sherow goes into deep detail about conversations and communication about the dam between locals, politicians, and federal agency officials. This explanation is thus only a recapitulation of the important moments within the larger story.

45. Sherow, "Discord in the 'Valley of Content,'" 329–31.

46. Sherow, *Watering the Valley*, 141–67.

47. Sherow, "Discord in the 'Valley of Content,'" 332, 335–36.

48. *Congressional Record*, 77th Cong., 1st sess., folder 15 Arkansas Valley Authority Act 1941–HR. 18223 + S. 280–Cong. Record Statements, Resolutions, etc., box 73, Chenoweth Collection.

49. The information about Johnson comes from a brief introduction to his records that are held at the Colorado State Archives. See "Colorado Governors," accessed August 25, 2017, http://www.colorado.gov/dpa/doit/archives/govs/johnson.html.

50. Edwin Johnson, *Congressional Record*, 6, folder 15 Arkansas Valley Authority Act 1941–HR. 18223 + S. 280–Cong. Record Statements, Resolutions, etc., box 73, Chenoweth Collection.

51. Farley, "Colorado and the Arkansas Valley Authority," 226

52. Governor Ralph Carr speech, n.d., 6, FD 15 Arkansas Valley Authority Act 1941–HR. 18223 + S. 280–Cong. Record Statements, Resolutions, etc., box 73, Chenoweth Collection.

53. Governor Ralph Carr speech, 8.

54. Clifford H. Stone to J. Edgar Chenoweth, February 11, 1941, FD 14 Water–Arkansas Valley Authority Act H.R. 1823 1941–Correspondance/General, box 73, Chenoweth Collection.

55. Chenoweth to Wilbur B. Foshay, January 24, 1941, folder 14 Water–Arkansas Valley Authority Act H.R. 1823 1941–Correspondance/General, box 73, Chenoweth Collection.

56. Foshay to Chenoweth, February 13, 1941, folder 14 Water–Arkansas Valley Authority Act H.R. 1823 1941–Correspondance/General, box 73, Chenoweth Collection.

57. Floyd M. Wilson to Alva B. Adams, February 8, 1941, emphasis in original, folder 14 Water–Arkansas Valley Authority Act H.R. 1823 1941–Correspondance/General, box 73, Chenoweth Collection.

58. Governor Ralph Carr speech, Washington DC, April 1941, folder 15 Arkansas Valley Authority Act 1941–HR. 18223 + S. 280–Cong. Record, Statements, Resolutions, etc., box 73, Chenoweth Collection.

59. "Resolutions Adopted by Conference of Governors of Western States Respecting Proposed Arkansas Valley Authority Act of 1941," February 7, 1941, folder 15 Arkansas Valley Authority Act 1941–HR. 18223 + S. 280–Cong. Record, Statements, Resolutions, etc., box 73, Chenoweth Collection.

60. *Lamar Daily News*, January 13, 1941.

61. Farley, "Colorado and the Arkansas Valley Authority," quotes from 234 and 232, respectively.

62. Welsh, *U.S. Army Corps of Engineers*, 60–61.

63. Welsh, *U.S. Army Corps of Engineers,* 69.

64. J. Edgar Chenoweth to Vera Pointer, May 10, 1944, box 73, folder 23 Water–Caddoa Dam + John Martin Reservoir on Arkansas River 1941–1944, box 73, Chenoweth Collection.

65. See Welsh, *U.S. Army Corps of Engineers*, 70–71, and Sherow, *Watering the Valley*, 154–65.

66. Max Mills, "Annual Report, Extension Service, Prowers County, December 1, 1944, to December 1, 1945," Prowers County Annual Reports, 1944–1945, 100, folder 16, box 67, Records of the Colorado Cooperative Extension.

67. Gutman, Great Plains Population and Environment Data.

68. A. J. Hamman, "Annual Report, Extension Service, Prowers County, November 1, 1935, to October 31, 1936," 1, folder 7, box 67.

69. A. J. Hamman, "Annual Report, Extension Service, Prowers County, November 1, 1936, to November 30, 1937," 1, folder 8, box 67.

70. These numbers come from calculations I made using census data from the 1930s and 1954 to check long-term trends. See *United States Census of Agriculture 1935*; *Sixteenth Census of the United States, 1940: Agriculture*; and *United States Census of Agriculture: 1954*.

71. *Sixteenth Census of the United States, 1940: Agriculture*.

72. *Sixteenth Census of the United States: 1940*.

73. Doherty, "Effects on Farmers of Change from Dryland to Irrigation in Baca County," 49–56.

74. Hamman, *My Long Journey*, 146.

75. Jack French, "Annual Report, Extension Service, Prowers County, November 30, 1939 to November 30, 1940," 148, folder 11, box 67. Also see the Water Commissioner's Field Records, reproduced in *Lamar Daily News,* June 22, 1940.

76. *Lamar Daily News,* January 31, 1939. See also *Lamar Daily News,* February 8, 1940.

77. Jack French, "Annual Report, Extension Service, Prowers County, November 30, 1940, to November 30, 1941," folder 12, box 8.

78. Jack French, "Annual Report, Extension Service, Prowers County, November 30, 1939, to November 30, 1940," 19, folder 11, box 8.

79. Max B. Mills, "Annual Report, Extension Service, Prowers County, June 12, 1944, to December 1, 1944," 24, folder 15, box 8.

80. A. J. Hamman, "Annual Report, Extension Service, Prowers County, November 1, 1936, to November 30, 1937," 92, folder 8, box 8.

81. A comparison of the narrative and the statistical summaries from agents in the two counties expose the difference.

82. *Lamar Daily News,* December 8, 1941.

83. *Lamar Daily News,* July 27, 1933.

5. ON THE MOVE

1. Claude E. Gausman, "Annual Report, Extension Service, Baca County, November 1, 1939 to November 1, 1940," folder 52, box 8, in Records of the Colorado Cooperative Extension.

2. Neil Maher discusses the way that community formation failed to abide by the grid patterns the federal government employed when trying to organize the West. See Maher, "'Crazy Quilt Farming on Round Land.'"

3. *Sixteenth Census of the United States: 1940.*

4. The best example is Gregory, *American Exodus.* Various works on the Dust Bowl also deal with the issue of migration, including Riney-Kehrberg, *Rooted in Dust* and Fearon, *Kansas in the Great Depression.*

5. Cunfer, *On the Great Plains,* 5–6, 19–35.

6. *Sixteenth Census of the United States, 1940: Agriculture*; Truedell, "Farm Tenancy Moves West."

7. A number of historians and New Deal contemporaries indicted the tenant system for neglecting to help tenants during the 1930s. See Volanto, "Leaving

the Land"; Cotton, "Regulations of Farm Landlord-Tenant Relationships";
Maddox, "Bankhead-Jones Farm Tenant Act."

8. S-C Lee, "Theory of the Agricultural Ladder," 53; Harris, "New Agricultural
Ladder," 258–59.

9. C. Rasmussen, "'Never a Landlord for the Good of the Land,'" 71. See also
Baldwin, *Poverty and Politics*, 22–46.

10. Gray et al., "Farm Tenancy," iii.

11. Gray et al., "Farm Tenancy," 5.

12. Gray et al., "Farm Tenancy," 10–23.

13. Robert T. McMillan, "Social Problems of Farm Families in Baca County,
Colorado," Report I of Sociology Section, Land Use Planning Division,
Resettlement Administration, Region XII (Amarillo TX, 1937), 43–44, folder
Baca County Sociological Study, box 2 General, SP 10 Records Relating to
Land Utilization 1936–1939, Records of the SCS.

14. McMillan, "Social Problems of Farm Families," 52, 63.

15. Flip F. Higbee, *Springfield Democrat Herald,* December 29, 1938.

16. McMillan, "Social Problems of Farm Families," 45.

17. Gray et al., "Farm Tenancy," 7.

18. Gray et al., "Farm Tenancy," 7.

19. *Lamar Daily News*, January 18, 1939

20. Harbaugh, "Twentieth-Century Tenancy," 97–101; Worster, *Dust Bowl,*
189. Harbaugh suggests that such reputations were often unwarranted and
unearned, given to tenants because they often challenged the idyllic notion
that agriculture was the safest path to competency. He argues that the evidence collected during the 1910s and 1920s promoted as proof that tenants
disparaged owners did not take an adequate survey of the country, nor did
it reflect variables that could adversely impact the soil such as wind, climate,
or crop choices.

21. Harbaugh, "Twentieth-Century Tenancy," 107; McMillan, "Social Problems
of Farm Families," 52.

22. Claude E. Gausman, "Annual Report, Extension Service, Baca County,
November 30, 1940, to November 30, 1941," 27–32, folder 52, box 8. .

23. Baldwin, *Poverty and Politics*, 94–95, 157–61. Other policies like the second
AAA and the Soil Conservation and Domestic Allotment Act tried to protect tenants more than the original by mandating that owners pay tenants

for their part in conservation or land retirement, but there was no serious mechanism to force owners to share their subsidies.

24. Maddox, "Bankhead-Jones Farm Tenant Act," 446.

25. Maddox, "Bankhead-Jones Farm Tenant Act," 447–48.

26. Jack N. French, "Annual Report, Extension Service, Prowers County, November 30, 1939 to November 30, 1940," folder 10, box 67; French, "Annual Report, Extension Service, Prowers County, November 30, 1940, to November 30, 1941," folder 11, box 67; *Lamar Daily News*, January 18, 1939.

27. Gausman, "Annual Report, Extension Service, Baca County, November 30, 1940 to November 30, 1941," 38, folder 52, box 8.

28. J. E. Morrison, "An Evaluation of the Farmers Home Administration," 1–3, folder Reports 1935–1950 (1), box 155.

29. Markoff, "Sugar Beet Industry," 70–73.

30. Hendrickson, "Sugar-Beet Laborer and the Federal Government," 45–46.

31. A. J. Hamman, "Annual Report of A. J. Hamman, State Supervisor Emergency Farm Labor Program, February 15, 1943 to December 31, 1943" (Fort Collins: Colorado State College of A. & M.A. and United States Department of Agriculture in Cooperation with Office of Labor, War Food Administration, 1944), 4.

32. A. J. Hamman, "Annual Report," 1. Hamman is the epitome of increased state involvement after 1933, having served as County Extension agent in Prowers County and then as head of the Irrigation Division in southeast Colorado before taking over management of the agricultural labor division of wartime employment.

33. Deutsch, *No Separate Refuge*, 108.

34. Deutsch, *No Separate Refuge,* 120.

35. Hamman, "Annual Report," 2.

36. Deutsch, *No Separate Refuge*.

37. Valdés, "Settlers, Sojourners, and Proletarians," 113.

38. Hamman, "Annual Report ," 2.

39. *Fifteenth Census of the United States: 1930*.

40. *Lamar Daily News*, September 9, 1933.

41. *Lamar Daily News*, May 6, 1938.

42. On the Wiley notice, see *Lamar Daily News*, November 13, 1933; for Lamar information, see Hamman, "Annual Report, Extension Service, Prowers County, November 1, 1935, to October 31, 1936," folder 6, box 67.

43. *Lamar Daily News*, January 12, 1934.

44. Historians have looked in two distinct ways at the *colonias*, the name often given to the Latino neighborhoods that sprouted up in sugar beet areas and other places where immigrant labor emerged. On the one hand, some have argued that such neighborhoods indicate ethnic solidarity and migrants' concerted effort made by migrants to maintain cultural independence while also taking advantage of economic opportunities. Conversely, other scholars have perceived such demographic patterns as an indication of racism and discrimination. These folks were effectively "colonized," and neither the recruiters who shuttled them to the area or the white residents living near them believed that they could—or should—amount to more than simple manual laborers. Moreover, Donato postulates that the migrant workers-turned-residents were exploited; they were underemployed, lived in relative squalor, and worked under deplorable conditions. See Donato, "Sugar Beets, Segregation, and Schools," 69–88. For further support of Donato's claims that the labor system was especially exploitive in northern Colorado fields, see *Denver Catholic Register*, May 23, 1929; Thos F. Mahoney, chairman of Longmont Council of Mexican Welfare Committee, Knights of Columbus, to John E. Gross, secretary-treasurer Colorado State Federation of Labor, January 25, 1930, folder Beet Field Workers' Conference–Greeley March 22, 1936–Correspondence with Locals August 1, 1936, box 46, Colorado State Federation of Labor, University of Colorado Archives.

45. Hendrickson, "Sugar-Beet Laborer and the Federal Government," 49.

46. It does seem that the AFL had some trepidation about the potential for extremism among agricultural workers. A letter from the AFL president, William Green, to the secretary-treasurer of the Colorado State Federation of Labor, John E. Gross, suggests that there could be some concern for incoming labor leader Paul Arias, who took over as president of the Colorado State Federation Agricultural Workers Unions. William Green to John E. Gross, n.d., Colorado State Federation of Labor, folder Beet Field Workers' Conference–Greeley March 22, 1936–Correspondence with Locals August 1, 1936, box 46.

47. "Constitution and By-Laws of the Federated Beet Workers of Colorado," October 1935, folder Beet Field Workers' Conference–Greeley March 22, 1936–Correspondence with Locals August 1, 1936, box 46.

48. Hendrickson, "Sugar-Beet Laborer and the Federal Government," 50–51.

49. Ham, "Sugar Beet Field Labor under the AAA," 643–44.

50. Hyde and McClelland, *History of the Extension Service*," 106; see also *Lamar Daily News*, January 24, 1934, and May 11, 1934.

51. Two of the better examples are L. Cohen, *Making a New Deal* and Denning, *Cultural Front*.

52. Balderrama and Rodríguez, *Decade of Betrayal*, 2–4, 75–76. Additional coverage of the repatriation saga can be found in Reisler, *By the Sweat of Their Brow* and Kirstein, *Anglo over Bracero*.

53. Deutsch, *No Separate Refuge*, 163–67.

54. Oppenheimer, "Acculturation or Assimilation," 165–67; Kulkosky, "Mexican Migrant Workers," 128.

55. Kulkosky, "Mexican Migrant Workers," 125. Eric Meeks exposes similar arguments made in favor of repatriation in Arizona. See Meeks, "Protecting the 'White Citizen Worker.'"

56. Kulkosky, "Mexican Migrant Workers," 127. Oppenheimer rightly points out that many of the reasons for this reluctance spread from white attempts to segregate and discriminate against the immigrants.

57. Deutsch, *No Separate Refuge*, 16.

58. Kulkosky, "Mexican Migrant Workers," 129–30. Conversely, John Gross complained to AFL president William Green and Harry Hopkins, head of the WPA, that the agency had done too little to help sugar beet workers and had "immediately and arbitrarily" taken workers off the WPA relief rolls. See Gross to Green, March 23, 1936, folder Beet Field Workers' Conference-Greeley March 22, 1936–Correspondence with Locals August 1, 1936, box 46. See also *Pueblo Star Journal*, April 24, 1936.

59. Francisco E. Balderrama and Raymond Rodríguez suggest that a more accurate assessment, based on Mexican documents, is closer to one million people sent home during the 1920s and 1930s. See Balderrama and Rodríguez, *Decade of Betrayal*, 122.

60. Deutsch, *No Separate Refuge*, 165–66.

61. *Lamar Daily News*, May 7, 1935.

62. *Lamar Daily News*, April 20, 1936; *Springfield Democrat Herald*, April 23, 1936.

63. "Alien Labor Found with Rail Passes," *Denver Post*, April 22, 1936.

64. *Lamar Daily News*, April 21, 1936.

65. *Lamar Daily News*, April 21, 1936.

66. *Lamar Daily News*, April 20, 1936

67. Deutsch, *No Separate Refuge*, 166.

68. *Lamar Daily News*, April 30, 1936.

69. *Lamar Daily News*, May 26, 1936. Prowers County agent A. J. Hamman noted widespread resistance to the Works Progress Administration for the same reason. See Hamman, "Annual Report, Extension Service, Prowers County, November 1, 1935 to October 31, 1936," folder 6, box 67.

70. See, e.g., *Springfield Democrat Herald*, July 1, 1937; *Springfield Democrat Herald*, August 25, 1938; *Lamar Daily News*, August 17, 1940

71. Ubbelohde, Benson, and Smith, *A Colorado History*, 321–25; Abbott, Leonard, and Noel, *Colorado*, 297–307. Quote reprinted from Abbott et al., 306.

72. Claude E. Gausman, "Annual Report, Extension Service, Baca County, November 1, 1941, to December 30, 1941," folder 53, box 8; Claude E. Gausman, "Annual Report, Extension Service, Baca County, January 1, 1942, to December 1, 1942," 14, folder 54, box 8.

6. FOOD FOR VICTORY

1. Gutmann, Great Plains Population and Environment Data.

2. *Lamar Daily News*, January 9, 1939.

3. *Lamar Daily News*, February 26, 1939.

4. *Springfield Democrat Herald*, April 20, 1939.

5. *Lamar Daily News*, September 11, 1939.

6. *Lamar Daily News*, October 26, 1939.

7. *Springfield Plainsman Herald*, January 30, 1941.

8. Wickard, "Food Will Win the War and Write the Peace."

9. *Plainsman Herald*, December 10, 1942.

10. H. H. Finnell, quoted in *Lamar Daily News*, July 14, 1941.

11. *Lamar Daily News*, December 31, 1942.

12. *Lamar Daily News*, January 8, 1941.

13. *Plainsman Herald*, September 14, 1944; *Lamar Daily News*, March 5, 1942.

14. Claude E. Gausman, "Annual Report, Extension Service, Baca County, November 1, 1941, to December 1, 1942," folder 53, box 8, 15. *Lamar Daily News* covered the debate for much of spring 1942; see, e.g., *Lamar Daily News*, May 6, 1942.

15. *Lamar Daily News*, February 16, 1944.

16. *Lamar Daily News*, June 2, 1942.

17. Jack N. French, "Annual Report, Extension Service, Prowers County, November 30, 1940, to November 30, 1941," folder 11, box 67.

18. Fiset, "Thinning, Topping, and Loading," 126. See also stories in *Lamar Daily News*, February 5, 1942, and February 19, 1942.

19. *Lamar Daily News*, March 4, 1942.

20. Anderson and Hamman, *Resume*, 2–3.

21. Hamman, *My Long Journey*, 136.

22. Anderson and Hamman, *Resume* , 13.

23. A. J. Hamman, "How to Fight on the Farm Front," folder Annual Report—Emergency Farm Labor 1943, box 137, Records of the Colorado Cooperative Extension.

24. A. J. Hamman, "Annual Report of A. J. Hamman, State Supervisor Emergency Farm Labor Program, February 15, 1943, to December 31, 1943," 49, folder Annual Report—Emergency Farm Labor 1943, box 137.

25. Carpenter, "'Regular Farm Girl,'" 163–64; "U.S. to Mobilize Land Army of over 3 Millions," *Denver Post*, January 25, 1943.

26. Hamman, "Annual Report, State Supervisor Emergency Farm Labor Program," 24–25, 61–62.

27. Hamman, "Annual Report, State Supervisor Emergency Farm Labor Program," 24–40.

28. A solid description of the early efforts at starting the program and its goals can be found in Galarza, *Merchants of Labor*, 45–48. On Colorado, see the brief account in Hamman, "Annual Report, State Supervisor Emergency Farm Labor Program" 28–29.

29. Anderson and Hamman, *Resume*, 24–26.

30. Three examples from Cindy Hahamovitch cover the importation of Jamaican workers but focus more closely on the eastern United States. See Hahamovitch, "'In America Life Is Given Away,'" 134–60; Hahamovitch, *Fruits of Their Labor*; Hahamovitch, *No Man's Land*.

31. Max Mills, "Annual Report, Extension Service, Prowers County, December 1, 1944, to December 1, 1945," 28, folder 16, box 67. Hamman noted that "we" called Jamaicans "Jakes." See Hamman, *My Long Journey*, 142.

32. Hamman, "Annual Report, State Supervisor Emergency Farm Labor Program," 33, folder Annual Report Emergency Farm Labor 1944, box 137.

33. Hamman, *My Long Journey*, 142.

34. Hamman, "Annual Report, State Supervisor Emergency Farm Labor Program," 33.

35. *Lamar Daily News*, May 22, 1945.

36. There is little attention to the program in Colorado, with most of the literature focusing on California. See, e.g., Mitchell, *They Saved the Crops*; Galarza, *Merchants of Labor*; Kirstein, *Anglo over Bracero*; Driscoll, *Tracks North*. An important corrective is Hurt, *Great Plains during World War II*, 217–24.

37. *Lamar Daily News*, March 24, 1944, and May 22, 1945.

38. Hamman, "Annual Report, State Supervisor Emergency Farm Labor Program," 28–29.

39. Hamman, "Annual Report, State Supervisor Emergency Farm Labor Program," 29.

40. W. Rasmussen, *Emergency Farm Labor Supply Program, 1943–1947*, 229.

41. W. Rasmussen, *Emergency Farm Labor Supply Program, 1943–1947*, 264.

42. W. Rasmussen, *Emergency Farm Labor Supply Program, 1943–1947*, 232.

43. W. Rasmussen, *Emergency Farm Labor Supply Program, 1943–1947*, 258–60; Hahamovitch, *No Man's Land*, 67–85.

44. Hamman, "Annual Report, State Supervisor Emergency Farm Labor Program, 1944," 20–22.

45. W. Rasmussen, *Emergency Farm Labor Supply Program, 1943–1947*, 219–25.

46. Hamman, "Annual Report, State Supervisor Emergency Farm Labor Program" 40.

47. Anderson and Hamman, *Resume*, 32–34

48. Anderson and Hamman, *Resume*, 30–34.

49. On German POWs broadly, see Gansberg, *Stalag, U.S.A*; G. Thompson, *Prisoners on the Plains*; A. Thompson, *Men in German Uniform*. On German POWs in Colorado, see Paschal, "Enemy in Colorado"; Jepson, "Camp Carson, Colorado," 32–53.

50. Landsberger, *Prisoners of War at Camp Trinidad*, 7–12. Landsberger worked as a translator at the camp. His recollection of the early days, combined with his research in local newspapers, grounds this description. His work is one of the only examples to deal extensively with the situation at Camp Trinidad.

51. Landsberger, *Prisoners of War at Camp Trinidad*, 45–60.

52. Hamman, "Annual Report, State Supervisor Emergency Farm Labor Program, 1944," 26–28.

53. Anderson and Hamman, *Resume*, 3, 20.

54. Hamman, *My Long Journey*, 141.

55. FDR's decision to do so is still a point of debate in some circles. There are a number of works devoted to the order, the philosophy that led to its creation, and the first stages of evacuation. Roger Daniels is at the forefront of scholars of internment. See Daniels, *Prisoners without Trial*. Analysis of the decision and its impact on Colorado can be found in Harvey, *Amache*; Schrager, *Principled Politician*. Karl Lillquist's recent work best addresses internee agricultural labor even though he focuses on camp production rather than internees as paid labor outside the camps. Lillquist, "Farming the Desert."

56. For a good overview of the labor element of internment, see Heimburger, "Life beyond Barbed Wire."

57. See brief explanations of the population pre–World War II in Abbott, Leonard, and Noel, *Colorado*, 302–4; Ubbelohde, Benson, and Smith, *A Colorado History*, 324–26. A more complex view of the population's development up to and during the early stages of WWII can be found in chaps. 8–10 in Hosokawa, *Colorado's Japanese Americans*. The details on the Prowers County families come from *Lamar Daily News*, January 24, 1936.

58. The *Lamar Daily News* claimed that the population of Issei and Nisei increased by about 30 percent, to roughly 5,000 total in the state. See *Lamar Daily News* May 5, 1942.

59. Schrager, *Principled Politician*, 166.

60. Hosokawa, *Colorado's Japanese Americans*, 87–91.

61. See Harvey, *Amache*, 25–48; Schrager, *Principled Politician*, 140–46; *Lamar Daily News*, March 30, 1942.

62. Schrager, *Principled Politician*, 187, 284.

63. Hosokawa, *Colorado's Japanese Americans*, 90–94.

64. J. Smith, "New Deal Public Works at War," 67

65. See M. Johnson, "At Home in Amache," 5–8, and Wei, "'The Strangest City in Colorado,'" 4.

66. Robert Harvey covers the debates in Denver papers well. See Harvey, *Amache*, 118–24; Takahara, *Off the Fat of the Land*, 36–45.

67. *Plainsman Herald*, September 3, 1942.

68. *Lamar Daily News* September 19, 1942.

69. *Plainsman Herald*, September 10, 1942.

70. *Lamar Daily News*, June 4–6, 1942. The paper published the WRB's statements over a three-day period.

71. *Granada Bulletin*, October 17, 1942.

72. Joseph McClelland interview with Louise Bashford, April 8, 1981. Transcript and audio tape available in Joseph McClelland Collection, University of Colorado–Denver archives.

73. *Granada Pioneer*, October 31, 1942.

74. *Granada Bulletin*, October 24, 1942.

75. Heimburger, "Life beyond Barbed Wire," 33n92.

76. Mildred Garrison interview with Louise Bachford, March 26, 1981, 2–3. Transcript and audio tape available in McClelland Collection.

77. Anderson and Hamman, *Resume*, 50.

78. Claude Gausman, "Annual Report, November 1, 1941, to December 1, 1942," 12, folder 54, box 8.

79. Jack N. French, "Annual Report, December 1, 1941, to September 01, 1942," 5, folder 13, box 67.

80. Colorado Preservation, Baca County Survey, accessed August 25, 2017, http:// www.coloradopreservation.org/crsurvey/rural/baca/sites/baca_resources _agriculture.html.

81. Martin Eriksen, "Annual Report, Extension Service, Baca County, April 16, 1943, to November 30, 1943," folder 55, box 8; *Plainsman Herald*, September 30, 1943.

82. Martin Eriksen, "Annual Report, Extension Service, Baca County, December 1, 1943 to December 1, 1944," folder 56, box 8.

83. Anderson and Hamman, *Resume*, 17, 25.

84. Martin Eriksen, "Annual Report, Extension Service, Baca County, December 1, 1944, to November 30, 1945," folder 57, box 8.

85. Two examples of this optimism and boosterism are found in *Plainsman Herald*, December 13 and 27, 1945.

86. *Lamar Daily News*, March 25, 1943, and April 22, 1943.

87. "Governor Vivian Renews Demands Army Farmers Be Released," *Denver Post*, April 16, 1943; Hurt, *Great Plains during World War II*, 203–4.

88. Digital Collections of Colorado show Gausman as a member of sports teams and a member of the Sigma Phi Epsilon Fraternity during his days in Fort Collins. See http://lib.colostate.edu/digital-collections/.

89. Claude Gausman, "Annual Report, Extension Service, Prowers County, December 1, 1942, to December 1, 1943," 18–21, folder 14, box 67.

90. *Lamar Daily News*, June 29, 1942.

91. *Lamar Daily News*, September 10, 1942

92. *Granada Pioneer*, November 14, 1942.

93. *Granada Pioneer*, December 24, 1942.

94. *Granada Pioneer*, December 24, 1942.

95. *Granada Pioneer*, November 3, 1943.

96. Heimburger, "Life beyond Barbed Wire," 32n56.

97. McClelland interview with Bashford, April 8, 1981, 13.

98. Heimburger, "Life beyond Barbed Wire," 15.

99. Lillquist, "Farming the Desert," 83–95; Wei, "'The Strangest City in Colorado,'" 6.

100. Lillquist, "Farming the Desert," 85.

101. Heimburger, "Life beyond Barbed Wire," 21.

102. Claude Gausman, "Annual Report, Extension Service, Prowers County, December 1, 1943, to June 1, 1944," 26, folder 15, box 67.

103. *Lamar Daily News*, March 24, 1944.

104. Anderson and Hamman, *Resume*, 24–33.

105. Hamman, "Annual Report, State Supervisor Emergency Farm Labor Program," 33; Max Mills, "Annual Report, Extension Service, Prowers County, December 1, 1944, to December 1, 1945," 28, folder 16, box 67.

106. Anderson and Hamman, *Resume*, 4–5.

107. Anderson and Hamman, *Resume*, 35–36.

108. Anderson and Hamman, *Resume*, 37.

109. Hamman, "Annual Report, State Supervisor Emergency Farm Labor Program, 1944," 13; Anderson and Hamman, *Resume*, 40–41.

110. Anderson and Hamman, *Resume*, 38–40.

111. Anderson and Hamman, *Resume*, 38–39. See also the description of Jamaican workers offered in Max Mills, "Annual Report, Extension Service, Prowers County, December 1, 1944, to December 1, 1945," 28, folder 16, box 67.

112. Landsberger, *Prisoners of War at Camp Trinidad*, 181.

113. Landsberger, *Prisoners of War at Camp Trinidad*, 38–40.

114. *Lamar Daily News*, March 18, 1939.

115. *Lamar Daily News*, January 7, 1939.

116. A. J. Hamman, "Annual Report, Extension Service, Prowers County, November 1, 1936, to November 30, 1937," 51, folder 8, box 67.

117. Anderson and Hamman, *Resume*, 38–39.

118. This counters the still popular "transformation thesis" that suggests the war, in and of itself, was transformative for the West. To appreciate the role of the New Deal and these connections, however, shows that the war was a contributing factor to big changes but not the exclusive reason. See Nash, *American West Transformed*. Nearly every book published about the war after this work had to deal with the thesis in one way or another; most often the subsequent literature investigated one part of Nash's argument to consider his general position. Also see Lotchin, *Bad City in the Good War*.

119. Cochrane, *Development of American Agriculture*, 127.

120. Brief explanations of this shift can be found in Cochrane, *Development of American Agriculture*, 132–36; Dimitri, Effland, and Conklin, *20th Century Transformation*, 4–6.

121. Dimitri, Effland, and Conklin, *20th Century Transformation*, 9.

122. *Plainsman Herald*, October 9, 1941.

123. *Lamar Daily News*, December 8, 1941.

124. *Lamar Daily News*, October 6, 1943.

7. AN UNQUENCHABLE THIRST

1. "Financial Campaign to Support Frying Pan Project Gets Underway in Prowers County," *Lamar Daily News*, January 14, 1955. See also Milenski, "Short and Early History of Fry Ark Project," 1

2. A. J. Hamman[?], "Soil Conservation Specialist Report," box 127, folder 6.

3. "The Editor Speaking," *Lamar Daily News*, July 6, 1953.

4. Wiener, Pulwarty, and War, "Bite without Bark."

5. Milenski, "Short and Early History of Fry Ark Project," 4.

6. "High Wind and Dust Stops Traffic Lamar–Springfield," *Plainsman Herald*, January 3, 1952.

7. "Springfield Men Have Pneumonia," *Plainsman Herald*, March 4, 1954.

8. "Heavy Winds Spread Dust over County," *Plainsman Herald*, February 25, 1954.

9. "Rain Breaks the Protracted Drouth; 1 to 2 Inches Fell," *Plainsman Herald*, April 20, 1954.

10. "State's Crop Production Reported Smallest since 1935 Drought Year," *Plainsman Herald*, July 22, 1954.

11. "The Editor Speaking," *Lamar Daily News*, January 26, 1950.

12. "Public Forum—Little Dust Not Fatal," *Lamar Daily News*, July 16, 1953.

13. Chester Fithian, "Annual Report, Extension Service, Baca County, December 1, 1953 to November 30, 1954," 46, folder 52, box 8.

14. Nace and Pluhowski, "Drought of the 1950's," 50, 53–54.

15. "Governor Assures Area Drouth Aide, 'This Is No Funeral,'" *Lamar Daily News*, June 30, 1953.

16. "State Drouth Disaster Committee to Meet Here," *Lamar Daily News*, June 3, 1953; "Colorado Delegation Moves to Aid Governor in Fight against Drouth," *Lamar Daily News*, June 6, 1953.

17. "Baca County Included in Drouth Area," *Plainsman Herald*, August 7, 1952.

18. "Secretary Benson and Aides Stop Here on Drought Tour," *Plainsman Herald*, April 28, 1955.

19. "Rain Brings Hope to Dust Bowl Counties, Southeastern Colo.," *Plainsman Herald*, March 25, 1954.

20. "Rain Falls on Western Tri-State Area," *Lamar Daily News*, March 31, 1953; "Million Dollar Rain, Snow Soak State Aids Wheat," *Lamar Daily News*, March 31, 1953.

21. "Million Dollar Rain in County, Everybody Happy; Grass and Row Crops Grow," *Plainsman Herald*, July 25, 1954.

22. See Anderson, *Industrializing the Corn Belt*.

23. See Courtwright, "On the Edge of the Possible"; Spence, *Rainmakers*; Whitaker, "Making War on Jupiter Pluvius."

24. "Rainmaker Active in Baca County, Paid-off Monday," *Plainsman Herald*, April 6, 1950.

25. "C of C Endorses Rain Increase Project in Valley," *Lamar Daily News*, September 6, 1950.

26. "Valley Rain-Increasing Association Nears Goal of $45,000 for Full Year's Program," *Lamar Daily News*, September 26, 1950.

27. "Rainmakers Sign Contract with Farmers-Ranchers," *Plainsman Herald*, November 2, 1950.

28. "Rain-Maker Says Saved S.E. Colo. from Bad Drouth," *Lamar Daily News*, September 18, 1950.

29. "Did Cloud-Seeding Increase the Snow Fall?" *Plainsman Herald*, January 18, 1951.

30. "Plainsman Ads Get Results, It Helped Get Rain Machine," *Plainsman Herald*, February 1, 1951.

31. "The Editor Speaking," *Lamer Daily News*, September 18, 1954 and October 7, 1954.

32. "High Interest Shown in Weather Modification Project," *Lamar Daily News*, August 1, 1964; "Denver 'Cloud Master' Hired for Rain Project," *Lamar Daily News*, August 14, 1964.

33. "Cloud Seeding in Baca County Off; Can't Raise Funds," *Plainsman Herald*, August 9, 1951.

34. "Baca County's Come-back Is Like a Fairy Tale, but True," *Plainsman Herald*, September 7, 1950.

35. "Million Dollar Snow Fell Here Saturday AM," *Plainsman Herald*, January 18, 1951.

36. Riney-Kehrberg, "From the Horse's Mouth," 150.

37. "New Water Supply Might Make Irrig. of Drylands Soon," *Plainsman Herald*, December 2, 1954.

38. *Plainsman Herald*, February 17, 1955.

39. *Holly Chieftain*, January 10, 1958.

40. "Gov. Thornton Favors Research for Underground Water Sources," *Plainsman Herald*, April 1, 1954.

41. See, e.g., "Thornton Urges Quick Action on Drought Relief Program," *Plainsman Herald*, and "Farmers Discuss Erosion Control for Baca County," *Plainsman Herald*, both from February 25, 1954.

42. Arwin M. Bolin, "Annual Report, Extension Service, Baca County, December 1, 1955, to October 31, 1956," 65, box 9, folder 2.

43. Arwin M. Bolin, "Annual Report, Extension Service, Baca County, December 1, 1956, to October 31, 1957," 43, box 9, folder 3.

44. Colorado Division of Water Resources, "History of Water Rights" http://water.state.co.us/SurfaceWater/swrights/Pages/wrhistory.aspx.

45. "State Farm Group Proposes Ground Water Law," *Lamar Daily News*, December 15, 1956.

46. "Pump Irrigation Triples in 6 Years," *Lamar Daily News*, March 2, 1953.

47. This trend had started during the 1930s but briefly reversed during some of the 1940s when population seemed to stabilize. The drought of the 1950s was the most likely reason for another round of out-migration. See Arwin M. Bolin, "Annual Report, Extension Service, Baca County, December 1, 1956, to October 31 1957," 43, box 9, folder 3; Thomas J. Doherty, "Annual Report, Extension Service, Baca County, November 1, 1962, to October 31, 1963," 43, box 9, folder 9.

48. Greg Hobbs, "An Overview of Colorado Groundwater Law," *Colorado Ground-Water Association Newsletter*, Summer 2007, 9.

49. Thomas J. Doherty, "Annual Report, Extension Service, Baca County, November 1, 1962, to October 31 1963," 43, box 9, folder 9.

50. Fred Fitzsimmons, "Annual Report, Extension Service, Prowers County, November 1, 1963, to October 31, 1964," 3, box 68, folder 10.

51. Bolin, "Annual Report, December 1, 1956, to October 31, 1957," 68, box 9, folder 3.

52. "To Discuss Sugar Beet Planting," *Plainsman Herald*, December 20, 1956.

53. "Sign Up Sugar Beet Acreage," *Plainsman Herald*, January 10, 1957.

54. "Sugar Companies Interested in Baca Beet Growing," *Plainsman Herald*, January 4, 1957.

55. "Beet Harvest Underway Here," *Plainsman Herald*, October 24, 1957.

56. See, e.g., "Beets Are Beneficial to Community," *Plainsman Herald*, October 11, 1958.

57. "Irrigated Wheat Important to County Economy," *Plainsman Herald*, June 7, 1962.

58. "Tony Ochoa Named Baca Champion in Sugar Beet 'Sugar per Acre' Contest," *Plainsman Herald*, January 26, 1961.

59. "Irrigation Wells Boost Economy," *Plainsman Herald*, June 5, 1958.

60. A good explanation of how the Newlands Act influenced and was influenced by the balance of independence and dependence is found in Campbell, "Newlands, Old Lands."

61. Water Conservation District Southeastern Colorado, "Fryingpan-Arkansas Project," FD Ark-Fryingpan Materials Including "Rotary Spokes" (1962–71).

62. Water Conservation District Southeastern Colorado, "Fryingpan-Arkansas Project" FD Ark-Fryingpan Materials Including "Rotary Spokes" (1962–71).

63. No obvious direct connection between the two projects exists, however. It seems likely that proponents of each knew of the other project, but because of differing goals, organizational methods, and timelines, the two projects did not develop in a parallel fashion.

64. James Ogilvie, ephemera, n.d., 7, box 2, folder Fryingpan–Arkansas Information [Maps, Correp, etc 1959–1969], Papers of James L. Ogilvie, Water Resources Archive, Colorado State University; Milenski, *In Quest of Water*, 2.

65. Southeastern Colorado Water Conservancy District, "Fryingpan Arkansas Project."

66. Petersen, "Fryingpan Arkansas Project," 36–43.

67. Terence Brace, "History of the Fryingpan-Arkansas Project," n.d., 28–29.

68. Milenski, *In Quest of Water*, 14.

69. *Hearing before the Subcommittee on Irrigation and Reclamation of the Committee on Interior and Insular Affairs*, 87th Cong., 2nd sess., 30.

70. *Hearing*, 49.

71. *Hearing*, 49.

72. Burris, *Fourteen Statements*, 15–20.

73. Milenski, *In Quest of Water*, 10.

74. Burris, *Fourteen Statements*, 51.

75. Petersen, "Fryingpan Arkansas Project," 35–36.

76. Brace, "History of the Fryingpan-Arkansas Project," 28–30.

77. Milenski, *In Quest of Water*, 16.

78. Helmer Reenberg, "August 17, 1962–President John F. Kennedy's Remarks at the High School Stadium in Pueblo, Colorado," accessed August 25, 2017 https://www.youtube.com/watch?v=t5bshypn9wQ.

79. Two good books on this subject are Philpott, *Vacationland* and Childers, *Colorado Powder Keg*.

8. BACK TO WORK

1. *Springfield Plainsman Herald*, November 1, 1945.

2. *Lamar Daily News*, August 14, 1945.

3. "The Editor Speaking," *Lamar Daily News*, August 15, 1945.

4. "The Editor Speaking," *Lamar Daily News*, September 7, 1945.

5. *Plainsman Herald*, December 13, 1945.

6. *Plainsman Herald*, December 27, 1945.

7. "Dust Storms Predicted Here," *Plainsman Herald*, September 26, 1946.

8. "Prediction of Dust Hazards," *Plainsman Herald*, August 29, 1946; "Dust Storms Predicted Here," *Plainsman Herald*, September 26, 1946.

9. "Prediction of Dust Hazards," *Plainsman Herald*, August 29, 1946.

10. "Great Plains Hit Jackpot," *Plainsman Herald*, August 28, 1947.

11. "Conservation Is Urged This Fall," *Plainsman Herald*, September 26, 1946.

12. Jett Hardware and Lumber Company advertisement, *Plainsman Herald*, May 30, 1946.

13. *Plainsman Herald*, July 4, 1946.

14. See *Plainsman Herald*, September 26 and October 24, 1946.

15. Guy Robbins, "Annual Report, Extension Service, Baca County, December 1, 1945, to November 15, 1947," n.p., box 8, folder 58.

16. Stanley L. Boyes, "Annual Report, Extension Service, Prowers County, December 1, 1947, to December 1, 1948," box 67, folder 19.

17. Chester Fithian, "Annual Report, Extension Service, Baca County, December 1, 1952 to November 30, 1953," 8, box 9, folder 1.

18. Chester R. Fithian, "Annual Report, Extension Service, Baca County, December 1, 1952, to November 30, 1953," 96–99, box 9, folder 1.

19. Fithian, "Annual Report, Extension Service, Baca County, December 1, 1953, to November 30, 1954," 8, box 9, folder 2.

20. Vernon Carter and Bruce E. Young, "Annual Report, Extension Service, Prowers County,, December 1, 1952, to November 39, 1953," 146, box 67, folder 23.

21. References to such involvement are common in these years. See Records of the Colorado Cooperative Extension, Baca County Annual Reports, 1929.

22. "Baca County Conservation District," accessed August 25, 2017, http://www .freewebs.com/bacaccd/hist/hist01.html.

23. "Plans to Organize Conservancy to Be Discussed Here," *Lamar Daily News*, November 30, 1954.

24. James W. Read, "Annual Report, Extension Service, Prowers County, November 1, 1960, to October 31, 1961," 17, box 68, folder 8.

25. See, e.g., "How Soil Districts Operate," *Holly Chieftain*, January 29, 1959, and "Are Soil Conservation Districts Worthwhile?" *Holly Chieftain*, February 5, 1959.

26. *Holly Chieftain*, February 1, 1958.

27. *Holly Chieftain*, February 21, 1958.

28. Several articles from local newspapers speak to these programs. See, e.g., "800,000 Acres of Baca Land Now Tilled for Wind Erosion," *Plainsman Herald*, April 22, 1954; "Baca County Farmers Eligible For Emergency Aid Program," *Plainsman Herald*, May 13, 1954; "To Get $150,000 for Emergency Practices," *Lamar Daily News*, May 14, 1954.

29. "Farmers May Get Tax Refunds for Soil Practices," *Plainsman Herald*, July 15, 1954.

30. Vernon Carter, "Annual Report, Extension Service, Prowers County, December 1, 1951, to November 30, 1952, n.p., box 67, folder 22.

31. "Colorado Department of Agriculture Responsibilities to Conservation," accessed August 25, 2017, https://www.colorado.gov/pacific/sites/default /files/Colorado%20department%20of%20agriculture%20responsibilities %20to%20conservation%20(11–2010).pdf.

32. Chester Fithian, "Annual Report, Extension Service, Baca County, December 1, 1954, to November 30, 1955," 19, box 9, folder 2.

33. House Document, Public Law 1021, August 7, 1956 (Washington DC: Government Printing Office, 1956), 1115, accessed August 25, 2017, https://www .gpo.gov/fdsys/pkg/statute-70/pdf/statute-70-Pg1115.pdf.

34. Helms, "Great Plains Conservation Program."

35. Clem H. Dodson, "Roy Hanes Signs First Great Plains Conservation Contract in S E Colorado," *Plainsman Herald*, August 20, 1958.

36. Helms, "Great Plains Conservation Program."

37. "Conservation High in Arkansas Valley Area," *Plainsman Herald*, August 3, 1961.

38. "New High in Baca Conservation," *Plainsman Herald*, August 10, 1961.

39. Helms, "Brief History of the USDA Soil Bank Program."

40. "Soil Bank Benefits Are Many, Says Chairman," *Holly Chieftain*, October 10, 1958.

41. "Baca County Farmers Will Soon Get $2,371,925 in Soil Bank Payments," *Plainsman Herald*, June 6, 1957.

42. "Nearly 200 Baca County Farmers Ask ASC to Establish Rates for Soil Bank," *Plainsman Herald*, October 2, 1958.

43. "193 Applying for Soil Bank Program," *Lamar Daily News*, October 30, 1958.

44. Hurt, "National Grasslands," 259.

45. Hurt, "National Grasslands," 259.

46. Lewis, "National Grasslands in the Old Dust Bowl," 42–45.

47. Lewis, "National Grasslands in the Old Dust Bowl," 142.

48. "Comanche National Grassland FAQ," accessed August 25, 2017, https://www
.fs.usda.gov/wps/portal/fsinternet/cs/detail.

49. Kent Brooks, "Challenge of the Broomcorn Harvesting Machine," accessed
August 25, 2017, http://bacacountyhistory.com/?p=463.

50. For a few examples of these views, see Kirstein, *Anglo over Bracero*; D. Cohen,
"Caught in the Middle."

51. "Mexico Refuses Send Nationals Do Farm Work," *Lamar Daily News*, February
16, 1946.

52. "Mexican Labor Cut until Living Conditions Better," *Lamar Daily News*,
March 18, 1953.

53. "Mexican Worker Influx Not Bad, Official States," *Lamar Daily News*, April
10, 1958.

54. "Mexican Worker Influx Not Bad, Official States," *Lamar Daily News*, April
10, 1958.

55. Potts, "Social Profile."

56. Karraker, *Agricultural Seasonal Laborers of Colorado and California*, 6. Cyrus Karraker was president of the Pennsylvania Citizens Committee on Migrant Labor.

57. Potts, "School Bells for Children Who Follow the Crops," 441

58. Vest, *The Teachers Say*, 27.

59. Potts, "Social Profile," 4, 5.

60. Potts, "Social Profile," 7, 9–15.

61. Potts, "Social Profile," 19–21.

62. Potts, "Social Profile," 22–27.

63. Rossi, *Colorado Migrant Education Program Summary Report*," 1, Tables 2–4.

64. McCracken, *Migrant Farmworkers in Colorado*, 4–5, 70–71.

65. McCracken, *Migrant Farmworkers in Colorado*, 88–89.

66. "Set Trial of Broom Corn Workers for Assault with Intent to Kill," *Lamar
Daily News*, October 5, 1955.

67. "Workers Arriving and Going 'Broke,'" *Plainsman Herald*, September 8, 1949.

68. See, e.g., "No Charges Filed Yet in Injury of Broomcorn Worker Run Over
by Car," *Plainsman Herald*, September 13, 1962.

69. "Labor Migration Will Meet Harvest Needs," *Lamar Daily News*, June
18, 1962.

70. "Transient Who Died Here Is Identified," *Plainsman Herald*, September 21, 1961.

71. "Two Broom Corn Cutters Seriously Injured Wednesday," *Plainsman Herald*, September 24, 1953.

72. "Vilas Farm Worker Dies Sunday Morning in Baca Car Wreck," *Lamar Daily News*, June 2, 1958.

73. "2-Day Vaqueros Fiesta Opens Saturday," *Lamar Daily News*, May 21, 1962.

74. James W. Read, "Annual Report, Extension Service, Prowers County, November 1, 1958, to October 31, 1959," 3, box 68, folder 6.

CONCLUSION

1. *Denver Post*, July 20, 2012, accessed August 25, 2017, http://www.denverpost.com/news/ci_21116040/first-drought-now-pests-are-descending-parched-colorado.

2. See, e.g., "Drought Puts Baca on Edge," *Pueblo Chieftain*, April 25, 2011.

3. See Hurt, "National Grasslands," 246–59; Lewis, "National Grasslands in the Dust Bowl," 161–71.

4. Ashworth, *Ogallala Blue*, 10.

5. Guru and Horne, *Ogallala Aquifer*.

6. Danbom, *Born in the Country*, 244.

7. "John Martin Reservoir Drying Up," June 25, 2006, accessed August 25, 2017, http://www.kktv.com/home/headlines/3423251.html.

8. "John Martin to Add Water," *Lamar Ledger*, May 21, 2009, accessed August 25, 2017, http://www.lamarledger.com.

9. Pictures of people commiserating around the piles of dead fish at the bottom of the reservoir and an explanation of the push for a permanent pool can be found in the J. Edgar Chenoweth Collection, folder 28 Water–John Martin Reservoir 1948–1965, box 78.

10. "Drought Puts Baca on Edge," *Pueblo Chieftain*, April 25, 2011, accessed August 25, 2017, http://www.chieftain.com/.

11. Farm Service Agency, "Program Fact Sheets: Conservation Reserve Program," accessed August 25, 2017, http://www.fsa.usda.gov.

12. USDA, "Dust Settles on Baca County Thanks to CRP," accessed February 17, 2019, https://www.fsa.usda.gov/FSA/printapp?fileName=ss_Co_artid_613.html&newsType=crpsuccessstories.

BIBLIOGRAPHY

ARCHIVES/MANUSCRIPT MATERIALS

Allot, Gordon L. Papers. Western History Collection. University of Colorado, Norlin Library, Boulder.

Arkansas Valley, ephemera, 1894–1976. Stephen H. Hart Library and Research Center, Denver.

The Arkansas Valley Truth: Devoted to the Development of the Arkansas Valley in Colorado. Chicago: Rand McNally, 1898.

Blinn, Philo K. *Development of the Rockyford Cantaloupe Industry.* Bulletin 108, March 1906. Fort Collins: Agricultural Experiment Station, 1906.

Bureau of the Census. *Thirteenth Census of the United States Taken in the Year 1910: Population.* Washington DC: Government Printing Office, 1913.

———. *Thirteenth Census of the United States Taken in the Year 1910: Agriculture.* Washington DC: Government Printing Office, 1914.

———. *Fourteenth Census of the United States, 1920: Population.* Washington DC: Government Printing Office, 1922.

———. *Fifteenth Census of the United States, 1930: Agriculture.* Washington DC: Government Printing Office, 1932.

———. *Fifteenth Census of the United States, 1930: Population.* Washington DC: Government Printing Office, 1932.

———. *Sixteenth Census of the United States, 1940.* Washington DC: Government Printing Office, 1943.

———. *Sixteenth Census of the United States, 1940: Agriculture.* Washington DC: Government Printing Office, 1943.

———. *United States Census of Agriculture, 1925.* Washington DC: Government Printing Office, 1928.

———. *United States Census of Agriculture, 1935.* Washington DC: Government Printing Office, 1936.

———. *United States Census of Agriculture, 1945.* Washington DC: Government Printing Office, 1946.

Chenoweth, J. Edgar. Papers. Western History Collection. University of Colorado, Norlin Library, Boulder.

Civil Works Administration. Pioneer Interviews Index. Colorado Historical Society, Denver CO.

Code, W. E. "Construction of Irrigation Wells in Colorado." Fort Collins: Colorado Experiment Station, 1935.

Colorado Soil Conservation Act, Soil Erosion Districts, Colorado House Bill 258.

Colorado State Federation of Labor. Labor Collection. University of Colorado, Norlin Library, Boulder.

Congressional Record, 63rd Cong., 2nd sess. Washington DC: Government Printing Office, 1914.

Costigan, Edward P. Papers. Western History Collection. University of Colorado, Norlin Library, Boulder.

Cottrell, H. M. *Dry Land Farming in Eastern Colorado.* Bulletin 145. Fort Collins: Agricultural Experiment Station of the Colorado Agricultural College, December 1909.

Digital Collections of Colorado. http://digitool.library.colostate.edu.

Dimitri, Carolyn, Anne Effland, and Neilson Conklin. *The 20th Century Transformation of U.S. Agriculture and Farm Policy.* Economic Information Bulletin 3. Washington DC: USDA, 2005. http://www.ers.usda.gov/media/259572/eib3_1_.pdf.

Douglas, M. R. "Migratory Labor in Colorado: Colorado Legislative Council Report to the Colorado General Assembly." Denver: December, 1962.

Gaer, Joseph. *Toward Farm Security: The Problem of Rural Poverty and the Work of the Farm Security Administration, Prepared under the Direction of the FSA Personnel Training Committee for FSA Employees.* Washington DC: Government Printing Office, 1941.

Garrison, Lloyd. Collection. Auraria Library Special Collections, Denver.

Geographical and Geological Survey of the Rocky Mountain Region (US), John Wesley Powell, Grove Karl Gilbert, Clarence E. Dutton, A. H. Thompson, and

Willis Drummond. *Report on the Lands of the Arid Region of the United States, with a More Detailed Account of the Lands of Utah. With maps.* Washington: Government Printing Office, 1879.

Gray, Lewis C., et al. *Farm Tenancy: Message from the President of the United States Transmitting the Report of the Special Committee on Farm Tenancy.* Washington DC: Government Printing Office, 1937.

Hyde, Blanche, and J. H. McClelland. *History of the Extension Service of Colorado State College, 1912 to 1941.* Fort Collins: Extension Service, Colorado State College of Agricultural and Mechanic Arts, 1941.

McClelland, Joseph. Collection. Auraria Library Special Collections, Denver.

Miles, Donald L. "Salinity in the Arkansas Valley of Colorado." Fort Collins: Colorado Extension Service, 1977.

National Farmers Union. Labor Collection. University of Colorado, Norlin Library, Boulder.

Odum, Kathy. Collection. Auraria Library Special Collections, Denver.

Office of Board of Supervisors. "The Western Baca County Soil Erosion District." Springfield CO: 1938.

Office of the State Engineer. *Fifteenth Biennial Report of the State Engineer to the Governor of Colorado for the Years 1909–10.* Denver: Smith-Brooks, 1911.

Records of the Bureau of Agricultural Economics. Record Group 83. National Archives and Records Administration–Rocky Mountain Region, Denver.

Records of the Office of the Chief of Engineers. Record Group 77. National Archives and Records Administration–Rocky Mountain Region, Denver.

Records of the Civilian Conservation Corps, Colorado State Archives, Denver.

Records of the Colorado Cooperative Extension, Colorado Agricultural Archives, Colorado State University, Fort Collins.

Records of the Great Plains Agricultural Council, Colorado Agricultural Archive, Colorado State University, Fort Collins.

Records of the Natural Resources Conservation Service. Record Group 114. National Archives and Records Administration–Rocky Mountain Region, Denver.

Records of the Natural Resources Conservation Service. Record Group 114. National Archives and Records Administration–Southwest Region, Fort Worth.

Records of the Soil Conservation Service. Record Group 114. National Archives and Records Administration–Southwest Region, Fort Worth.

Records of the War Manpower Commission. Record Group 211. National Archives and Records Administration–Rocky Mountain Region, Denver.

Soil Conservation Service. *Soil Survey of Baca County, Colorado*. Washington DC: Department of Agriculture, 1973.

Tiggs, Ernest. Collection. Auraria Library Special Collections, Denver.

Underwood, John J. "Physical Land Conditions in the Western and Southeastern Baca County Soil Conservation Districts." Washington DC: Department of Agriculture, 1944.

U.S. Congress, 44th, 2nd sess. "An Act to Provide for the Sale of Desert Lands in Certain States and Territories." *U.S. Statutes at Large* 19:377. Washington DC: Washington Government Printing Office, 1877.

U.S. Congress, 57th, 1st sess. "An Act Appropriating the Receipts from the Sale and Disposal of Public Lands in Certain States and Territories to the Construction of Irrigation Works for the Reclamation of Arid Lands." *U.S. Statutes at Large* 32:388–90. Washington DC: Government Printing Office, 1902.

U.S. Congress, 87th, 2nd sess. *Hearing before the Subcommittee on Irrigation and Reclamation of the Committee on Interior and Insular Affairs*. Washington DC: Government Printing Office, 1962.

U.S. Senate. *Report of the Country Life Commission: Special Message from the President of the United States Transmitting the Report of the Country Life Commission*. Washington DC: Government Printing Office, February 9, 1909.

Wickard, Claude. "Food Will Win the War and Write the Peace." September 8, 1941. *Vital Speeches of the Day*, 7:764–66.

Writers' Program of the Work Projects Administration. *The WPA Guide to 1930s Colorado*. Introduction by Thomas J. Noel. Lawrence: University Press of Kansas, 1987.

Writers' Program of the Work Projects Administration, Colorado. *Agriculture— Sugar Beets*. Denver: Colorado Historical Society, 1940.

PUBLISHED WORKS

Abbott, Carl, Stephen J. Leonard, and Thomas J. Noel. *Colorado: A History of the Centennial State*. 4th ed. Boulder: University Press of Colorado, 2005.

Abert, James William. *Expedition to the Southwest: An 1845 Reconnaissance of Colorado, New Mexico, Texas, and Oklahoma*. Lincoln: University of Nebraska Press, 1999.

Adelman, Jeremy, and Stephen Aron. "From Borderlands to Borders: Empires, Nation-States, and the Peoples in Between in North American History." *American Historical Review*, no. 104 (June 1999): 814–41.

Anderson, F. A., and A. J. Hamman. *A Resume of the Emergency Farm Labor Program in Colorado, 1943 to 1947.* Fort Collins: Extension Service, Colorado A&M College, 1947.

Anderson, Joseph. *Industrializing the Corn Belt: Agriculture, Technology, and Environment, 1945–1972.* DeKalb: Northern Illinois University Press, 2009.

Andrews, Thomas. *Killing for Coal: America's Deadliest Labor War.* Cambridge: Harvard University Press, 2008.

Ankli, Robert E. "Farm Income on the Great Plains and Canadian Prairies, 1920–1940." *Agricultural History* 51, no. 1 (1977): 92–103.

Arrington, Leonard J. "Science, Government, and Enterprise in Economic Development: The Western Beet Sugar Industry." *Agricultural History* 41, no. 1 (January 1967): 1–18.

Ashworth, William. *Ogallala Blue: Water and Life on the Great Plains.* New York: Norton, 2006.

Athearn, Frederic J. *Land of Contrast: A History of Southeast Colorado.* BLM Cultural Resources series, Colorado no. 17, chap. 8. www.nps.gov/history/history/online_books/blm/co/17/chap8.htm.

Atkins, James A. *Human Relations in Colorado, 1858–1959.* Denver: Colorado State Department of Education, 1961.

Baca County Historical Society. *Baca County.* Lubbock TX: Specialty, 1983.

Balderrama, Francisco E., and Raymond Rodríguez. *Decade of Betrayal: Mexican Repatriation in the 1930s.* Albuquerque: University of New Mexico Press, 1995.

Baldwin, Sidney. *Poverty and Politics: The Rise and Decline of the Farm Security Administration.* Chapel Hill: University of North Carolina Press, 1968.

Baltensperger, Bradley H. "Larger and Fewer Farms: Patterns and Causes of Farm Enlargement on the Central Great Plains, 1930–1978." *Journal of Historical Geography* 19, no. 3 (1993): 299–313.

Baskin, O. L., & Co. *History of the Arkansas Valley, Colorado.* Chicago: O. L. Baskin, Historical, 1881.

Beidleman, Richard G. "The 1820 Long Expedition." *American Zoologist* 26, no. 2 (1986): 307–13.

Berwanger, Eugene H. *The Rise of the Centennial State: Colorado Territory, 1861–1876*. Urbana: University of Illinois Press, 2007.

Betz, Ava. *Prowers County, Colorado: A Prowers County History*. Lamar CO: Prowers County Historical Society, 1986.

Black, John D. "Notes on 'Poor Land' and 'Submarginal Land.'" *Journal of Farm Economics* 27, no. 2 (May 1945): 345–74.

Bonner, Robert E. "Elwood Mead, Buffalo Bill Cody, and the Carey Act in Wyoming." *Montana: The Magazine of Western History*, April 1, 2005, 36–51.

Bonnifield, Paul. *The Dust Bowl: Men, Dirt, and Depression*. Albuquerque: University of New Mexico Press, 1979.

Borneman, Walter R. "Black Smoke among the Clouds." *Colorado Magazine* 52, no. 4 (1975): 317–38.

Bowman, Charles W. "The History of Bent County." In *History of the Arkansas Valley, Colorado*, ed. W. B. Vickers and O. L. Baskin. Evansville IN: Unigraphics, 1971.

Brinkley, Alan. *The End of Reform: New Deal Liberalism in Recession and War*. New York: Knopf, 1995.

Burns, Ken. *The Dust Bowl*. San Francisco: Kanopy Streaming, 2015.

Burris, E. Duvoid. *Fourteen Statements: History of Fryingpan-Arkansas Project and Southeastern Colorado Water Conservancy District*. Pueblo: Southeastern Colorado Water Conservancy District, 1975.

Campbell, Robert B. "Newlands, Old Lands: Native American Labor, Agrarian Ideology, and the Progressive-Era State in the Making of the Newlands Reclamation Project, 1902–1926." *Pacific Historical Review* 71, no. 2 (May 2002): 203–38.

Cannon, Brian Q. *Remaking the Agrarian Dream: New Deal Rural Resettlement in the Mountain West*. Albuquerque: University of New Mexico Press, 1996.

Carpenter, Stephanie Ann. "'Regular Farm Girl': The Women's Land Army in World War II." *Agricultural History* 71, no. 2 (Spring 1997): 162–85.

Causley, Fred. "Agricultural Experiment Stations." In *The Encyclopedia of Oklahoma History and Culture*. www.okhistory.org.

Chiang, Connie. "Imprisoned Nature: Toward an Environmental History of the World War II Japanese American Incarceration," *Environmental History* 15 (April 2010): 236–67.

Childers, Michael W. *Colorado Powder Keg: Ski Resorts and the Environmental Movement*. Lawrence: University Press of Kansas, 2012.

Choate, Jean. *Disputed Ground: Farm Groups That Opposed the New Deal Agricultural Program.* Jefferson NC: McFarland, 2002.

Christman, Abigail D. *The Legacy of the New Deal on Colorado's Eastern Plains.* Denver: Colorado Preservation, 2008.

Clements, Kendrick A. *Hoover, Conservation, and Consumerism: Engineering the Good Life.* Lawrence: University Press of Kansas, 2000.

Cochrane, Willard W. *The Development of American Agriculture: A Historical Analysis.* Minneapolis: University of Minnesota Press, 1979.

Cohen, Deborah. "Caught in the Middle: The Mexican State's Relationship with the United States and Its Own Citizen-Workers, 1942–1954." *Journal of Ethnic History* 20, no. 3 Migration and the Making of North America (Spring 2001): 110–32.

Cohen, Lizabeth. *Making a New Deal: Industrial Workers in Chicago, 1919–1939.* New York: Cambridge University Press, 1990.

Colorado Preservation. Baca County Survey. Accessed online August 25, 2017. http://www.coloradopreservation.org/crsurvey/rural/baca/sites/baca_resources _agriculture.html.

———. "Depression and Dust Bowl." Accessed on March 10, 2018. http:// www.coloradopreservation.org/crsurvey/rural/baca/sites/baca_resources _depression.html.

Colorado State College, Extension Service. *Keeping the Farm at Home.* Fort Collins, July 1935.

Cornebise, Alfred E. *The CCC Chronicles: Camp Newspapers of the Civilian Conservation Corps, 1933–1942.* Jefferson NC: McFarland, 2004.

Cotton, Albert H. "Regulations of Farm Landlord-Tenant Relationships." *Law and Contemporary Problems* 4, no. 4, Farm Tenancy (October 1937): 508–38.

Coues, Elliott, ed. *The Expeditions of Zebulon Montgomery Pike.* Vol. 2. New York: Dover, 1987.

Courtwright, Julie. "On the Edge of the Possible: Artificial Rainmaking and the Extension of Hope on the Great Plains." *Agricultural History* 89, no. 4 (Fall 2015): 536–58.

Cronon, William. "A Place for Stories: Nature, History, and Narrative." *Journal of American History* 78 (1992): 1347–76.

———, ed. *Uncommon Ground: Toward Reinventing Nature.* New York: Norton, 1995.

Cunfer, Geoff. *On the Great Plains: Agriculture and Environment*. College Station: Texas A&M University Press, 2005.

Cutler, Phoebe. *The Public Landscape of the New Deal*. New Haven: Yale University Press, 1985.

Danbom, David B. *Born in the Country: A History of Rural America*. 2nd ed. Baltimore: Johns Hopkins University Press, 2006.

Daniels, Roger. *Prisoners without Trial: Japanese Americans in World War II*. New York: Hill and Wang, 1993.

De Baca, Vincent C. *La Gente: Hispano History and Life in Colorado*. Denver: Colorado Historical Society, 1998.

Denning, Michael. *The Cultural Front: The Laboring of American Culture in the Twentieth Century*. New York: Verso, 1997.

Deutsch, Sarah. *No Separate Refuge: Culture, Class, and Gender on an Anglo-Hispanic Frontier in the American Southwest, 1880–1940*. New York: Oxford University Press, 1987.

Dillon, Richard H. "Stephen Long's Great American Desert." *Proceedings of the American Philosophical Society* 111, no. 2 (April 14, 1967): 93–108.

Dilsaver, Lary M., and Craig E. Colten, eds. *The American Environment: Interpretations of Past Geographies*. Lanham MD: Rowman & Littlefield, 1992.

Doherty, Thomas J. "Effects on Farmers of Change from Dryland to Irrigation in Baca County." Master's thesis, Colorado State University, 1964.

Donato, Rubén. "Sugar Beets, Segregation, and Schools: Mexican Americans in a Northern Colorado Community, 1920–1960." *Journal of Latinos and Education* 2, no. 2 (2003): 69–88.

Driscoll, Barbara A. *The Tracks North: The Railroad Bracero Program of World War II*. Austin: CMAS Books, Center for Mexican American Studies, University of Texas at Austin, 1999.

Durrell, Glen R. "Homesteading in Colorado." *Colorado Magazine* 51, no. 2 (Spring 1974): 93–114.

Egan, Timothy. *The Worst Hard Time: The Untold Story of Those Who Survived the Great American Dust Bowl*. New York: Mariner Books, 2006.

Emmons, David M. *Garden in the Grasslands: Boomer Literature of the Central Great Plains*. Lincoln: University of Nebraska Press, 1971.

Environmental Working Group. Farming and Farm Subsidies. Available online at http://www.ewg.org/farmsubsidies.

Farley, Mary. "Colorado and the Arkansas Valley Authority." *Colorado Magazine* 48, no. 3 (1971): 221–34.

Fausold, Martin L. "President Hoover's Farm Policies, 1929–1933." *Agricultural History* 51, no. 2 (April 1977): 362–77.

Fearon, Peter. *Kansas in the Great Depression: Work Relief, the Dole, and Rehabilitation*. Columbia: University of Missouri Press, 2007.

Fernandez, Marilyn, and Stephen S. Fugita. *Altered Lives, Enduring Community: Japanese Americans Remember Their World War II Incarceration*. Seattle: University of Washington Press, 2004.

Fiege, Mark. *Irrigated Eden: The Making of an Agricultural Landscape in the American West*. Seattle: University of Washington Press, 1999.

———. "The Weedy West: Mobile Nature, Boundaries, and Common Space in the Montana Landscape." *Western Historical Quarterly* 36, no. 1 (Spring 2005): 23–47.

Firkus, Angela. "The Agricultural Extension Service and Non-Whites in California, 1910–1932." *Agricultural History* 84, no. 4 (Fall 2010): 506–30.

Fiset, Louis. "Thinning, Topping, and Loading: Japanese Americans and Beet Sugar in World War II." *Pacific Northwest Quarterly* 90 (Summer 1999): 123–29.

Fitzgerald, Deborah. *Every Farm a Factory: The Industrial Ideal in American Agriculture*. New Haven: Yale University Press, 2003.

"Floyd M. Wilson and the Alfalfa Milling Industry." *Colorado Magazine* 21, no. 2 (March 1944): 100.

Foley, Neil. T*he White Scourge: Mexicans, Blacks, and Poor Whites in Texas Cotton Culture*. Berkeley: University of California Press, 1997.

Forrest, Suzanne. *The Preservation of the Village: New Mexico's Hispanics and the New Deal*. Albuquerque: University of New Mexico Press, 1989.

Galarza, Ernesto. *Merchants of Labor: The Mexican Bracero Story: An Account of the Managed Migration of Mexican Farm Workers in California, 1942–1960*. Charlotte/Santa Barbara: McNally & Loftin, 1964.

Gamboa, Erasmo. *Mexican Labor and World War II: Braceros in the Pacific Northwest, 1942–1947*. Austin: University of Texas Press, 1990.

Gansberg, Judith M. *Stalag, U.S.A.: The Remarkable Story of German POWs in America*. New York: Crowell, 1977.

Garcia, Alephonso. "Beet Seasons in Wyoming: Mexican-American Family Life on a Sugar Beet Farm near Wheatland during World War II." *Annals of Wyoming* 73, no. 2 (2001): 14–17.

Gates, Paul W. "Homesteading in the High Plains." *Agricultural History* 51, no. 1 (January 1977): 109–33.

Gease, Derly V. "William N. Byers and the Case for Federal Aid to Irrigation in the Arid West." *Colorado Magazine* 45, no. 4 (1968): 340–45.

Gillis, Earle A. "'We Traded Your High Chair for a Quarter of Beef': Two Years on the Colorado Flats." *Colorado Heritage*, Spring 2005, 38–47.

Gordon, Linda. *Dorothea Lange: A Life beyond Limits.* New York: Norton, 2009.

Grant, Michael Johnston. *Down and Out on the Family Farm: Rural Rehabilitation in the Great Plains, 1929–1945.* Lincoln: University of Nebraska Press, 2002.

Gray, Lewis C. "Federal Purchase and Administration of Submarginal Land in the Great Plains." *Journal of Farm Economics* 21, no.1, Proceedings Number (February 1939): 123–31.

———. "Research Relating to Policies for Submarginal Areas." *Journal of Farm Economics* 16, no. 2 (April 1934): 298–303.

Great Plains Drought Area Committee. *Report of the Great Plains Drought Area Committee, August 1936.* Washington DC, 1936.

Gregg, Sara M. *Managing the Mountains: Land Use Planning, the New Deal, and the Creation of a Federal Landscape in Appalachia.* New Haven: Yale University Press, 2010.

Gregory, James N. *American Exodus: The Dust Bowl Migration and Okie Culture in California.* New York: Oxford University Press, 1989.

Griffin, Ronald C., and John R. Stoll. "Evolutionary Processes in Soil Conservation Policy." *Land Economics* 60, no. 1 (February 1984): 30–39.

Grove, Wayne A. "The Mexican Farm Labor Program, 1942–1964: Government-Administered Labor Market Insurance for Farmers." *Agricultural History* 70, no. 2 (1996): 302–20.

Guglielmo, Thomas A. "Fighting for Caucasian Rights: Mexicans, Mexican Americans, and the Transnational Struggle for Civil Rights in World War II Texas." *Journal of American History* 92, no. 4 (2006): 1212–37.

Gundy, William Davis. "Snapshots from Old Soddy: A Farm Girl's Life on the Eastern Colorado Plains in the 1930s and '40s." *Colorado Heritage*, Autumn 2004, 2–15.

Guru, Manjula V., and James E. Horne. *The Ogallala Aquifer.* Kerr Center for Sustainable Agriculture, 2000. http://www.kerrcenter.com/publications/ogallala _aquife.pdf.

Gutmann, Myron P. *Great Plains Population and Environment Data: On-Line Extraction System*. Computer file. Ann Arbor: University of Michigan, 2005.

Hafen, LeRoy R., ed. *To the Pike's Peak Gold Fields, 1859*. Lincoln: University of Nebraska Press, 2004.

Hahamovitch, Cindy. *The Fruits of Their Labor: Atlantic Coast Farmworkers and the Making of Migrant Poverty*. Chapel Hill: University of North Carolina Press, 1997.

———. "'In America Life Is Given Away': Jamaican Farmworkers and the Making of Agricultural Immigration Policy." In *The Countryside in the Age of the Modern State: Political Histories of Rural America*, 134–216. Ithaca NY: Cornell University Press, 2001.

———. *No Man's Land: Jamaican Guestworkers in America and the Global History of Deportable Labor*. Princeton: Princeton University Press, 2011.

Hall, Frank. *History of the State of Colorado*. Chicago: Blakely Printing, 1895.

Ham, William T. "Sugar Beet Field Labor under the AAA." *Journal of Farm Economics* 19, no. 2 (May 1937): 643–47.

Hamilton, David E. *From New Day to New Deal: American Farm Policy from Hoover to Roosevelt, 1928–1933*. Chapel Hill: University of North Carolina Press, 1991.

Hamman, A. J. *My Long Journey*. Marjorie J. Miller, 1989.

Harbaugh, William H. "Twentieth-Century Tenancy and Soil Conservation: Some Comparisons and Questions." *Agricultural History* 66, no. 2 (Spring 1992): 95–119.

Hargreaves, Mary W. M. "Land-Use Planning in Response to Drought: The Experience of the Thirties." *Agricultural History* 50, no. 4 (1976): 561–82.

Harris, Marshall. "A New Agricultural Ladder." *Land Economics* 26, no. 3 (August 1950): 258–67.

Hart, John Mason, ed. *Border Crossings: Mexican and Mexican-American Workers*. Wilmington DE: Scholarly Resources, 1998.

Hart, Stephen Harding, and Archer Butler Hulbert. *Zebulon Pike's Arkansaw Journal*. Westport CN: Greenwood Press, 1972.

Harvey, Robert. *Amache: The Story of Japanese Internment in Colorado during World War II*. Lanham MD: Taylor Trade, 2004.

Hays, Samuel P. *Conservation and the Gospel of Efficiency: The Progressive Conservation Movement, 1890–1920*. Cambridge: Harvard University Press, 1959.

Headley, J. C. "Soil Conservation and Cooperative Extension." *Agricultural History* 59, no. 2 (April 1985): 290–306.

Heimburger, Christan. "Life beyond Barbed Wire: Japanese American Labor during Internment at Amache and Topaz." Available at http://centerwest.org /wp-content/uploads/2011/01/heimburger2008.pdf.

Heimburger, Christian. "Life Beyond Barbed Wire: The Significance of Japanese American Labor in the Mountain West, 1942-1944." Ph.D diss., University of Colorado at Boulder, 2013.

Heisler, Barbara Schmitter. "The 'Other Braceros': Temporary Labor and German Prisoners of War in the United States, 1943–1946. *Social Science History* 31, no 1 (Summer 2007): 239–71.

Helms, Douglas. "Brief History of the USDA Soil Bank Program." *Historical Insights* 1 (January, 1985), 1.

———. "Conserving the Plains: The Soil Conservation Service in the Great Plains." *Agricultural History* 54, no. 2 (1990): 58–73.

———. "The Development of the Land Capability Classification." In *Readings in the History of the Soil Conservation Service,* 60–73. Washington DC, 1992. https://www.nrcs.usda.gov/wps/portal/nrcs/detail/national/about/history/ ?cid=nrcs143_021436.

———. "Great Plains Conservation Program, 1956–1981: A Short Administrative and Legal History," *Great Plains Conservation Program: 25 Years of Accomplishment.* SCS National Bulletin no. 300-2-7. November 24, 1981.

———. "Hugh Hammond Bennett and the Creation of the Soil Erosion Service." *Journal of Soil and Water Conservation* 64, no. 2 (March/April 2009)): 68A–74A.

———, ed. *Readings in the History of the Soil Conservation Service.* Historical Notes no. 1. Washington DC: USDA, 1992.

Henderson, Henry L., and David B. Woolner, eds. *FDR and the Environment.* New York: Palgrave Macmillan, 2005.

Hendrickson, Kent. "The Sugar-Beet Laborer and the Federal Government: An Episode in the History of the Great Plains in the 1930s." *Great Plains Journal* 3, no. 2 (1964): 44–59.

Hewes, Leslie. "Early Suitcase Farming in the Central Great Plains." *Agricultural History* 51, no. 1 (January 1977): 23–37.

Hewitt, William L. "Mexican Workers in Wyoming during World War II: Necessity, Discrimination, and Protest." *Annals of Wyoming* 54, no. 2 (1982): 20–33.

Hill, James H. "A History of Baca County." Master's thesis, Colorado State College of Education, 1941.

Hoffman, Abraham. *Unwanted Mexican Americans in the Great Depression*. Tucson: University of Arizona Press, 1974.

Hoig, Stan. *The Sand Creek Massacre*. Norman: University of Oklahoma Press, 1961.

Holleran, Michael. "Historic Context for Irrigation and Water Supply Ditches and Canals in Colorado." Denver: Colorado Center for Preservation Research, 2005. Available at http://cospl.coalliance.org/fez/eserv/co:3740/ucdh612d632005internet.pdf.

Hosokawa, Bill. *Colorado's Japanese Americans: From 1886 to the Present*. Boulder: University Press of Colorado, 2005.

Hurt, R. Douglas. *The Dust Bowl: An Agricultural and Social History*. Chicago: Nelson-Hall, 1981.

———. "Federal Land Reclamation in the Dust Bowl." *Great Plains Quarterly* 6, no. 2 (1986): 94–106.

———. *The Great Plains during World War II*. Lincoln: University of Nebraska Press, 2008.

———. "The National Grasslands: Origin and Development in the Dust Bowl." *Agricultural History* 59, no. 2, The History of Soil and Water Conservation: A Symposium (April 1985): 256.

Jepson, Daniel A. "Camp Carson, Colorado: European Prisoners of War in the American West during World War II." *Midwest Review* 13 (1991): 32–53.

Johnson, Melyn. "At Home in Amache: A Japanese-American Relocation Camp in Colorado." *Colorado Heritage* 1 (1989): 2–11.

Johnson, Roger T. "Part-Time Leader: Senator Charles L. McNary and the McNary-Haugen Bill." *Agricultural History* 54, no. 4 (October 1980): 527–41.

Johnson, Vernon Webster, and Raleigh Barlowe. *Land Problems and Policies*. New York: Arno Press, 1979.

Karraker, Cyrus. *Agricultural Seasonal Laborers of Colorado and California*. Lewisburg: Pennsylvania, 1962.

Kelsey, Lincoln David, and Cannon Chiles Hearne. *Cooperative Extension Work*. Ithaca NY: Comstock, 1949.

Kennedy, David. *Freedom from Fear: The American People in Depression and War, 1929–1945*. New York: Oxford University Press, 1999.

Kirstein, Peter N. *Anglo over Bracero: A History of the Mexican Worker in the United States from Roosevelt to Nixon*. San Francisco: R and E Research Associates, 1977.

Kulkosky, Tanya W. "Mexican Migrant Workers in Depression-Era Colorado." In *La Gente: Hispano History and Life in Colorado*, edited by Vincent C. De Baca. Denver: Colorado Historical Society, 1998.

Landsberger, Kurt. *Prisoners of War at Camp Trinidad, Colorado, 1943–1946: Internment, Intimidation, Incompetence, and Country Club Living.* New York: Arbor Books, 2007.

Lawson, Merle P., and Charles W. Stockton. "Desert Myth and Climatic Reality." *Annals of the Association of American Geographers* 71, no. 4 (December 1981): 527–35.

Leckie, William H. *The Military Conquest of the Southern Plains.* Norman: University of Oklahoma Press, 1963.

Lecompte, Janet. *Pueblo, Hardscrabble, Greenhorn: The Upper Arkansas, 1832–1856.* Norman: University of Oklahoma Press, 1978.

Lee, R. Alton. "Drought and Depression on the Great Plains: The Kansas Transition from New Deal Work Relief to Old Age Pensions." *Heritage of the Great Plains* 39, no. 1 (2006): 5–29.

Lee, Shu-Ching. "The Theory of the Agricultural Ladder." *Agricultural History* 21, no. 1 (January 1947): 53–61.

Leonard, Stephen J. *Trials and Triumphs: A Colorado Portrait of the Great Depression, with FSA Photographs.* Niwot: University Press of Colorado, 1993.

Leuchtenburg, William. *Franklin D. Roosevelt and the New Deal, 1932–1940.* New York: Harper & Row, 1963.

Lewis, Michael E. "National Grasslands in the Dust Bowl." *Geographical Review* 79, no. 2 (April 1989): 161–71.

———. "The National Grasslands in the Old Dust Bowl: A Long-term Evaluation of Agricultural Adjustment through Land Use Change." PhD diss., University of Oklahoma, 1988.

Lillquist, Karl. "Farming the Desert: Agriculture in the World War II–Era Japanese-American Relocation Centers." *Agricultural History* 84, no. 1 (Winter 2010): 74–104.

Limerick, Patricia Nelson. *The Legacy of Conquest: The Unbroken Past of the American West.* New York: Norton, 1987.

Lockwood, Jeffrey. *Locust: The Devastating Rise and Mysterious Disappearance of the Insect That Shaped the American Frontier.* New York: Basic Books, 2004.

Loeffler, M. John. "Beet-Sugar Production on the Colorado Piedmont." *Annals of the Association of American Geographers* 53, no. 3 (September 1963): 364–90.

Lookingbill, Brad D. *Dust Bowl, USA: Depression America and the Ecological Imagination, 1929–1941.* Athens: Ohio University Press, 2001.

Lotchin, Roger W. *The Bad City in the Good War: San Francisco, Los Angeles, Oakland, and San Diego.* Bloomington: Indiana University Press, 2003.

Lowitt, Richard. *The New Deal and the West.* Norman: University of Oklahoma Press, 1984.

Lowitt, Richard, and Maurine Beasley, eds. *One Third of a Nation: Lorena Hickok Reports on the Great Depression.* Urbana: University of Illinois Press, 1981.

Mackey, Mike. *Remembering Heart Mountain: Essays on Japanese American Internment in Wyoming.* Powell WY: Western History, 1998.

Maddox, James G. "The Bankhead-Jones Farm Tenant Act." *Law and Contemporary Problems* 4, no. 4, Farm Tenancy (October 1937): 434–55

Maher, Neil. "'Crazy Quilt Farming on Round Land': The Great Depression, the Soil Conservation Service, and the Politics of Landscape Change on the Great Plains during the New Deal Era." *Western Historical Quarterly* 31, no. 3 (2000): 319–39.

———. *Nature's New Deal: The Civilian Conservation Corps and the Roots of the American Environmental Movement.* New York: Oxford University Press, 2007.

Malin, James. *The Grassland of North America: Prolegomena to Its History.* 4th ed. Gloucester MA: Peter Smith, 1967.

Markoff, Dena Sabin. "The Beet Sugar Industry in Microcosm: The National Sugar Manufacturing Company, 1899 to 1967." PhD diss., University of Colorado, 1980.

Matsumoto, Valerie J. *Farming the Home Place: A Japanese American Community in California, 1919–1982.* Ithaca NY: Cornell University Press, 1994.

May, William John, Jr. *The Great Western Sugarlands: The History of the Great Western Sugar Company and the Economic Development of the Great Plains.* New York: Garland, 1989.

McCracken, Robert D. *Migrant Farmworkers in Colorado.* Timnath CO: Human Resources Press, 1979.

McDean, Harry. "Federal Farm Policy and the Dust Bowl: The Half-Right Solution." *North Dakota History* 47, no. 3 (1980): 21–31.

———. "Western Thought in Planning Rural America: The Subsistence Homesteads Program, 1933–1935." *Journal of the West* 31, no. 4 (1992): 15–25.

McHendrie, A. W. "The Early History of Irrigation in Colorado and the Doctrine of Appropriation." In *A Hundred Years of Irrigation in Colorado*. Denver: Colorado Water Conservation Board, 1952.

McWilliams, Carey. *Factories in the Field: The Story of Migratory Farm Labor in California*. Boston: Little, Brown, 1940.

Meeks, Eric V. "Protecting the 'White Citizen Worker': Race, Labor, and Citizenship in South-Central Arizona, 1929–1945." *Journal of the Southwest* 48, no. 1 (Spring 2006): 91–113.

Milenski, Frank. *In Quest of Water: A History of the Southeastern Colorado Water Conservancy District and the Fryingpan-Arkansas Project*. Privately published, 1993.

———. "Short and Early History of Fry Ark Project." Available at https://dspace.library.colostate.edu/bitstream/10217/89699/1/WFMP03403.pdf.

Mitchell, Don. *They Saved the Crops: Labor, Landscape, and the Struggle over Industrial Farming in Bracero-Era California*. Athens: University of Georgia Press, 2012.

Momii, Dick, and Chizuko Momii. "Americans First: Colorado's Japanese American Community during World War II." Interview by William Wei. *Colorado Heritage*, Winter 2005, 18–20.

Montoya, Fawn-Amber. "From Mexicans to Citizens: Colorado Fuel and Iron's Representation of Nuevo Mexicans, 1901–1919." *Journal of the West* 45, no. 4 (2006): 29–35.

Murray, Alice. *Historical Memories of the Japanese American Internment and the Struggle for Redress (Asian America)*. Stanford: Stanford University Press, 2007.

———. *What Did the Internment of Japanese Americans Mean?* Boston: Bedford/St. Martin's, 2000.

Nace, R. L., and E. J. Pluhowski. "Drought of the 1950's with Special Reference to the Midcontinent." Geological Survey Water-Supply Paper 1804. Washington DC: Government Printing Office, 1965.

Nash, Gerald. *The American West Transformed: The Impact of the Second World War*. Bloomington: Indiana University Press, 1985.

Nelson, Paula M. *The Prairie Winnows Out Its Own: The West River Country of South Dakota in the Years of Depression and Dust*. Iowa City: University of Iowa Press, 1996.

Nelson, Peter. Tenancy: A Major Factor in Soil Conservation. *Journal of Land & Public Utility Economics* 14, no. 1 (February 1938): 88–91.

Ng, Wendy. *Japanese American Internment during World War II: A History and Reference Guide*. New York: Greenwood Press, 2002.

Nixon, Edgar B., ed. *Franklin D. Roosevelt and Conservation, 1911–1945*. Hyde Park NY: General Services Administration, National Archives and Records Service, Franklin D. Roosevelt Library, 1957.

Oliva, Leo E. *Soldiers on the Santa Fe Trail*. Norman: University of Oklahoma Press, 1967.

Opie, John. "Moral Geography in High Plains History." *Geographical Review* 88, no. 2 (1998): 241–58.

———. *Ogallala: Water for a Dry Land*. 2nd ed. Lincoln: University of Nebraska Press, 2000.

Oppenheimer, Robert. "Acculturation or Assimilation: Mexican Immigrants in Kansas, 1900 to World War II." *Western Historical Quarterly* 16, no. 4 (October 1985): 429–48.

Osteen, Ike. *A Place Called Baca*. Chicago: Adams Press, 1979.

Paschal, Allan W. "The Enemy in Colorado: German Prisoners of War, 1943–1946." *Colorado Magazine* 56 (Summer/Fall 1979): 119–42.

Peters, Scott J. "'Every Farmer Should Be Awakened': Liberty Hyde Bailey's Vision of Agricultural Extension Work." *Agricultural History* 80, no. 2 (2006): 190–219.

Petersen, Brian D. "The Fryingpan Arkansas Project: A Political, Economic, and Environmental History." Master's thesis, University of Montana, 2005.

Phillips, Sarah T. "FDR, Hoover, and the New Rural Conservation, 1920–1932." In *FDR and the Environment*, edited by Henry L. Henderson and David B. Woolner, 107–52. New York: Palgrave Macmillan, 2005.

———. *This Land, This Nation: Conservation, Rural America, and the New Deal*. New York: Cambridge University Press, 2007

Philpott, William. *Vacationland: Tourism and Environment in the Colorado High Country*. Seattle: University of Washington Press, 2013.

Postel, Charles. *The Populist Vision*. New York: Oxford University Press, 2007.

Potts, Alfred M. "School Bells for Children Who Follow the Crops." *Elementary School Journal* 60, no. 8 (May 1960): 441.

———. "A Social Profile of Agricultural Migratory People in Colorado. Denver: Office of Instructional Services, 1959.

Putnam, Dan, et al. "The Importance of Western Alfalfa Production." Las Vegas: 29th National Alfalfa Symposium Proceedings, Alfalfa Council and UC Cooperative Extension. Accessed via http://ag.arizona.edu/crop/counties/yuma /farmnotes/fn 1101 alfalfaprod.pdf.

Rasmussen, Chris. "'Never a Landlord for the Good of the Land': Farm Tenancy, Soil Conservation, and the New Deal in Iowa." *Agricultural History* 73, no. 1 (Winter 1999): 393–410.

Rasmussen, Wayne D. *Farmers, Cooperatives, and USDA: A History of Agricultural Cooperative Service*. Washington DC: USDA, 1991.

———. *A History of the Emergency Farm Labor Supply Program, 1943–1947*. Agricultural Monograph No. 13. Washington DC: USDA Bureau of Economics, 1951.

———. *Taking the University to the People: Seventy-Five Years of Cooperative Extension*. Ames: Iowa State University Press, 1989.

Reid, Debra A. *Reaping a Greater Harvest: African Americans, the Extension Service, and Rural Reform in Jim Crow Texas*. College Station: Texas A&M University Press, 2007.

Reisler, Mark. *By the Sweat of Their Brow: Mexican Immigrant Labor in the United States, 1900–1940*. Westport CN: Greenwood Press, 1976.

Riney-Kehrberg, Pamela. "From the Horse's Mouth: Dust Bowl Farmers and Their Solutions to the Problem of Aridity." *Agricultural History* 66, no. 2 (1992): 137–50.

———. *Rooted in Dust: Surviving Drought and Depression in Southwestern Kansas*. Lawrence: University Press of Kansas, 1997.

———, ed. *Waiting on the Bounty: The Dust Bowl Diary of Mary Knackstedt Dyck*. Iowa City: University of Iowa Press, 1999.

Rollins, Peter, and Harris J. Elder. "Environmental History in Two New Deal Documentaries." *Film and History* 3, no. 3 (1973): 1–7.

Rossi, Nick. *Colorado Migrant Education Program Summary Report, September 1, 1968, through August 31, 1969*. Denver: Colorado Department of Education, 1970.

Ruíz, Vicki. *Cannery Women, Cannery Lives: Mexican Women, Unionization, and the California Food Processing Industry, 1930–1950*. Albuquerque: University of New Mexico Press, 1987.

Saloutos, Theodore. *The American Farmer and the New Deal*. Ames: Iowa State University Press, 1982.

———. "New Deal Agricultural Policy: An Evaluation." *Journal of American History* 61, no. 2 (September 1974): 394–416.

Sanchez, Virginia. *Forgotten Cuchareños of the Lower Valley*. Charleston SC: History Press, 2010.

Sanders, H. C., ed. *The Cooperative Extension Service*. Englewood Cliffs NJ: Prentice-Hall, 1966.

Schrager, Adam. *The Principled Politician: The Ralph Carr Story*. Golden CO: Fulcrum, 2008

Schulten, Susan. "How to See Colorado: The Federal Writers' Project, American Regionalism, and the 'Old New Western History.'" *Western Historical Quarterly* 36, no. 1 (Spring 2005): 49–70.

Schuyler, Michael W. *The Dread of Plenty: Agricultural Relief Activities of the Federal Government in the Middle West, 1933–1939*. Manhattan KS: Sunflower University Press, 1989.

———. "New Deal Farm Policy in the Middle West: A Retrospective View." *Journal of the West* 33, no. 4 (1994): 52–63.

Scott, Roy V. *The Reluctant Farmer: The Rise of Agricultural Extension to 1914*. Urbana: University of Illinois Press, 1970.

Secrest, Clark. "How Colorado Blew into Kansas." *Colorado Heritage*, Winter 1994, 14–17.

Seeley, Charles Livingstone. *Pioneer Days in the Arkansas Valley in Southern Colorado and History of Bent's Fort*. Denver: Charles Livingstone Seeley, 1932.

Sheflin, Douglas. "The Environment and the New Deal." In *Interpreting American History: The New Deal and the Great Depression*, edited by Aaron D. Purcell, 65–80. Kent OH: Kent State University Press, 2012.

Sherow, James Earl. "Discord in the 'Valley of Content': Strife over Natural Resources in a Changing Environment on the Arkansas River Valley of the High Plains." PhD diss., University of Colorado, 1987.

———. "Utopia, Reality, and Irrigation: The Plight of the Fort Lyon Canal Company in the Arkansas River Valley." *Western Historical Quarterly* 20, no. 2 (May 1989): 162–84.

———. *Watering the Valley: Development along the High Plains Arkansas River, 1870–1950*. Lawrence: University Press of Kansas, 1990.

Shideler, James H. "Herbert Hoover and the Federal Farm Board Project, 1921–1925." *Mississippi Valley Historical Review* 42, no. 4 (March 1956): 710–29.

Smith, Henry Nash. *Virgin Land: The American West as Symbol and Myth*. Cambridge: Harvard University Press, 1950.

Smith, Jason Scott. *Building New Deal Liberalism: The Political Economy of Public Works, 1933–1956*. New York: Cambridge University Press, 2006.

———. "New Deal Policy Works at War: The WPA and Japanese American Internment." *Pacific Historical Review* 72, no. 1 (2003): 63–92.

Smith, Michael M. "Beyond the Borderlands: Mexican Labor in the Central Plains, 1900–1930." *Great Plains Quarterly* 1, no. 4 (1981): 239–151.

Smith, Page. *Democracy on Trial: The Japanese-American Evacuation and Relocation in World War II*. New York: Simon & Schuster, 1995.

Spence, Clark C. *The Rainmakers: American "Pluviculture" to World War II*. Lincoln: University of Nebraska Press, 1980.

Steinbeck, John. *The Grapes of Wrath*. New York: Penguin Classics, 2006.

Stock, Catherine McNicol. *Main Street in Crisis: The Great Depression and the Old Middle Class on the Northern Plains*. Chapel Hill: University of North Carolina Press, 1992.

Stock, Catherine McNicol, and Robert D. Johnston, eds. *The Countryside in the Age of the Modern State*. Ithaca NY: Cornell University Press, 2001.

Stoll, Steven. *Larding the Lean Earth: Soil and Society in Nineteenth-Century America*. New York: Hill and Wang, 2002.

Sutter, Paul S. "Terra Incognita: The Neglected History of Interwar Environmental Thought and Politics." *Reviews in American History* 29, no. 2 (2001): 289–97.

Svaldi, David P. "The Rocky Mountain News and the Indians." *Journal of the West* 27, no. 3 (July 1988): 85–94.

Syndicate Land and Irrigation Company. *Valuable Data*. Denver, n.d.

Takahara, Kumiko. *Off the Fat of the Land: The Denver Post's Story of the Japanese American Internment during World War II*. Powell WY: Western History, 2003.

Takaki, Ronald. *Double Victory: A Multicultural History of America in World War II*. Boston: Little, Brown, 2000.

Taylor, Morris F. "The Town Boom in Las Animas and Baca Counties." *Colorado Magazine* 55, no. 2/3 (Spring/Summer 1978): 111–32.

Taylor, Sandra C. *Jewel of the Desert: Japanese American Internment at Topaz*. Berkeley: University of California Press, 1993.

Thompson, Antonio. *Men in German Uniform: POWs in America during World War II*. Knoxville: University of Tennessee Press, 2010.

Thompson, Glenn. *Prisoners on the Plains: German POWs in America*. Holdrege NE: Phelps County Historical Society, 1993.

Thompson, William Takamatsu. "Amache: A Working Bibliography on One Japanese American Concentration Camp." *Amerasia Journal* 19, no. 1 (1993): 153–59.

Truedell, Leon E. "Farm Tenancy Moves West." *Journal of Farm Economics* 8, no. 4 (October 1926): 443–50.

Tweton, D. Jerome. *The New Deal at the Grass Roots: Programs for the People in Otter Tail County, Minnesota.* St. Paul: Minnesota Historical Society Press, 1988.

Ubbelohde, Carl, Maxine Benson, and Duane A. Smith. *A Colorado History.* 9th ed. Boulder: Pruett, 2006.

Valdés, Dennis Nodín. "Settlers, Sojourners, and Proletarians: Social Formation in the Great Plains Sugar Beet Industry, 1890–1940." *Great Plains Quarterly* 10 (1990): 110–23.

Van Hook, Joseph Orlando. "Development of Irrigation in the Arkansas Valley." *Colorado Magazine* 10 (1933): 11.

———. "Settlement and Economic Development of the Arkansas Valley from Pueblo to the Colorado-Kansas Line, 1860–1900." PhD diss., University of Colorado, 1933.

Vest, H. Grant. *The Teachers Say–1959 Colorado Schools for Migratory Children.* Denver: Colorado State Department of Education, 1959.

Volanto, Keith J. "Leaving the Land: Tenant and Sharecropper Displacement in Texas during the New Deal." *Social Science History* 20, no. 4 (Winter 1996): 533–51.

Webb, Walter Prescott. *The Great Plains.* Lincoln: University of Nebraska Press, 1981.

Weglyn, Michi. *Years of Infamy: Untold Story of America's Concentration Camps.* New York: William Morrow, 1976.

Wei, William. "The Strangest City in Colorado: The Amache Concentration Camp." *Colorado Heritage,* Winter 2005, 2–17.

Weisiger, Marsha. *Dreaming of Sheep in Navajo Country.* Seattle: University of Washington Press, 2009.

Welsh, Michael. "Deserts, Gardens, and Cities: Rethinking Colorado's Arkansas Basin in the 20th Century." *Heritage of the Great Plains* 37, no. 2 (2004): 33–47.

———. *U.S. Army Corps of Engineers, Albuquerque District, 1935–1985.* Albuquerque: University of New Mexico Press, 1987.

West, Elliott. *The Contested Plains: Indians, Goldseekers, and the Rush to Colorado.* Lawrence: University of Kansas Press, 1998.

Westerlund, John S. *Arizona's War Town: Flagstaff, Navajo Ordinance Depot, and World War II.* Tucson: University of Arizona Press, 2003.

Wickens, James F. *Colorado in the Great Depression*. New York: Garland, 1979.

Wiener, John D., Roger S. Pulwarty, and David War. "Bite without Bark: How the Socioeconomic Context of the 1950s U.S. Drought Minimized Responses to a Multiyear Extreme Climate Event." *Weather and Climate Extremes* 11 (2016): 80–94.

Wilson, M. L. "Problem of Poverty in Agriculture." *Journal of Farm Economics* 22, no. 1 (February 1940): 10–29.

Wilson, Robert. "Landscapes of Promise and Betrayal: Reclamation, Homesteading, and Japanese American Incarceration." *Annals of the Association of American Geographers* 101, no. 2 (March 2011): 424–44.

Wishart, David J. *The Last Days of the Rainbelt*. Lincoln: University of Nebraska Press, 2013.

———, ed. *Encyclopedia of the Great Plains*. http://plainshumanities.unl.edu /encyclopedia/doc/egp.asam.002.

Wixson, Douglas, ed. *On the Dirty Plate Trail: Remembering the Dust Bowl Refugee Camps*. Austin: University of Texas Press, 2007.

Whitaker, Michael R. "Making War on Jupiter Pluvius: The Culture and Science of Rainmaking in the Southern Great Plains, 1870–1913." *Great Plains Quarterly* 33, no. 4 (Fall 2013): 207–19.

Worster, Donald. "The Dirty Thirties: A Study in Agricultural Capitalism." *Great Plains Quarterly* 6, no. 2 (1986): 107–16.

———. *Dust Bowl: The Southern Plains in the 1930s*. New York: Oxford University Press, 1979.

———. *Rivers of Empire: Water, Aridity, and the Growth of the American West*. New York: Pantheon Books, 1985.

Wunder, John R., Frances W. Kaye, and Vernon Carstensen, eds. *Americans View Their Dust Bowl Experience*. Niwot: University Press of Colorado, 1999.

Yoshino, Ronald W. "Barbed Wire and Beyond: A Sojourn through Internment—A Personal Recollection." *Journal of the West* 35, no. 1 (1996): 34–43.

INDEX

Baca County Emergency Drouth
Committee, 287
Baca County Labor Association, 228
Bailey, Liberty Hyde, 50–51
Balderrama, Francisco E., 348n59
Bankhead, John, 177
Bankhead-Jones Farm Tenant Act
(1937), 170, 177–79
Bennett, Hugh Hammond, 92, 95–
96, 111–13
Benson, Ezra, 254
Bent brothers, 127
Bent County, 115, *149*
Betz, Ava, 331n19
Betz, Fred, 69, 253, 257
Betz, Fred, Sr., 249
Big Thompson Project, 250
Bird, John, 284
Bitner, Rosalie, 323
Black, John D., 82
Black Sunday, 1, *2*, 248
Blinn, Philo K., 35
Bolin, Arwin, 261, 263
Boston MA, 35, 73
Boyes, Stanley, 285–86
braceros, 6, 11, 15, 16, 211–15, 299–302,
306. *See also* Latinos; Mexicans
Branom (chairman), 295
Brighton CO, 220
broomcorn: advances in production of,
244; ideal conditions for cultivation,
44, 227; laborers for, 192, 216, 227–29,
238, 241, 300, 311, 314; production in
Baca County, 9, 40, 227–29; wartime
production of, 205, 243. *See also* corn

Brown, Floyd, 162
Brown, Gerald E., 257
Brown, W. O., 113
Buckner, C. D., 292
Bureau of Agricultural Economics
(BAE), 87–91, 121
Bureau of Reclamation, 147, 148, 157–
58, 267, 269

Caddoa Dam and Reservoir. *See* John
Martin Dam and Reservoir
Caddoa Dam Board, 158
Caddoa Reservoir Association, 142. *See
also* Arkansas Basin Committee (ABC)
California, 1–2, 76, 124, 256–57, 272
Camp, Silas E., 260
Camp Carson, 217–18
Canadians, 26
cantaloupes. *See* melons
Carey Act (1894), 134. *See also* Desert
Land Act (1877)
Carr, Ralph, 152, 153, 155–56, 221–22
Carranza, Venustiano, 181
Carter, Vernon, 287, 292
Chalmers, K. W., 112
Chenoweth, J. Edgar, 153–54, 158,
217–18, 272, 285
Cheyenne Indians, 222. *See also*
American Indians
Chicago IL, 35, 73, 128, 248
children: of Japanese immigrants,
220; as laborers, *183*, 186, 187, 207,
216; of migrant laborers, 190, 303,
306, 307; outreach to, 65; of tenant
farmers, 175

Christy, Harold, 273
Christy, R. L., 222
Civilian Conservation Corps (CCC):
camp of, 228, 229, 314; for soil
conservation, 95, 114–15; in Springfield, 68, 333n46; in wartime, 199,
210; water projects of, 162
Civil Works Administration (CWA),
68, 136
Code, W. E., 133–34
colonias, 189, 347n44
Colorado: agricultural laborers in,
5–6, 9–11, 179–85, 190–91, 226,
244, 311–12; agricultural policy
in, 3–8; CCC workers from, 115;
crops grown in, 32–35, 39, 251;
drought in, 248, 252; economy in,
268; employment in, 68, 138, 143,
145, 146, 168, 193–95, 216, 238,
306; farming challenges in, 21–22,
27–30, 38–39, 42–44, 124–25, 317,
325–26; geopolitical importance of,
23; Japanese population in, 352n58;
map of southeastern, 10; migration in, 21–28, 31, 42–43, 75–76,
167–68, 179, 181–84, 189–91, 194,
215–16, 220–21, 235, 244, 302, 306;
negligent landowners in, 111, 119;
rainfall in, 256; as territory, 24–25;
urban development in, 275–77, 315;
water disputes in, 146–58. See also
Baca County; Prowers County
Colorado A&M, 50, 58, 230
Colorado Agricultural College. See
Colorado A&M

Colorado Agricultural Extension
Agency, 79
Colorado Arkansas Valley, Inc., 26
Colorado–Big Thompson Project,
267–68
Colorado Cooperative Extension Service: A. J. Hamman with, 346n32;
on Arkansas River flow, 144; in
Baca County, 41–42; farmers' attitudes toward, 58–64, 70–71, 111,
161, 165, 282, 285–88; and irrigation
techniques, 162, 261; labor management by, 15, 200, 206–11, 214–16,
218, 219, 226–30, 234–43; and
land purchasing program, 119; in
post-WWII economy, 281, 313, 314;
role of, 7, 12–13, 46–51, 53, 62–68,
236–37, 244; and soil conservation,
19–20, 53–54, 70, 73–74, 85–86,
91, 95–101, 104–9, 113, 164, 246,
251, 284–88, 290–93; and tenant
farmers, 178; in wartime, 199, 205–
6, 246. See also Emergency Farm
Labor Program
Colorado Department of Education,
302–4, 314
Colorado Division of Wildlife, 320
Colorado Fish, Game, and Parks
Service, 135
Colorado Fuel and Iron Co., 194
Colorado Green, 320
Colorado Ground Water Law, 261–62
Colorado Highway Patrol, 221
Colorado Migrant Council, 308–9
Colorado National Guard, 190–91

Territory, 24–25; and conservation efforts, 289–90; during droughts, 248; effect on farmers, 22, 129, 167–68, 186–87, 192–93, 318, 323–25; farmers' role in, 55, 82, 85; and farming methods, 19–21, 174, 312–13, 328n2; and farm subsidies, 321; governments' role in agricultural, 70, 77–82, 177–78, 204, 253, 254, 281–85, 296, 314–15, 334n16; internment camp's impact on, 224–25; and labor system, 186, 187, 192–93, 300, 302, 312; and landownership, 171–76; and rainmaking, 258; in Trinidad, 218; war's impact on, 198, 201, 205, 236; and water projects, 137–39, 142, 145–47, 150, 154–56, 268, 272, 275. *See also* Great Crash of 1929; Great Depression

Egan, Timothy, 1

Eisenhower, Milton, 221

Ellis, Clyde, 149–52, 154, 155

Emergency Farm Labor Program, 15, 181, 200, 206, 209, 237, 239, 243–44. *See also* Colorado Cooperative Extension Service

Emick family, 320

Environmental Working Group, 321

erosion control. *See* soil conservation

Executive Order 9066, 220, 222

Expanded Homestead Act, 120

Farley, Mary, 156

farm board, 56–57

Farm Bureau, 64, 204, 288

farmers: challenges to on Great Plains, 2–4, 116–17, 168–69, 317, 323–26; distrust of government, 111–16, 282–83; Dust Bowl's impact on, 11, 249–52; economic well-being of, 15, 19–22, 40–44, 46, 55–59, 63–70, 76–79, 82–87, 96, 105–6, 118–22, 127, 135, 142, 161, 171–72, 175–78, 186, 198–201, 204, 247, 266, 291, 312, 334n16; government assistance to, 5–7, 12–14, 16, 25, 46–47, 57–58, 60–61, 76, 77, 86, 92, 95, 97, 101, 105–6, 109, 119–22, 135–36, 172, 173, 314, 321–22; laborers preferred by, 211–13, 225–29; labor management by, 234–35, 240, 281; migration of, 26–29, 42–43, 120, 168, 174–75, 179; number of post-WWII, 245, 299; outreach to, 50–54, 57–59, 62–65, 70–71, 205; voluntary conservation by, 8, 13, 92–111, 114, 161–65, 198, 202, 286, 291–99, 313, 321, 323, 332n39; and wartime labor shortage, 199–200, 205

farmers, tenant: in Baca County, 88, 89, 90, 94; criticism of, 175–76, 345n23; decline in number of, 192–93; increase in Colorado, 39–40, 169–70; and landownership, 171–73, 177–79; migration of, 120, 168, 174–75, 179; resettlement of, 86–87, 178; rights of, 170; and soil conservation, 84–85, 94, 102, 104, 119, 121, 173–74

279; on irrigation wells, 259–61, 264; on laborers, 310–11; on land-ownership in Baca County, 297; on weather, 254, 256–59, 282

squash, 32

Stalford, Calvin, 285

Standard Soil Conservation Districts Law, 97

Stansbury, F. R., 114

State Agricultural College, 38

Steinbeck, John, 2, 248

Stewart, T. G., 73–74, 117–19

Stimson, Henry, 230

Stinson, J. A., 132

St. Louis MO, 35

Stock Raising Homestead Act (1916), 29

Stone, Clifford H., 153

Stonington CO, 95

strip cropping, 96, 106, 284, 294

Strohauer, Harry, 317

sugar beets: advances in production of, 244; and cantaloupe production, 36; conditions for cultivation of, 44, 251, 263; laborers for, 14, 179–81, *183*, 189–92, 213–15, 220, 229–32, 238, 240, 241, 300, 311, 314, 347n44, 347n46, 348n58; legislation regarding, 186–87; marketability of, 40; production during Great Depression, 22, 160; production in Colorado, 9, 15, 30, 33–35, 41, 170, 179–82; wartime production of, 205, 229–30, 241–43

sugar cane, 186

Sugar City CO, 34

sugar companies, 179–88, 193, 206, 209–10, 235–36, 241, 302

supply and demand. *See* price controls; production control

Swink, George, 32–33

Swink, Ralph, 175

Switzerland, 26

Syndicate Land and Irrigation Co., 27

tariffs, 33, 35, 56–57

Tennessee Valley Authority (TVA), 69, 124, 137–38, 149, 150, 156

terracing, 94, 96, 97, *98*, 105, 108, 291, 294

Texas: construction company from, 222; dust storms in, 267; farmers from, 26, 84; housing of migrant laborers in, 307; laborers from, 11, 181, 190, 191, 216, 235, 236, 299, 301, 305, 306, 309, 310

Thomas, Cyrus, 27

Thompson, Florence Owens, 1–2

Thompson, Ross, 223

Thornton, Dan, 254, 260, 292

Tigges, Ernest, 233

Timber Culture Act (1873), 29–30

Title III lands. *See* land, classification of submarginal

tree planting, *98*; act pertaining to, 29–30; in Baca County, 106; government assistance for, 105, 108, 294, 313; in Prowers County, 108, 109, 164; voluntary, 70

319; and weather modification, 257–58; and westward expansion, 24–30, 42, 140–41. *See also* Civilian Conservation Corps (CCC); Farm Security Administration (FSA); New Deal; Soil Conservation Service (SCS); War Relocation Authority (WRA); Works Progress Administration (WPA)

U.S. military, 194, 218, 219, 220, 229, 230

U.S. Office of Education, 303

U.S. Postal Service, 223

U.S. Supreme Court, 146, 158

U.S. War Department, 198, 230, 237

Utah, 190, 306

Vahles brothers, *50*

"Valley of Content," 21, 124, 328n3

Victory Farm Volunteers, 207

Vilas School, 68

Vivian, John, 230, 301

von Wechmar, Rüdiger Freiherr, 240–41

Wallace, Henry A., 92, *107*, 115, 175–76

Walsh, Lauriston, 110

Walsh CO, 228–29, 264, 300

War Food Administration, 206

War Manpower Commission, 219

War Relocation Authority (WRA), 221–24, 232, 233

War Relocation Board (WRB), 224

Washington DC, 73, 248

water: availability in Baca County, 37–38; availability on Great Plains, 26–32, 43, 161–62, 259, 281; CCC

projects involving, 115, 162; ground sources of, 132–34, 251, 259; local control of, 136, 149–50, 263, 266, 275; pumping of, 132, 251, 260–63, 288; rights to, 129, 131–35, 144, 146, 149–58, 161–62, 165, 233, 265–66, 271, 272, 278; soil's ability to hold, 93. *See also* irrigation

water conservation: farmers' attitudes toward, 162–65, 251, 282, 286; in post-WWII economy, 313; programs and policies for, 53, 70, 275, 288, 289, 293, 298, 321; by tenant farmers, 173; in wartime, 198, 200, 202, 205

watermelons. *See* melons

water use: Dust Bowl's effects on, 4, 5; by irrigated farmers, 161–62, 272; and soil health, 126; in southeastern Colorado, 9, 13–16, 136, 317, 319. *See also* irrigation

Water Users Association, 288

Watson, W. R., 114

weather, 21, 26–27, 30, 45, 135, 252–54, 282. *See also* drought; rainfall

Weather Modification Committee, 257

Weather Research Association, 256, 258

Webb Soil Erosion District, 163

Weld County, 317

West Baca County Farm Labor Association, 229

West Baca Soil Conservation District, 109–10

Western Baca County Soil Erosion District, 103–5, 116–17, 282–83, 286

Western Slope Farms, 86

West Prowers County Farm Labor Association, 236

West Slope, 181, 248, 250, 266, 273, 274

wheat: conditions for cultivation of, 44; at internment camps, 233; laborers for, 14, 169, 170, 179, 227, 229; marketability of, 35, 40, 56; production during Great Depression, 22; production in Baca County, 9, 20, 21, 36, 39, 41, 56, 78, 94, 160, 204, 263, 264; production in Prowers County, 159, 160; and soil quality, 81, 94, 117, 118, 292, 295; subsidies for, 61, 66, 321; wartime production of, 205, 229, 243, 283

whites, 9, 14, 24, 182, 187–89, 214, 220, 305–6, 347n44. *See also* labor, racism toward

Wickard, Claude, 201

Wiggins CO, 303

Wilderness Society, 269

wildlife, 109, 269–71, 289, 294, 295, 298, 320, 321, 363n9

Wiley CO, 185

Wilson, Floyd, 32, 154, 155

Wilson, M. L., 82–83, *107*

wind erosion. *See* soil conservation

Wind Erosion Control Program, 108–9, 113

windmills, 319–21

Women's Land Army, 207

Woods, Ralph R., 300

Works Progress Administration (WPA), 68, 136–37, 189, 191, 199, 222, 348n58, 349n69

World's Columbian Exposition, 26

World War I, 21, 36, 39, 129, 201–2, 210, 279, 313, 328n2

World War II: agricultural production during, 165, 197–202, 229, 243, 279–80; conservation efforts during, 237, 246; effects on agriculture, 3, 9, 13, 15, 244, 355n118; and farming technology, 255; labor shortage during, 6, 9–11, 169, 193–95, 206, 231–34, 300–301; news about, 164, 279; water projects during, 157

Wyoming, 221, 242

Xcel Energy, 320

Yarborough, A. A., 90

Young, Bruce, 287

Young, Mrs. M. H., 252

CPSIA information can be obtained
at www.ICGtesting.com
Printed in the USA
LVHW031132020721
691689LV00004B/217